T0200704

Construction Safety

About the Author

Dr. M. Rashad Islam is an Associate Professor and an undergraduate program coordinator at the Colorado State University Pueblo. He is also a registered Professional Engineer (PE), and an ABET Program Evaluator for civil and construction engineering undergraduate programs. He teaches civil and construction engineering major courses. Dr. Islam received a Ph.D. degree with the grade of "distinction" in the research, and a GPA of 4.0 out of 4.0 in the coursework from the University of New Mexico, USA. He has a master's degree jointly from the University of Minho (Portugal) and the Technical University of Catalonia (Barcelona, Spain). Dr. Islam has more than 100 publications including several reference books, several textbooks, over 60 scholarly journal articles, several technical reports, and conference papers. His other major textbooks are *Pavement Design—Materials, Analysis and Highways*, McGraw Hill, ISBN: 9781260458916, *Civil Engineering Materials—Introduction and Laboratory Testing*, ISBN-13: 978-0367224820, and *Engineering Statics*, ISBN: 9781003098157. Dr. Islam is a member of the American Society of Civil Engineers (ASCE) and is associated with the American General Contractors (AGC).

Construction
Safety
Health, Practices, and OSHA

M. Rashad Islam

New York Chicago San Francisco
Athens London Madrid
Mexico City Milan New Delhi
Singapore Sydney Toronto

Library of Congress Cataloging-in-Publication Data

Names: Islam, M. Rashad, author.
Title: Construction safety : health, practices, and OSHA / M. Rashad Islam.
Description: New York : McGraw Hill, 2021. | Includes index. | Summary:
  "This textbook for Construction Safety and Safety Management courses
  takes a student-friendly approach to the topic, providing a
  comprehensive overview, up to date information on OSHA standards, and
  discussing the OSHA.com construction safety training"—Provided by
  publisher.
Identifiers: LCCN 2021021205 | ISBN 9781264257829 (acid-free paper) | ISBN
  9781264257836 (ebook)
Subjects: LCSH: Building—United States—Safety measures. | United States.
  Occupational Safety and Health Administration.
Classification: LCC TH443 .I75 2021 | DDC 690/.22—dc23
LC record available at https://lccn.loc.gov/2021021205

McGraw Hill books are available at special quantity discounts to use as premiums and sales promotions or for use in corporate training programs. To contact a representative, please visit the Contact Us page at www.mhprofessional.com.

Construction Safety: Health, Practices, and OSHA

Copyright © 2022 by McGraw Hill. All rights reserved. Printed in the United States of America. Except as permitted under the United States Copyright Act of 1976, no part of this publication may be reproduced or distributed in any form or by any means, or stored in a data base or retrieval system, without the prior written permission of the publisher.

1 2 3 4 5 6 7 8 9   CCD   25 24 23 22 21

ISBN 978-1-264-25782-9
MHID 1-264-25782-1

This book is printed on acid-free paper.

Sponsoring Editor Ania Levinson	**Proofreader** Kirti Dogra, MPS Limited
Editing Supervisor Donna M. Martone	**Indexer** May Hasso
Acquisitions Coordinator Elizabeth M. Houde	**Production Supervisor** Lynn M. Messina
Project Manager Rishabh Gupta, MPS Limited	**Composition** MPS Limited
Copy Editor Bindu Singh, MPS Limited	**Art Director, Cover** Jeff Weeks

Information contained in this work has been obtained by McGraw Hill from sources believed to be reliable. However, neither McGraw Hill nor its authors guarantee the accuracy or completeness of any information published herein, and neither McGraw Hill nor its authors shall be responsible for any errors, omissions, or damages arising out of use of this information. This work is published with the understanding that McGraw Hill and its authors are supplying information but are not attempting to render engineering or other professional services. If such services are required, the assistance of an appropriate professional should be sought.

Contents

Preface

Due to the rapid increase in construction activities, construction safety became a very important topic in the construction industry. Construction education is also increasing at the university level. However, there is a shortage in construction safety educational materials. The author is thus motivated to write this construction safety textbook for classroom learning. The primary audience of the book is undergraduate and graduate students interested in the construction industry and academia. The book is written based on the Occupational Safety and Health Administration (OSHA), Part 1926—Safety and Health Regulations for Construction (https://www.osha.gov/). Numerous examples and practice activities are provided for effective classroom learning.

The materials presented in this book are entirely for educational purposes only. It should not be used to provide guidance to your customers, used as OSHA standards, or for clients in lieu of competent, certified legal advice. Readers are always encouraged to see the updated regulations by OSHA (https://www.osha.gov/). All parties involved in the publishing of this textbook will not be liable for any inappropriate use of this information beyond educational purposes.

After studying this book, readers should seek out online certification by OSHA.com at https://www.osha.com/. OSHA.com's outreach training courses are accepted by OSHA and powered by 360training.

The book has been thoroughly inspected with the help of professional editors to fix typos, editorial issues, and poor sentence structure. Despite this, if there is any issue, please excuse the author and report it at islamunm@gmail.com. Any suggestion to improve this book, or any issue reported, will be fixed in the next edition with proper acknowledgment. Thank you.

M. Rashad Islam, PhD, PE
Associate Professor
Colorado State University—Pueblo

CHAPTER 1

Introduction

(Photo courtesy of 123RF.)

1.1   Nature of Construction Operations

Construction project sites are the most hazardous workplaces other than mining and agriculture. The construction industry accounts for about 10% of injury accidents and 20% of industrial workforce deaths, even though construction workers account for just about 8% of industrial man-hours worked in the United States. The construction industry encompasses a wide range of companies, specialized crafts, and types of projects. Projects range from single houses to multibillion dollar major infrastructure projects.

Fatality cases reported by the United States Bureau of Labor Statistics (BLS 2019) for the construction industry from 2011 to 2018 are provided in Table 1.1. It shows that 1008 construction deaths occurred in the private sector last year, compared to 971 in 2017. The industry's fatal accident rate was at 9.5 per 100,000 full-time equivalent workers. The rate of the last year represents a decrease from 2016 to 2015, when the figure was 10.1. Deaths in the heavy and construction sectors increased by 18% in 2018, to 180 from 152, and fatalities in buildings construction increased slightly, from 196 in the previous year to 200 in 2018. According to BLS figures, specialty trade contractors had 609 fatalities last year, one fewer than in 2017. The Bureau stated that the fatality rates for individual construction sectors were not available. In an encouraging result, BLS reported that the number of deaths caused by falls, slips, and trips across all sectors decreased by 11% to 791 in 2018. That compares with a high for this BLS data series of 887 in 2017. The Bureau traces the decline in falls to a decrease of 14% from 713 fatal falls to a lower level to 615. The total of 615 is the lowest annual mark for this type of fall since 2013, noted by BLS. Construction safety is, therefore, a very important topic. The ability of a construction company to eliminate or mitigate the risk of accidents is essential for the successful execution of a construction project. The implementation of effective safety measures reduces project costs and demonstrates concern for the welfare of people working on or passing on the project. Safety records are often considered by project owners when selecting construction companies to build their projects. There are two key aspects of the safety of the project site:

1. Safety of persons working on the site
2. Safety of the general public who may be near the project site

Both aspects must be addressed when developing project-specific safety plans. These plans describe the hazards to be faced by construction workers and the general public during the different phases of construction and the measures to be taken to reduce the risk of injury to workers or the public. Examples of construction workers' safety measures include the use of personal protective equipment (PPE) and the installation of barricades around floor openings. Examples of public safety measures include perimeter fences and warning signs to prevent unauthorized individuals from entering the project site.

Year	Construction Fatalities	Fatality Rates[a]
2011	738	9.1
2012	806	9.9
2013	828	9.7
2014	899	9.8
2015	937	10.1
2016	991	10.1
2017	971	9.5
2018	1008	9.5

[a]Fatalities per 100,000 full-time equivalent workers. This statistics is for private-sector construction industry. (Source: Bureau of Labor Statistics)

TABLE 1.1   Fatality Cases within the Construction Industry

The primary causes of construction jobsite injuries are:

- Falling from an elevation
- Being struck by something
- Trenching and excavation cave in
- Being caught between two objects
- Electrical shock

Many hazards exist on all construction sites: sharp edges, falling objects, openings in floors, chemicals, noise, and a myriad of other potentially dangerous situations. Mitigation measures are required to reduce the risk for injury, and further training is required to ensure that the entire workplace has a healthy working attitude. Most of the construction projects are special and implemented in a variety of work environments.

Construction workers are, therefore, constantly expected to become familiar with new situations that could potentially be hazardous. Besides, the composition of construction project teams varies from project to project, and many craft workers may work for different employers, resulting in a lack of conformity and continuity. Craft workers may only work during some phases of the work on a project site and then move to another project site. The ongoing change in the composition of the work force on a project provides the project manager, director, and field supervisors with significant leadership challenges. Another big safety problem for construction site supervisors is the increased employment of workers whose second language is English. Not only do these workers have difficulty reading and understanding safety signage, but they may be unwilling to report unsafe jobsite conditions or working practices. It is critical that supervisors be able to implement good safety standards on all individuals working on a jobsite. This may require safety signage to be published in multiple languages and safety guidance to be provided in multiple languages. When construction projects are under way, there is a constant sequence of circumstances in which construction workers and/or the general public can be exposed to the risk of injury. Recognizing such situations and taking action to control or mitigate these jobsite hazards are extremely important for construction leaders. Many construction projects are carried out in excavations below the ground or in the air above the ground. Construction work is subjected in many instances to natural forces such as rain, snow, wind, or other climatic conditions. The best way of mitigating the risk of injury is to introduce measures to protect workers and the public.

The complexity and potential hazard environments in a construction site in Kuala Lumpur city are shown in Fig. 1.1. Considering different groups, construction people working simultaneously could have very high accident potential.

The creation of a safe work environment depends on the physical conditions of the working environment and the behavior or working attitude of the people working on the site. Safety planning must begin during the initial planning of a construction project, together with the development of cost estimates and the schedule of projects. The initial safety plan needs to detail how project safety will be handled, including the roles and responsibilities of project members, available resources, potential hazards and mitigation measures, training requirements, and safety equipment requirements. Requiring everyone on the project site to wear appropriate PPE may have an impact on the productivity of workers, and the purchase of appropriate safety equipment may have an impact on project costs.

FIGURE 1.1 Complexity of a construction site with the potential of accidents.
(Photo courtesy of Zukiman Mohamad from Pexels.)

1.2 Introduction to OSHA

The Occupational Safety and Health Administration (OSHA) is an agency of the United States Department of Labor. Congress established the agency under the Occupational Safety and Health Act (OSH Act) on December 29, 1970. OSHA's mission is to "assure safe and healthy working conditions for working men and women by setting and enforcing standards and by providing training, outreach, education, and assistance." The Agency is also charged with the enforcement of a variety of whistleblower statutes and regulations.

Federal OSHA is a small agency; with our state partners, we have approximately 2100 inspectors responsible for the health and safety of 130 million workers, employed at more than 8 million worksites around the nation—which translates to about one compliance officer for every 59,000 workers.

OSHA workplace safety inspections have shown that injury rates and injury costs have been reduced without adverse effects on employment, sales, credit ratings, or firm survival (Levine et al., 2012). For more than four decades, OSHA and our state partners, along with employers' contributions, health and safety practitioners, unions and advocates, have had a significant impact on safety at work. Some milestones of OSHA impacts are listed below:

- Worker deaths in America were down—on average, from about 38 worker deaths a day in 1970 to 14 a day in 2017.
- Worker injuries and illnesses were down—from 10.9 incidents per 100 workers in 1972 to 2.8 per 100 in 2017.
- The work-related fatality rate was cut to historic low for 2002 to 2004.
- From 2003 to 2004, the number of workplace injuries and illnesses reduced by 4% and the rate of lost workday cases dropped by 5.8%.
- In 2005, OSHA conducted close to 39,000 inspections and issued just over 85,000 citations for violations.
- In 2004, the Consultation Program made over 31,000 visits to employers.

OSHA regional and area offices employ inspectors whose duties include visits to construction projects to ensure compliance with mandated safety and health procedures and to assess significant fines for failure to comply with the required procedures. OSHA job safety and health standards typically consist of guidelines to avoid hazards that have been established by research and experience to be detrimental to personal safety and health. The Act allowed individual states to lay down their own occupational health and safety standards as long as the state standards are at least as stringent as the federal requirements. Several states have adopted their own occupational safety and health laws and employ inspectors to ensure enforcement of state-owned construction projects. Failure to comply with regulatory criteria usually results in significant citations and fines. Rules created by all federal regulatory agencies are collected into a multipart document called the Code of Federal Regulations (CFR). OSHA 29 CFR 1926 standards focus on the construction industry, and identify the specific work-related risks associated with it.

Example 1.1

OSHA _____ are rules that describe the methods that employers must use to protect workers from hazards.

A. Website

B. Trainings

C. Quick takes

D. Standards

Answer: D

1.3 Importance of Safe Practices

Disabling or catastrophic damage to a construction project can have a major negative effect on the execution of construction operations. Accidents cost money, have an adverse impact on worker morale and productivity, and lead to adverse publicity about the project, the construction company, and the project owner. It is the responsibility of the construction company to provide a safe working environment for all construction workers on the site, including those employed by subcontractors, and to protect the public from harm. This is a major concern when significant construction work is taking place inside a facility, such as a hospital, which is in operation. The primary factors that motivate safe practices on construction sites are:

- Humanitarian concern for workers and the public
- Economic cost of accidents
- Regulatory requirements for worksite safety

It is a normally accepted principle that a person should not suffer an injury while working for an employer. This is based on a humanitarian concern for every individual's wellbeing. In addition to the humanitarian concern, there is a significant adverse economic impact if an accident occurs. Accidents are costly, as discussed in the next section, and often lead to uncompensable delays in completing the construction project.

Most successful construction companies have recognized the importance of safety management and have developed effective company safety programs that include:

- New employee orientation
- Safety training
- Project-specific accident prevention plans
- Jobsite surveillance

Good safety practices reduce business expenses as they reduce workers' compensation and liability insurance rates and mitigate the expenses of accidents and injuries at work. Construction companies often pay 10 to 20% of their direct labor costs for workers' compensation insurance premiums, depending on the type of craft labor they employ, which is a significant cost of doing business. The quality of the safety system of a construction company is also a key factor in the company's ability to be prequalified and allowed to submit a proposal on a project. Project owners do not want unsafe contractors working on their projects because the owners do not want negative publicity related to construction accidents. Also, unsafe project sites often result in citations and fines from state or federal inspections of occupational safety and health. Implementing and enforcing a strong safety program also demonstrates the interest of the company management in the welfare of people working on the jobsite. A construction company's ability to deliver a quality project is directly affected by the ability and motivation of the individuals working on the jobsite, whether they work for the general contractor or for one of the subcontractors. Providing a safe working environment shows the commitment of the management to the welfare of the workers, which ensures that the workers continue to work on the jobsite to make the project a success.

1.4   Workers' Rights

OSHA provides workers the right to a safe and healthful workplace. OSH Act states, "Each employer shall furnish to each of his employees employment and a place of employment which are free from recognized hazards that are causing or are likely to cause death or serious physical harm to his employees." A safe and healthful workplace means that hazards are removed and workers are trained. If a hazard cannot be removed completely, protection such as PPE must be provided to workers free of cost to them. In summary, workers have the right to:

- Working conditions that do not pose a risk of serious harm. Figure 1.2 is an example of risky work practice where a construction worker is walking very close to an excavator.
- Receive information and training (in a language and vocabulary the worker understands) about workplace hazards, methods to prevent them, and the OSHA standards that apply to their workplace.
- Review records of work-related injuries and illnesses for the past 5 years. The employer must provide this by the end of the next workday.
- File a complaint asking OSHA to inspect their workplace if they believe there is a serious hazard or that their employer is not following OSHA's rules. OSHA will keep all identities confidential.
- Exercise their legal rights without retaliation, including reporting an injury to their employer or OSHA or raising health and safety concerns. If a worker has been retaliated for using his or her right, he or she must file a complaint with OSHA as soon as possible, but no later than 30 days.
- Be free from retaliation for exercising safety and health practices.
- Participate in an OSHA inspection directly. You have the right to talk to the inspector privately.
- Complain or request hazard correction from employer.

Under federal law, you are entitled to a safe workplace. Your employer must ensure that the workplace is free of known health and safety hazards. If you have any concerns, you have the right to speak out about them without fear of retaliation. You also have the right to:

- Be trained in a language you understand
- Work on machines that are safe
- Be provided required safety gear such as gloves or a harness and lifeline for falls
- Be protected from toxic chemicals
- Request an OSHA inspection, and speak to the inspector
- Report an injury or illness, and get copies of your medical records
- See copies of the workplace injury and illness log
- Review records of work-related injuries and illnesses
- Get copies of test results done to find hazards in the workplace

FIGURE 1.2   A construction worker close to an excavator.
(Photo courtesy of Nicholas Lim from Pexels.)

Another important workers' right is to know about the hazardous chemicals present in the workplace. Employers must have a written complete hazard communication program that includes information on:

- Container labeling
- Safety Data Sheets (SDSs)

Worker training must include the physical and health hazards of the chemicals and their protective measures, including specific procedures that the employer has implemented to protect workers such as work practice, emergency procedures, and PPE.

Workers have the right to examine and copy exposure and medical records, including records of workplace monitoring or measuring a toxic substance. This is important if you have been exposed to toxic substances or harmful physical agents in the workplace. Examples of toxic substances and harmful physical agents are:

- Some metal and dusts such as lead, cadmium, silica, etc.
- Biological agents such as bacterial, viruses, fungi, etc.
- Physical stresses such as noise, heat, cold, vibration, repetitive motion, ionizing and nonionizing radiation, etc.

Example 1.2
Under OSHA, workers have specific rights:
[Select all that apply]

   A. Right to examine exposure and medical records including monitoring or measuring toxic substances

   B. Right to file an OSHA complaint and participate in an OSHA inspection

   C. Right to information about the hazards they face including Safety Data Sheets (SDSs)

D. Right to hazard and overtime pay for hazardous tasks

E. Right to information about injuries and illness in the workplace

F. Right to modify OSHA safety programs and enforce OSHA in the workplace

G. Right to training from employers on health and safety hazards and standards

H. Right to set permissible exposure levels for all workers who may encounter toxic substance

Answers: A, B, C, E, G

Example 1.3
SDS stands for:

A. Safety Data Sheet

B. Safe Data System

C. System Data Safety

D. System Data Sheet

Answer: A

1.5 Summary

Construction site is a very complex and dynamic workplace consisting of workers from different background and expertise. Accident in construction is thus very common. After Occupational Safety and Health Act was established in 1971 to give structure to the worker protection activities, accident in construction industry and other areas decreased remarkably. OSHA holds the employer responsible for providing a workplace that is free from recognized hazards. The current mission of OSHA is "to save lives, prevent injuries, and protect the health of America's workers." It is the employer's responsibility to keep the workplace free from any known or recognized hazard that is likely to cause injury or illness to their workers. Workers have the right to know what hazardous materials they are required to work around. Workers must have a way to report hazardous conditions, injuries, and illnesses. Employers must provide training for the workers to inform them on hazard recognition and hazard control. In addition to training, the employers must have medical screening and monitoring when employees are exposed to certain hazards.

1.6 Multiple-Choice Questions

1.1 Which government agency enforces the standards set out in the Occupational Safety and Health Act?

A. Department of Health and Human Services

B. Department of Agriculture

C. Department of Commerce

D. Department of Labor

1.2 After OSHA was established, the number of injuries and illnesses:

A.   Decreased
B.   Stayed the same
C.   Increased
D.   Decreased only for large employers

1.3 In a construction project, safety planning must begin:

A.   At the initial cost estimate phase of a project
B.   After submitting the bid but before being awarded
C.   After the job award but before the job starts
D.   At the time of job starting

1.4 Why is maintaining and monitoring safety a challenge for construction project?

A.   Continuously changing task in a jobsite
B.   English language is a second language for many workers
C.   Different shifts and rotations of workers and supervisors
D.   All of the above

1.5 BLS reported that in 2018 the number of deaths caused by falls, slips, and trips across all sectors decreased to:

A.   615
B.   887
C.   791
D.   971

1.6 If a worker has been retaliated for using his or her right, he or she must file a complaint with OSHA as soon as possible, but no later than:

A.   15 days
B.   30 days
C.   45 days
D.   60 days

1.7 Construction companies often pay _____ of their direct labor costs for workers' compensation insurance premiums.

A.   2 to 5%
B.   5 to 10%
C.   10 to 20%
D.   20 to 40%

1.8 Federal OSHA has approximately:

A.   1100 inspectors
B.   2100 inspectors
C.   4100 inspectors
D.   5100 inspectors

1.9 Federal OSHA has approximately 2100 inspectors responsible for the health and safety of:

A.   30 million workers
B.   60 million workers
C.   130 million workers
D.   190 million workers

1.10 Federal OSHA has approximately 2100 inspectors responsible for the health and safety of 130 million workers, employed at more than:

    A.   2 million worksites
    B.   4 million worksites
    C.   6 million worksites
    D.   8 million worksites

1.11 OSHA was established in the year:

    A.   1950
    B.   1960
    C.   1970
    D.   1980

1.7   Practice Problems

1.1 Why does a construction site have so much potential for accident to occur?

1.2 List the primary causes of construction jobsite injuries.

1.3 State OSHA's mission.

1.4 What is the importance of safe practices?

1.5 How can accident in construction be prevented a little bit?

1.8   Critical Thinking and Discussion Topic

Two construction undergraduate students are talking. One student says construction operation is very simple and straightforward. We see it almost every day of our life. Anybody works here knows the risk of working or safety issue related to construction activities. A university course is not required to gain knowledge on construction safety. Construction safety is all about common sense. The other student says there is a need for this course in academia as construction operation is versatile and dynamic in nature. People of different skills and orientation work here in various roles. Based on this discussion, give your opinion and the logics of your opinion.

(Photo courtesy of Pixabay from Pexels.)

CHAPTER 2
Personal Safety

(Photo courtesy of 123RF.)

2.1 Basics of Personal Safety

Personal safety is an individual's ability to go about their everyday life free from the threat or fear of psychological, emotional, or physical harm from others. It is often said that personal safety is a matter of common sense. However, common sense does not always result in common practice. For personal safety in the workplace to become common practice, it needs to be prioritized across the organization and championed by strong leadership. Without a clear organizational approach to personal safety, the response of each worker to a threatening or confrontational situation can differ dramatically and will rely heavily on individual judgments. In order to achieve consistent and effective personal safety at the workplace, each worker must know how to identify, assess, and reduce or manage the risk of violence and aggression. Employers' consideration and care of their employees strengthens not only the relationship between members of the public and employees but also the relationship between the organization and its staff.

2.2   Personal Protective Equipment (PPE)

In construction industry, personal protective equipment (PPE), shown in Fig. 2.1, is considered the last line of defense when it comes to protecting workers from injuries on the jobsite. Employers are expected to enforce engineering controls and other safety measures to prevent accidents and injuries. In the case that these steps fail or are not possible, the PPE is there to prevent injury when hazards occur. Employers are expected to pay for

Figure 2.1   Basic personal protective equipment (PPE).
(Photo courtesy of 123RF.)

and supply workers with all PPE and are responsible for ensuring their use where appropriate. All PPE should meet the American National Standards Institute (ANSI) specifications. Employers can allow workers to use their own PPE. If employee-owned equipment is allowed, employers are responsible for ensuring that the PPE is in good working condition, but they are not required to reimburse workers for PPE they provide themselves. In order to best provide protection to your workers, here are some tips for ensuring proper use of PPE. Your PPE program should outline hazards present and the proper PPE that is to be worn. The program should explain how to select and ensure the proper fit for PPE. The PPE program needs to cover proper inspection and maintenance of all PPE.

Example 2.1
PPE is required when _____ controls are not feasible for a worker or working condition.

    A. OSHA

    B. Engineering

    C. EPA

    D. Union

   Answer: B

2.3 Head Protection

2.3.1 Basics

Head injuries are commonly caused by falling or flying objects, or by bumping the head against a fixed object. Head injuries can be minor or deadly. Injuries can include minor abrasions, concussions, lacerations, fractures, burns, or even electrocution. Your employer must access the hazards in the workplace to determine if head protection is needed. You must receive training in the proper use of head protection and understand the following concepts:

- When head protection is necessary.
- What type of head protection is necessary.
- How to properly put on, take off, adjust, and wear head protection.
- Limitations of head protection.
- Proper care, maintenance, useful life, and disposal of head protection.

You must show that you understand the training and can use head protection properly before you will be allowed to perform work requiring its use.

2.3.2 Head Protective Equipment

Hard hats (Fig. 2.2) are designed to resist the penetration, absorb the shock from a blow, and provide protection from electrical shock and burn. Head protectors in the form of protective hats can resist penetration and absorb the shock of a blow. The shell of the protective hat is hard enough to resist many blows, and the suspension system keeps

FIGURE 2.2 Two construction workers with hard hats.
(Photo courtesy of Amanul Rezwan from Pexels.)

the shell away from the wearer's skull. Some protective hats can also protect against electrical shock. Protective hats are made in the following types and classes:

- Type 1—Helmets with a full brim.
- Type 2—Brimless helmets with a peak extending forward from the crown.
- Class G—General service, limited voltage. Intended for protection against impact hazards. Used in mining, construction, and manufacturing.
- Class E—Utility service, high voltage. Used by electrical workers.
- Class C—Special service, no voltage protection. Designed for lightweight comfort and impact protection. Used where there is a possibility of bumping the head against a fixed object.

Be sure to follow the manufacturer's directions for adjusting the suspension in the hard hat to ensure a comfortable and secure fit. Don't stuff anything in the hard hat while wearing it, because it could compromise the gap between the shell and the suspension.

Refer to the manufacturer's recommendation if you are considering painting or affixing stickers on hard hats. Most manufacturers don't recommend painting a hard hat because the paint can damage the shell, but they're generally good with affixing stickers on the hard hat.

Some hard hats can be worn backwards as long as the suspension can be reversed. The ball caps should not be worn under the hard hat. Winter liners and cooling head-wear are generally fine to wear with hard hats as long as they can be worn seated down on the head and do not interfere with the suspension bands. Hard hats, such as cracks and dents, should be inspected for damage before each use. Damaged hard hats should be withdrawn and replaced.

When you work in an area where there is a possible risk of head injury from falling or flying objects, or where there is a risk of electrical shock and burns, you must wear a hard hat. You should take good care of your hard hat to prolong its life and safety. The safety provided by a hard hat is limited if it is damaged or if it is not used as intended.

1. Check your hard hat every day for signs of dents, cracks, or penetration. If any of these signs of damage are found, do not use the helmet. This inspection should include the shell, suspension, headband, sweatband, and any accessories.

2. Do not store or carry your hard hat on the rear-window shelf of a car. Sunlight and high heat can degrade the helmet. Long periods of exposure to sunlight can lead to damage from ultraviolet rays. Signs of damage include dulling, chalking, crazing, or flaking on the shell surface.

3. Clean your hard hat with warm, soapy water. Scrub and rinse the shell with clean, hot water (about 140°F). Inspect the shell for damage after the shell has been cleaned.

4. Don't paint your hard hat. Some types of paints and thinners can damage the shell or weaken the helmet. Consult with the manufacturer of hard hats for recommendations on the use of any form of solvent to clean paint, tars, oils, or other helmet materials.

You might think the last thing you'd ever do would be to wear a hard hat at home, but there are some situations when wearing head protection is a good idea. A hard hat can protect you from being injured by falling objects or by hitting your head on fixed objects if you need to:

- Prune trees
- Work on the ground while someone is up doing roof repairs
- Work under someone who is doing home repairs or remodeling while up on a scaffold ladder
- Work in tight areas such as a crawlspace or attic

2.4 Hearing Protection
2.4.1 Basics
A sound source sends out vibrations into the air. These vibrations are called sound waves. The ear changes the energy in sound waves into nerve impulses that travel to the brain and are then interpreted. Sound is measured by its frequency and intensity. Frequency is the pitch (high or low) of a sound. High-frequency sound can be more damaging to your hearing than low-frequency sound. Intensity is the loudness of a sound. Loudness is measured in decibels (dB). Intensity that exceeds an average of 85 dB over an 8-hour day may cause hearing loss. According to OSHA standards, employers must provide hearing protection when workers are exposed to noise at levels of 85 dB or more averaged over an 8-hour period.

In general, there are three types of noise:

1. Wide band is a noise that is spread across a wide range of frequencies. Examples are the noise produced in most of the manufacturing settings and the operation of most internal combustion engines.

2. Narrow band noise is restricted to a narrow range of frequencies. Example includes noise from various kinds of power tools, circular saws, fans, and planers.

3. Impulse noise is composed of temporary "beats" that can occur on-and-off repeating patterns. Jack hammers, or power, or punch presses are good examples of tools that cause impulse noise.

How you're affected by noise depends on several things—loudness and frequency, length of exposure, and even your age and health. A temporary hearing loss can occur from a short exposure to loud noise, but your hearing soon recovers when the noise stops. If the level is high enough for long enough, gradually it can cause permanent hearing loss.

Hearing loss due to noise in the workplace is frequently ignored because it usually takes place over a long period of time and may not be readily apparent until the damage is done. There's no visible wound or other sign of injury.

But, too much noise can cause a variety of problems. It can make you tired and irritable from the strain of talking or trying to listen over loud sounds. You may not be able to hear important instructions about work or safety due to excessive noise.

Evidence exists that other physical damage may occur because of the way the body reacts to noise.

Studies link noise to high blood pressure, ulcers, headaches, and sleeping disorders. Add these potential threats to the obvious damage that noise can do to the hearing.

2.4.2   OSHA Regulation

Hearing protection must be provided and used when noise levels or duration cannot be reduced to those specified in limit by OSHA standard number 1926.52(d). The limits vary from 90 to 115 dB depending on the loading duration. Acceptable hearing protection included earmuffs and earplugs to be fitted to the ear. Earplugs must be individually fitted by a competent person. Plain cotton is not an acceptable protective device, according to OSHA.

2.4.3   Hearing Protective Devices

Various kinds of hearing protective devices are available for use in the workplace. Your employer will determine what types of hearing protection devices (HPDs) are needed. Hearing protectors do not completely block sound but reduce the amount of sound that enters the sensitive areas of the ear. By doing so, they offer some protection.

Enclosure.   The enclosure type of hearing protection completely surrounds the head like an astronaut's helmet. This type of protection is not too popular due to its cost and discomfort.

Earplugs.   Earplugs fit in the ear canal. They come in three forms:

1. **Custom-molded** earplugs are made for specific individuals, molded to the exact shape of that person's ear. Made from soft silicone rubber of plastic, they are reusable.

2. **Molded inserts**, often called premolded, are also made from soft silicone rubber or plastic. They are reusable and should be kept very clean. Use warm, soapy

water to clean them after each use, rinse off the soap, and store them in a clean carrying case.

3. **Formable** plugs are usually made of foam rubber; they are disposable.

Canal Caps.   Canal caps (also known as superaural) seal the external edge of the ear canal to reduce sound. The caps are made of a soft, rubber-like substance and are held in place by a headband. This type of ear protection is a good alternative for those who can't use earplugs or for workers who enter and leave high noise areas frequently during the course of their work day.

Earmuffs.   Earmuffs (also known as circumaural) fit over the whole ear to seal out noise. A typical muff is made up of three basic parts—cups, cushions, and headband as shown in Fig. 2.3. The cups are made of molded plastic and are filled with liquid, air, or foam. The headband simply holds the cups against the head. It may be worn over the head, behind the neck, or under the chin. There are also specialty earmuffs for different job requirements. Dielectric muffs have no metal parts for those workers exposed to high voltages. Electronic earmuffs reduce hazardous noise, but magnify wanted sounds like voices. Folding earmuffs are designed for use in situations where protection isn't required full-time but must be quickly available when needed. Cap-mounted muffs are attached directly to hard hats.

2.4.4   Effectiveness

In general, earplugs can reduce noise reaching the ear by 25–30 dB. Earmuffs can reduce noise 20–25 dB. Combinations of the two protectors can give 3–5 dB more protection. No matter what type of protection device you select, remember that the only effective hearing protector is the one that you wear!

Figure 2.3 Construction worker with earmuffs.
(Photo courtesy of 123RF.)

It is very important to keep track of your hearing by having it tested periodically. An audiometric test is a procedure for checking a person's hearing. Employers with facilities where noise exposure equals or exceeds an average of 85 dB over an 8-hour day are required to provide their employees with audiometric testing.

A trained technician uses an instrument (an audiometer) to send sounds (tones) through headphones. The person being tested responds to the test sounds when they are first heard. The chart that records responses to the test sounds is called an audiogram.

This test checks hearing ability so that any hearing loss can be identified and dealt with properly and promptly.

2.4.5   Selection of Hearing Protection

The selection of the right hearing protection depends on several factors:

- **The noise hazard**—what noise levels will you be dealing with?
- **Frequency of the noise**—will it be low pitch or high pitch? [Some earplugs or muffs reduce the force of noise (attenuate) better at lower frequencies than at the higher frequencies.]
- **Fit and comfort**—the protective devices must fit properly and be comfortable enough to wear as long as they are needed.
- **Noise Reduction Rating or NRR**—all hearing protectors carry a label indicating the NRR; a higher number on the label means more effectiveness.

2.4.6   Work at Working Safely

You are ultimately responsible for protecting your own hearing. You have the most to lose if you suffer hearing loss as a result of on-the-job noise hazards. Let's review a few important reminders about hearing conservation:

1. Disposable earplugs may be more convenient to use than long-term use plugs, but make sure they fit you correctly so that they will be effective.
2. Employees whose noise exposure equals or exceeds 85 dB over an 8-hour period are required to have an annual audiometric test to check their hearing.
3. Keep hearing protectors in good operational order with routine maintenance and replacement of defective parts.
4. Don't use homemade hearing protectors such as wadded cotton or tissue paper. They don't work.
5. Wear ear protection at home for any noisy job like operating a chainsaw or using various kinds of shop equipment. And watch the volume on your stereo headphones.

The sounds of everyday life—nature, music, the voices of family and friends—all add pleasure and meaning to our lives. Value them enough to protect your hearing.

Hearing loss can result from exposure to any source of excessive noise, even if you aren't at work. For example, you can be exposed to high noise levels when you:

- Use power tools
- Attend concerts and sporting events

- Listen to loud music on head phones
- Shoot targets at gun range

It is just as important to wear hearing protection during these noisy activities as it is to wear it in a noisy workplace. Even if you are not exposed to noise at work, consider having periodic hearing tests to identify any hearing loss that may develop.

Example 2.2
The OSHA noise standard requires that a hearing conservation program is used to help the employer get the 8-hour TWA noise level below:

A. 85 dB

B. 65 dB

C. 55 dB

D. 90 dB

Answer: A

Example 2.3
What does research show about hearing loss?

A. Instantaneous but temporary

B. Painful and permanent

C. Gradual but temporary

D. Gradual but permanent

Answer: C

2.5   Eye and Face Protection

2.5.1   Basics

Eye and face protection (Fig. 2.4) is needed when workers are exposed to hazards from flying particles, molten metal, liquid chemicals, acids, caustic liquids, chemical gases, vapors, and light radiation. Every day, an estimated of 1000 eye injuries occur in American workplaces including construction. The financial cost of these injuries is enormous—more than $300 million per year is lost in production time, medical expenses, and workers compensation. No dollar figure can adequately reflect the personal toll these accidents take on injured workers. The BLS reports that nearly three out of every five workers injured were not wearing eye protection at the time of their accidents. About 40% of the injured workers were wearing some form of eye protection when the accident occurred, but often, it was not the correct eye protection for the job being done.

In the case of workers wearing corrective lenses, prescriptions should be incorporated into the eye protection design or the eye protection should be designed to be worn over the prescription lens.

Figure 2.4   Construction worker with eye and face protection.
(Photo courtesy of Cleyder Duque from Pexels.)

Face masks, safety goggles, and safety glasses should be selected on the basis of the hazard they were designed to protect workers from injury. It should fit snugly and comfortably and not interfere with a worker's movement. Protective eyewear should be durable, cleanable, and capable of being disinfected.

When welding operations are carried out by workers, make sure that they are supplied with the proper filter lens shade number depending on the type of welding they will perform. Safety goggles with an appropriate optical density based on the laser wavelength should be used when operating with lasers.

2.5.2   Causes of Eye Hazards

Here are some of the causes of eye injuries:

- Flying objects or particles
- Chemicals such as injurious gases, vapors, and liquids
- Dusts or powders, fumes, and mists
- Splashing metal
- Thermal and radiation hazards such as heat, glare, ultraviolet, and infrared rays
- Electrical arcing and sparks

The BLS found that almost 70% of the eye injuries accidents studied resulted from flying or falling objects or sparks striking the eye. Injured workers estimated that nearly three-fifths of tile objects were smaller than pin heads. Most of the particles were said to be traveling faster than hand-thrown objects when accidents occurred. Chemicals caused one-fifth of the injuries.

Example 2.4

Which one of these conditions does not present a potential eye hazard?

    A. Toxic gases, vapors, and liquids

    B. Dusts, powders, fumes, and mists

    C. Harsh lighting from industrial lamps

    D. Molten metals

Answer: C

2.5.3   Location of Eye Hazards

Potential eye hazards can be found in nearly every industry, but BLS reported that more than 40% of injuries occurred among craft workers, like mechanics, repairers, car painters, and plumbers. Over a third of the 40% injured operated machinery, such as assemblers, sanders, and grinding machine operators. Laborers suffered about one-fifth of the eye injuries. Almost half the injured workers were employed in manufacturing; slightly more than 20% were in construction.

2.5.4   How to Protect Eyes

Your employer must identify hazards in your work area and determine how best to prevent eye injuries. Eye hazards can be minimized by the use of machine guards and shields. PPE is needed if the hazards aren't eliminated. Your employer must also provide eyewash facilities if they are exposed to corrosive materials.

Equipment Guards.   Be sure to use any guards, screens, and shields attached to equipment. If the guards do not completely remove the hazards of the eyes and face, wear eye and face protection. Movable screens may be used to distinguish workers at one machine from workers at nearby workstations. Portable welding screens may be placed around welding areas to protect other workers from lighting and radiation.

Eyewash Facilities.   Eyewash stations should be located within 10 seconds of a hazard area. If employees accidentally get something into their eyes, they must go directly to the eyewash station and flush their eyes with water for at least 15 minutes, then get medical attention. The employee should hold the eyelids open and "look" directly into the water streams. They should not rub their eyes. Rubbing the eyes may scratch or embed particles. Employees should seek medical attention immediately.

Personal Protective Equipment.   A wide variety of safety equipment is available. Protective eye and face equipment must comply with ANSI guidelines and be marked directly on the piece of equipment (e.g., glass frames and lenses).

Safety Glasses.   The most common type of protective equipment for the eyes is safety glasses. They have protection against flying chips or particles, and may have tinted lenses for radiation and laser hazards. They may look like normal street-wear glasses; they are made of glass, plastic, or polycarbonate. But they are made much stronger than street-wear lenses, are impact-resistant, and come in prescription or nonprescription

(Plano) forms. Some styles of Plano safety glasses are designed to fit over prescription glasses. Safety frames are stronger than street-wear frames. Different styles of frames are available for different jobs. Safety glasses also are available with side shield guards. Side shields provide protection for the sides of your eyes. Eye-cup side shields provide more thorough eye protection from hazards that come from the front, side, top, or bottom.

Goggles.   Goggles are very similar to safety glasses but fit closer to the eyes. They can provide additional protection in hazardous situations involving liquid splashes, fumes, vapors, and dust. Some models can be worn over prescription glasses.

Face Shields.   Full-face protection is often required to guard against dust, sprays, molten metal, and chemical splashes. Face shields do not protect you from impact loading. When you need to wear a face shield, always wear other eye protection such as goggles or safety glasses under it.

Cleaning.   You should keep your eye protection safe and clean daily. Use lens cleaners recommended by the equipment manufacturer or mild soap and warm water. Strong solvents can damage the lens. Dirty, scratched, or cracked lenses reduce vision and seriously reduce protection. Replace damaged eye and face protection immediately.

Contact Lenses.   Most workers can wear their contacts on the job safely. It's important to note that your contacts should be worn along with extra eye protection. It's always a good idea to have a spare pair of contacts or prescription glasses with you in case the pair you normally wear is missing or damaged when you're working. You might also want to make sure your supervisor or plant first aid personnel know that you wear contacts, in the event of any injury on the job.

Absorptive Lenses.   Absorptive lenses are used to absorb or screen out unwanted light and glare. Most ordinary sunglasses do not provide the right glare protection. For welding or work with torches, goggles or helmets are available with filter lenses to shield the eyes from radiation and glare.

Better Training and Education.   The BLS reported that most workers were hurt while doing their regular jobs. Workers injured while not wearing protective eyewear most often said they believed it was not required in the situation. Even though the vast majority of employers furnished eye protection at no cost to employees, about 40% of the workers received no information on when and what kind of eyewear should be used.

Example 2.5
While Shelby is working with acid, it splashes onto her goggles and drips through the vent holes into her right eye. If her employer is OSHA-compliant, Shelby should be able to reach an eyewash station in _____ seconds.

    A. 5

    B. 10

    C. 15

    D. 30

    Answer: B

2.5.5 OSHA Regulations

Regulations have been issued by the Occupational Safety and Health Administration (OSHA) on personal protective equipment (PPE) in general, and on eye protection in particular. You can find these regulations in 29 CFR 1910.132-133. Your employer must assess the hazards in the workplace to determine if eye and face protection is needed. You must receive training in the proper use of eye and face protection and understand:

- When eye and face protection is necessary.
- What eye and face protection is necessary.
- How to properly put on, take off, adjust, and wear goggles, face shields, etc.
- PPE limitations.
- Proper care, maintenance, useful life, and disposal of eye and face protection.

You must show that you understand the training and can use eye and face protection properly before you will be allowed to perform work requiring its use.

Let's review the following important rules about eye safety:

- Match safety equipment to the degree of hazard present.
- Know what protective devices are available on the job and how they can protect you.
- Make sure machine guards are in place.
- Know the location and operation of emergency eyewash.
- Street-wear eyeglasses are not designed to be safety glasses and should never be used as such.
- Face shields should not be used alone, but always with other eye protection such as goggles or glasses.
- Make sure any safety device you use fits properly.
- Safety equipment should be maintained in good condition and replaced when defective.

Flying objects, chips, and particles; dust; mist; and chemical splashes can injure your eyes during home projects as easily as they can hurt you at work. A good practice when selecting eye protection for use at home is to:

- Wear safety glasses with side shields or wear eyeglasses to protect your eyes from objects, particles, or chips that can fly away during tool use or fall into your eyes as you reach overhead to work.
- Wear safety goggles to protect your eyes from splashes of liquids, mists, or dust.

Example 2.6
Face shields do not protect the face from _____.

    A. Dust

    B. Impacts

C. Splashes

D. Sprays

Answer: B

2.6   Hand Protection

2.6.1   Basics

Human hand is the most commonly used tool in almost every workplace. Hand protection (Fig. 2.5) is essential because our hands are exposed to so many hazards in the workplace. A number of disabling accidents on the job involve the hands. Without your fingers or hands, your ability to work would be greatly reduced. Human hands are unique. No other species in the world has hands capable of grabbing, carrying, shifting, and manipulating objects like human hands. You are one of the greatest assets, and, as such, you need to be protected and cared for. Select the right glove for the job. The hand is the most commonly injured body part at the jobsite. Gloves should fit snugly and comfortably while still allowing for full dexterity of the fingers. You don't want workers removing gloves because they don't fit properly or because they can't feel what they are doing when working with tools or equipment.

Figure 2.5   Construction worker with hand gloves in addition to other protections. (Photo courtesy of Cleyder Duque from Pexels.)

2.6.2   Potential Hand Hazards

Some potential hazards to the hand are listed below:

- Traumatic Injuries. An employee can suffer a traumatic injury to his or her hands in many ways:
  o Tools and machines with sharp edges can cut hands.
  o Staples, screwdrivers, nails, chisels, and stiff wire can puncture hands.
  o Getting your hands caught in machinery can sprain, crush, or remove your hands and fingers.
- Contact Injuries. Coming into contact with caustic or toxic chemicals, biological substances, electrical sources, or extremely cold or hot objects can irritate or burn one's hands. Toxic substances are poisonous substances, some of which can be absorbed through one's skin and enter the body.
- Repetitive Motion Injuries. Whenever you repeat the same hand movement over a long period of time, you run the risk of repetitive motion problems. Repetitive motion problems can appear as a numb or tingling sensation, chronic or acute pain, loss of gripping power in your hands, or in many other ways.
- Preventative Measures. Poorly maintained machinery, tools, sloppy work areas, and cluttered aisles all contribute to hand injuries. Good hygiene includes hand washing. Proper washing helps remove germs and dirt from your hands. Clean hands are less susceptible to infection and other skin problems such as contact dermatitis.

2.6.3   OSHA Regulations

OSHA requires your employer to select and provide you with hand protection when you are exposed to hazards such as skin absorption of toxic chemicals, cuts or lacerations, abrasions, punctures, chemical burns, or severe temperature conditions. Your employer must assess the hazards in the workplace to determine if hand protection is needed. Your employer must train you in the proper use of hand protection. You must know:

- When hand protection is necessary.
- What type is necessary.
- How to properly put on, take off, adjust, and wear gloves, mitts, or other protection.
- Hand protection limitations.
- Proper care, maintenance, useful life, and disposal of hand protection.

You must show that you understand the training and can use hand protection properly before you will be allowed to perform work requiring its use.

2.6.4   Hand Protective Device

The first defense in the battle to minimize hand injuries is engineering controls designed for equipment during manufacturing or used to alter the working environment to make it safe. Machine guards protect hands and fingers from moving parts and should not be changed or removed.

PPE can help reduce the frequency and severity of hand and finger injuries.

Gloves.   Gloves are perhaps the most commonly used type of PPE. They provide protection to fingers, hands, wrists, and forearms. Gloves should be selected to protect against specific hazards. Types range from common canvas work gloves to highly specialized gloves used in specific industries. There are gloves designed to protect against a number of hazards in the workplace: gloves for protection against cuts and lacerations, welding gloves, heavy-duty rubber gloves for working with concrete, chemical-resistant gloves for protection against burns, and insulated gloves for electrical work.

Leather.   Leather gloves are useful for handling rough or abrasive materials. Leather gloves are also useful for handling hot objects.

Canvas.   Canvas or cloth gloves are worn for general light-duty protection for handling rough objects.

Metal Mesh.   Metal mesh gloves are worn by workers in the meat-packing industry who work with sharp knives.

Chemical-Resistant.   Rubber, vinyl, or neoprene gloves are used when handling caustic chemicals like acids, cleansers, or petroleum products. These gloves are rated as being safe for use with certain kinds of chemicals. For chemical-resistance information, read the glove manufacturer's chemical resistance charts. They rate each glove material and how it withstands specific chemicals.

Wear only gloves that fit your hand. Too small gloves can tire your hands, and too large gloves are clumsy to work with. With great care, gloves should be worn near moving equipment or machinery parts. The gloves could get caught and pull your fingers or your hand into the machine. Gloves should be properly cared for and cleaned. They should be regularly inspected for changes in shape, hardening, stretching, or tearing.

2.6.5   Safe Removal of Gloves
The instructions for the safe removal of contaminated gloves are listed below:

- Pull one glove near your wrist toward your fingertips until the glove folds over.
- Carefully grab the fold and pull toward your fingertips. As you pull, you are turning the inside of the glove outwards.
- Pull the fold until the glove is almost off.
- To avoid contamination of your environment, continue to hold the removed glove. Completely remove your hand from the glove.
- Slide your finger from your glove-free hand under the remaining glove. Continue to slide your finger toward your fingertips until almost half of your finger is under the glove.
- Turn your finger 180 degrees and pull the glove outwards and toward your fingertips. As you do this, the first glove will be encased in the second glove. The inside of the second glove will also be turned outwards.
- Grab the gloves firmly, by the uncontaminated surface (the side that was originally touching your hand). Release your grasp of the first glove you removed. Pull your second hand free from its glove. Dispose of the gloves properly.

2.6.6   Work at Working Safely

Because you use your hands every day on the job, they can easily be injured. Keep these points in mind to protect your hands as you work:

- Protective equipment selection depends upon the nature of the hazards in your workplace.
- Gloves should fit you properly and be maintained in the same careful way as other safety equipment.
- Do not wear gloves where they could get caught in moving machine parts.

You can probably think of plenty of reasons to wear gloves to protect your hands at home. Here are a few examples:

- When handling rough or sharp objects or using hand tools, you can help prevent abrasions, blisters, and cuts by wearing leather or canvas gloves.
- If your hands can get wet from cleaning products, paints, or other chemicals, you can avoid getting dry, cracked skin or chemical burns by wearing rubber gloves.
- Hand protection is even necessary for something as simple as taking a hot pan off of the stove. Using a pot holder prevents painful burns.

2.7   Foot Protection

2.7.1   Basics

The 26 bones in the foot are arch-shaped to provide a wide, strong support for the body's weight. Because of how valuable our feet obviously are to us, we want to protect them from the hazards of the workplace as shown in Fig. 2.6.

2.7.2   Potential Hazards

Potential hazards to the foot are listed below:

- **Impact Injuries.** If you have ever stubbed your toe, you know that impact injuries can hurt. At work, heavy objects can fall on your feet. If you work around sharp objects, you might step on something sharp and puncture your foot.
- **Injuries from Spills and Splashes.** Liquids such as acids, caustics, and molten metals can spill onto your shoes and boots. These hazardous materials can cause chemical and heat burns.
- **Compression Injuries.** Heavy machinery, equipment, and other objects can roll over your feet. The result of these types of accidents is often broken or crushed bones.
- **Electrical Shocks.** Accidents involving electricity can cause severe shocks and burns.
- **Extremes in Cold, Heat, and Moisture.** If not protected, your feet can suffer from frostbite if you must work in an extremely cold environment. Extreme heat, on the other hand, can blister and burn your feet. Finally, extreme moisture in your shoes or boots can lead to fungal infections.

Figure 2.6   Construction worker with foot protection.
(Photo courtesy of Aidan Jarrett from Pexels.)

- **Slipping.** Oil, water, soaps, wax, and other chemicals can cause you to slip and fall.
- **Chemicals.** Chemicals and solvents corrode ordinary shoes and can harm your feet.

Some preventative measures for foot safety are listed below:

- Avoid poorly maintained machinery, tools, sloppy work areas, and cluttered aisles; all these contribute to foot injuries.
- Select and use the right kind of footwear for the job you are going to be performing. Avoid footwear made of leather or cloth if you work around acids or caustics. These chemicals quickly eat through the leather or cloth and can injure your feet.
- Select footwear that fits properly.
- Inspect your footwear before you use it. Look for holes and cracks that might leak.
- Replace footwear that is worn or torn.
- After working with chemicals, cleanse your footwear appropriately to rinse away any chemicals or dirt before removing footwear.

- Avoid borrowing footwear; footwear is personal protective equipment.
- Store footwear in a clean, cool, dry, ventilated area.

When most people think of foot protection on the construction site, they think of steel or composite-toed boots. These protect workers from tools, materials, or equipment falling or dropping on the foot.

Composite-toed and steel-toed footwear both offer a high level of protection with steel-toed boots getting the slight edge. Composite-toed boots are lighter and don't conduct heat, which make them preferable if you are working outside in extreme heat or cold temperatures. They also don't conduct electricity, which is good if you are doing electrical work or work around live wires.

When selecting foot protection, you also want to choose slip-resistant shoes to protect against slip and fall, and to protect against sharp objects, such as misplaced nails, with puncture-resistant soles.

Example 2.7
Which one of these workplace conditions does not require employers to provide foot protection to their employees?

    A. Heavy objects that might roll onto feet

    B. Sharp objects that might penetrate shoes

    C. Molten metal that might splash on feet

    D. Loose surface that might cause slipping

Answer: D

2.7.3 OSHA Regulations

The OSHA has developed regulations that specify foot protection to keep your feet safe at work. These regulations are listed in 29 CFR 1910.132 and 136. Your employer must analyze the hazards at the workplace in order to decide whether foot protection is needed. Your employer needs to train you in the proper use of foot protection. You must know:

- When protective footwear is necessary.
- What footwear is necessary.
- How to properly put on, take off, adjust, and wear protective footwear.
- Protective footwear limitations.
- Proper care, maintenance, useful life, and disposal of protective footwear.

You must show you understand the training and can use safety shoes properly before you will be allowed to perform work requiring their use.

2.7.4 Foot Protective Shoe

Safety shoes are available in many varieties to suit very specific industrial applications. Here are the descriptions of some types of safety footwear.

Safety Shoes.   Standard safety shoes have toes that meet standard testing requirements. Steel, reinforced plastic, and hard rubber are used for safety toes, depending on their intended use. Such shoes are used by workers in many types of general industries.

Metatarsal Guards.   Shoes with metatarsal or instep guards protect the upper foot from impacts. For these shoes, the metal guards stretch over the foot rather than just over the toes.

Conductive Shoes.   Conductive shoes allow the static electricity which builds up in the wearer's body to drain to the ground harmlessly. By preventing the accumulation of static electricity, most conductive shoes keep electrostatic discharge from igniting sensitive explosive mixtures. Workers often wear these shoes in munitions facilities or refineries. Do not use these shoes if you work near exposed electrical circuits.

Chemical-Resistant Safety Boots.   Rubber or plastic safety boots offer protection against oil, water, acids, corrosives, and other industrial chemicals. They are also available with features such as steel-toe caps, puncture-resistant soles, and metal guards. Some rubber boots are designed to be pulled over regular safety shoes.

Electrical Hazard Shoes.   Electrical hazard shoes offer insulation from electrical shock hazards from contact with circuits of 600 Volts or less under dry conditions. These shoes are used in areas where employees work on live or potentially live electrical circuits. The toe box of the shoe is insulated, so there is no exposed metal. These shoes are most effective when dry and in good repair.

Sole Puncture-Resistant Footwear.   Puncture-resistant soles in safety shoes protect against hazards of stepping on sharp objects that can penetrate standard shoe soles. They are used primarily in construction work.

Static Dissipative Shoes.   Static dissipative footwear is designed to reduce the accumulation of excess static electricity by conducting the body charge to the ground while maintaining some resistance to electrical shock. Such shoes are worn by workers in the electronics industry, where static items can be harmful.

Foundry Shoes.   Foundry shoes are used by welders and molders in foundries or steel mills where there is a risk of hot splashes of molten metal or flying sparks. Instead of laces, they have elastic gores to hold the top of the shoe close to the ankle. They can then be removed quickly in case hot metal or sparks get inside.

Add-On Foot Protection.   Metatarsal guards and shoe covers can be attached to safety shoes or boots for greater protection from falling objects.

Rubber spats protect feet and ankles against chemicals. Puncture-proof inserts made of steel can be slipped into safety shoes or boots to protect against underfoot hazards. Strap-on cleats fastened to shoes can provide greater traction in ice or snow.

2.7.5   Work at Working Safely

Safety shoes can prevent serious, even disabling injuries at relatively low cost. As a review, let's look at some of the excuses that keep workers from using safety footwear. In each case, we'll "stamp out" the excuse with the facts!

They're ugly!—Some people are willing to sacrifice safety for style. However, safety shoes are now available in fashionable styles, ranging from running shoes to western boots.

They're too expensive!—When the cost is spread out over the life of the shoe, the price of safety is only pennies a day.

They're not comfortable!—Safety shoes should fit just like any other shoe.

They're too clumsy for climbing!—The key here is to find a shoe suitable for working on ladders or scaffolding. Anyone who works in those situations should wear a shoe with a defined heel and good traction.

Steel toe caps will cut off my toes if crushed!—Toe caps are designed to give a "buffer zone" of space over the toes in case they are crushed.

I don't know where to buy them!—Some stores specialize in the sale of safety shoes. Shoe mobiles, in-plant stores, and catalog order programs bring the shoe right to you on the job.

I'm a safe worker; I won't have an accident!—"It can't happen to me" is a dangerous myth that has been proven wrong again and again. So don't take a chance with your two good feet. Obtain proper safety footwear and wear it all the time on the job.

Consider wearing protective shoes when you are away from work if you engage in activities that could lead to foot injuries. For example, if you're doing home remodeling, moving stones for a landscaping project, or stacking firewood, you would benefit from the protection provided by safety shoes. And if you don't do something that requires exposure to falling or breaking objects, make sure to wear sturdy footwear while you're doing stuff like operating a lawn mower, climbing a ladder, or walking where sharp items can be placed on the floor.

Example 2.8

Which one of the following potential hazards to feet is most uncommon in the workplace?

    A. Frostbite

    B. Compression injuries

    C. Impact injuries

    D. Electrical shocks

Answer: A

2.8   Respirator

A respirator is a protective device that covers the worker's nose and mouth or the entire face and head to keep airborne contaminants out of the worker's respiratory system and to provide a safe air supply.

2.8.1   Types of Respirators

There are two major categories of respirators:

1. Air purifying respirators

2. Supplied air respirators

Air purifying respirators include:

- Air purifying disposable particulate masks
- Air purifying half mask respirators
- Air purifying full face mask respirators
- Gas masks
- Powered air purifying respirators

Supplied air respirators include:

- Airline respirators
- Emergency escape breathing apparatus
- Self-contained breathing apparatus (SCBA)

2.8.2 Selecting the Correct Respirator

The first step in selecting the correct respirator is to determine the level of hazard that is posed by the environment in which one will be working. To do this, one must be able to answer five basic questions:

1. What type of contaminant is present?
2. What is the form of the contaminant?
3. How toxic is the contaminant?
4. What is the concentration of the contaminant?
5. What will be the length and duration of the exposure?

Employee should always work with their supervisor or safety professional to determine the correct answers to these questions. Without the technical knowledge to make correct decisions, it's best to consult with an industrial hygienist or safety professional who is trained to provide professional guidance on proper respirator selection and use.

Example 2.9

What is the main purpose of a respirator?

A. To keep air contaminants out of a person's respiratory system
B. To enable a person to breath in a vacuous atmosphere
C. To enable a person to respire efficiently
D. To provide oxygen in hazardous gas environments

Answer: A

2.9   Personal Training

Employers must provide formal training (Fig. 2.7) to all employees who may be exposed to ergonomic hazards to inform them of the hazards associated with their work and of the tools, machinery, and equipment they use. Information that must be included in the training includes specific risk factors, their causes, recognizing and reporting symptoms, and the prevention of these occurrences.

2.9.1   Job-Specific Training

All new employees and those assigned new tasks must be made aware of the specific risks associated with a particular job before they start their work. A practical demonstration should be arranged in order to show the employees how to use all the tools and equipment properly and how to carry out all procedures efficiently. The initial training program should incorporate the following:

- How to use, handle, and maintain all tools, machinery, and equipment that have to be used as a part of the job.
- How to use the special tools, if any, associated with a particular job.
- How to use safety equipment and guards along with personal protective equipment to ensure safety.
- How to properly lift and the proper procedures to follow when an object is too heavy to lift safely without assistance.

FIGURE 2.7   Training operation in a workplace.
(Photo courtesy of Pixabay from Pexels.)

2.9.2 Training for Supervisors

In addition to training received by employees, supervisors should be provided with additional training to help them recognize early signs and symptoms of risk, hazardous work practices, how to correct such practices, and how to reinforce the ergonomic program.

2.9.3 Training for Managers

Apart from employees and supervisors, managers must also be made aware of their responsibilities to implement ergonomic principles that ensure the safety and health of all employees. They must also be familiar with the problems and risks associated with all tasks.

2.9.4 Training for Engineers and Maintenance Personnel

On-site engineers and maintenance workers must also be trained in order to recommend and enforce the best available machinery, equipment, designs, work practices, and tools to minimize the risk of injuries and bodily damage to staff.

2.9.5 Training for PPE

Workers ought to be trained on the proper selection and use of PPE and be aware that the PPE is essential for different conditions and tasks of the workplace. Workers assigned to wear PPE should know how to put it on and take it off, as well as how to adjust it to ensure proper fit. Let workers be aware of the capabilities and shortcomings of the PPE that they are required to wear.

Workers should be able to properly inspect PPE before each use to determine if it is damaged and needs to be replaced or repaired. They should also be taught the proper way to maintain and care for the equipment. PPE is an important aspect of keeping workers safe on the construction site. When engineering and administrative measures malfunction or are not feasible, PPE is essential to protecting employees from injuries in the event of accidents. The PPE should be comfortable and fit properly so that it will not conflict with the worker's ability to do his job comfortably. Safety managers and supervisors should routinely inspect the jobsite to make sure workers are wearing the assigned PPE and should determine if additional equipment is needed.

2.10 Summary

Personal protective equipment (PPE) protects against biological, chemical, or physical hazards on the job. The employer will assess the workplace to determine if hazards are present, or likely to be present, which may necessitate the use of PPE. This protective equipment is often the final solution for hazards that cannot be eliminated through the use of engineering controls. Also, wherever possible, PPE will be provided for the exclusive use of a single employee.

2.11   Multiple-Choice Questions

2.1   If your job exposes you to the risk for skin absorption of harmful substance:

A.   You need to wear hand protection.

B.   You can only work a limited number of houses.

C.   You should wash your hand only at the end of your work.

D.   OSHA classifies it as a high hazard occupation.

2.2   If you need to wear hand protection, you must know:

A.   The limitation of hand protection

B.   The type of hand protection needed

C.   When you need to wear the hand protection

D.   All of the above

2.3   You must show that you understand the limitation of hand protection:

A.   Whenever you tear a glove

B.   Every time you get a new pair of gloves

C.   Before you're allowed to do work that requires the use of hand protection

D.   All of the above

2.4   Which of the following ways can cause a traumatic injury to an employee's hands?

A.   Tools and machines with sharp edges can cut hands.

B.   Staples, screwdrivers, nails, chisels, and stiff wire can puncture hands.

C.   Getting your hands caught in machinery can sprain, crush, or remove your hands and fingers.

D.   None of the above.

2.5   Who is responsible for providing and paying for most PPE?

A.   OSHA

B.   The employee

C.   The employer

D.   The safety manager

2.6   Who must assess workplace hazards to determine if foot protection is needed?

A.   Each employee

B.   The employer

C.   OSHA

D.   The insurance company

2.7   You don't need to know about the limitations of protective footwear if you:

A.   Wore these safety shoes on a previous job

B.   Don't have any need to wear protective footwear

C.   Wear electrical safety protective footwear

D.   Aren't responsible for selecting protective footwear

2.8   Before you can be allowed to perform work requiring protective footwear, you must:

A.   Remove all of hazards from the work area

B.   Try on at least three different styles of safety shoes

C.   Show you understand your training and can properly use the protective footwear

D.   Test the effectiveness of the protective footwear

2.9 You must know how to properly put on, adjust, and wear eye and face protection:

    A. Only if you are new on the job

    B. Only after you have had an eye injury

    C. If you need to use eye and face protection on the job

    D. Only if you are the person assigned to clean eye and face protection

2.10 Before you are allowed to do the job that requires the use of eye protection:

    A. You must receive training

    B. You must show that you understand the training

    C. You must show that you can use the eye protection

    D. All of the above

2.12 Practice Problems

2.1 List the type of protective hats made.

2.2 List and discuss the popular types of hearing protection.

2.3 List some causes of eye injuries.

2.4 List some protective equipment of eye injuries.

2.5 List and discuss some potential hazards to the hand.

2.6 List the instructions for the safe removal of contaminated gloves.

2.7 List and discuss some potential hazards to the foot.

2.8 List some preventative measures for foot safety.

2.13 Critical Thinking and Discussion Topic

Two construction employers are discussing. One employer says construction safety is the responsibility of the employees. They should work safe while they work. The company may give guidelines on the first day of the employment. Soon after that it should be their responsibilities to follow the safety practices. The employer should not force safety practice on a regular basis. The other employer says the employer should force safety practice on a regular basis and it is the responsibility of the employer to make sure the employees are following it. Based on this discussion, give your opinion and the logics of your opinion.

CHAPTER **3**

Responsibility, Culture, and Ethics

(Photo courtesy of Pexels.)

3.1 General

Construction industry involves different types of personnel, such as laborers, owners, etc., of different backgrounds, motives, and attitudes. Factors involved include human behavior, different construction sites, the difficulties of works, unsafe safety culture, dangerous machinery and equipment being used, and noncompliance to the various set procedures. Study shows that an accident and injury at the worksite is often the result of workers' behavior, work practices or behavior, and work culture (Mahmood and Mohammed, 2006). Safety and health culture are more related to workers' safety practices. It is expected that an efficient safety management system and ethical practice will become a culture in the construction industry, involving all parties.

3.2   Company and Safety

The effectiveness of a company safety and health program is directly related to management's commitment to safety. Company leaders must establish the safety culture within their companies by stressing the significance of safety in meetings and their visits to project sites. We will also ensure that adequate resources are provided to implement a comprehensive company-wide safety program that mandates the creation of a detailed written injury-and-illness prevention plan for each project site. A company safety and health program should contain the following elements:

- Hazard analysis
- Hazard prevention—actions to be taken to keep workers safe
- Policies and procedures for working safely—rules to be followed by employees and subcontractors
- Employee training—type and frequency of training
- Continual workplace inspection—walk-around inspections of jobsites
- Enforcement of company safety policies and procedures—steps to be taken when violations occur

---

Example 3.1

The effectiveness of a company safety and health program is directly related to:

A. Management's commitment to safety

B. Management's commitment to workers

C. Workers' commitment to safety

D. Management's commitment to community

Answer: *A*

The best answer is the management's commitment to safety. The second best choice is (C)— workers' commitment to safety. Other two options are true as well.

OSHA requires that at least one person on each jobsite be designated as the *competent person* responsible for regular site inspections to conform to the required safety practices and procedures. In order to be considered competent, a person must be aware of the specific types of work to be carried out at the workplace, as well as of all required companies and legally mandated safety and health practices and procedures. The competent person may have other tasks; however, he or she is assigned specific responsibility for the safety of the project site. Everyone on a project from senior management to the newest employee has responsibility for safety:

- Project managers and superintendents are responsible for establishing and enforcing safety policies and procedures, providing necessary resources, and effectively communicating safety and health information to all people working on the site, both the contractor's employees and the subcontractors (Fig. 3.1).

FIGURE 3.1 Ensuring safety is a group work.
(Photo courtesy of 123RF.)

- Field supervisors implement and enforce these safety policies and procedures, as well as perform hazard analysis, employee training sessions, accident investigations, and safety inspections.
- Employees are responsible for following established safety procedures, reporting safety risks, and attending safety training and meetings.

Project leaders and field supervisors must set the standard regarding safety on their projects and enforce safety standards at all times. A continual safety awareness program focused on reducing accidents is essential. Frequent (at least weekly) jobsite safety inspections should be conducted to identify hazards and ensure compliance with job-specific safety rules. Safety will be addressed at every project meeting. Foremen should hold day-to-day safety meetings to review the safety aspect of the tasks to be carried on that day. They play a critical role in establishing and maintaining a safe jobsite. Safety is day-to-day, hands-on duty of all craft labor supervisors on the project. Supervisors must conduct training to ensure that their workers are knowledgeable of:

- The construction company's safety policies and procedures
- Specific accident prevention plans developed for the project
- Housekeeping procedures to be followed on the project
- Emergency procedures for the project site
- The proper use of all equipment to be employed on the project
- The identification of any hazardous materials to be used on the project and proper procedures for handling them

Safe construction procedures and techniques should be identified for each phase of the work to minimize the potential for accidents. Some of the ways to reduce the risk created by a hazard are as follows:

- Modify construction techniques to eliminate or minimize the hazard.
- Guard the hazard, for example, by fencing in the site.
- Provide a warning, such as back-up alarms on mobile equipment or warning signs.
- Provide special training.
- Equip workers with personal protective equipment (PPE), such as hard hats and hearing protection.

The hazards associated with each phase of work and selected mitigation strategies should be discussed with both contractor and subcontractor work crews prior to allowing them to start work. Many superintendents require daily safety meetings prior to allowing the workers to start work. These meetings address risks and mitigation strategies for the work to be performed that day. The meetings must address general housekeeping policies, emergency procedures, proper use of equipment, as well as any hazardous materials present and proper handling procedures for these materials.

In the event an accident does occur on the project site, a thorough investigation is needed to determine the cause of the accident. This information is needed to devise procedures to minimize the potential for a future recurrence. Supervisors generally are required to complete an accident report describing the results of their investigations. Most construction companies have standard formats for an accident report, but, in general, they require answering the following questions:

- What was the injured worker doing at the time of the accident?
- What were the jobsite conditions?
- What equipment was being used?
- Was the injured worker trained on the proper use of the equipment?
- Was the work being performed in accordance with company safety policies and procedures?
- Was the injured worker wearing all required PPE?
- What were the primary causes of the accident?

Many construction companies employ dedicated safety professionals who develop company safety policies and procedures, carry out training programs, and inspect project sites. These individuals, however, are technical advisers to the company leaders, who bear the ultimate responsibility for creating and maintaining an effective safety program. A company safety and health manager would have staff responsible for establishing the company's overall safety and health program. While the safety and health manager develops the program, it is the responsibility of company supervisors to implement it. In addition, company safety and health manager typically performs the following tasks:

- Conducting safety audits of construction sites and project offices
- Inspecting construction operations
- Conducting safety and health training

- Analyzing potential hazards
- Conducting accident investigations

Under the OSHA law, employers have a responsibility to provide a safe workplace. This is a short summary of key employer responsibilities:

- Provide a workplace free from serious recognized hazards and comply with standards, rules, and regulations issued under the OSH Act.
- Examine workplace conditions to make sure they conform to applicable OSHA standards.
- Make sure employees have and use safe tools and equipment and properly maintain this equipment.
- Use color codes, posters, labels, or signs to warn employees of potential hazards.
- Establish or update operating procedures and communicate them so that employees follow safety and health requirements.
- Employers must provide safety training in a language and vocabulary that workers can understand.
- Employers with hazardous chemicals in the workplace must develop and implement a written hazard communication program and train employees on the hazards they are exposed to and proper precautions (and a copy of safety data sheets must be readily available).
- Provide medical examinations and training when required by OSHA standards.
- Post, at a prominent location within the workplace, the OSHA poster (or the state-plan equivalent) informing employees of their rights and responsibilities.
- Report to the nearest OSHA office all work-related fatalities within 8 hours, and all work-related inpatient hospitalizations, all amputations, and all losses of an eye within 24 hours.
- Keep records of work-related injuries and illnesses. Employers with 10 or fewer employees and employers in certain low-hazard industries are exempt from this requirement.
- Provide employees, former employees, and their representatives access to the log of work-related injuries and illnesses.
- Provide access to employee medical records and exposure records to employees or their authorized representatives.
- Provide to the OSHA compliance officer the names of authorized employee representatives who may be asked to accompany the compliance officer during an inspection.
- Not discriminate against employees who exercise their rights under the Act, such as Whistler Protection.
- Post OSHA citations at or near the work area involved. Each citation must remain posted until the violation has been corrected, or for three working days, whichever is longer. Post abatement verification documents or tags.
- Correct cited violations by the deadline set in the OSHA citation and submit required abatement verification documentation.

OSHA encourages all employers to adopt a safety and health program. Safety and health programs, known by a variety of names, are universal interventions that can substantially reduce the number and severity of workplace injuries and alleviate the associated financial burdens on U.S. workplaces. Many states have requirements or voluntary guidelines for workplace safety and health programs. Also, numerous employers in the United States already manage safety using safety and health programs, and we believe that all employers can and should do the same. Most successful safety and health programs are based on a common set of key elements. These include management leadership, worker participation, and a systematic approach to finding and fixing hazards.

Example 3.2
To be competent, a person must be:

   A. Aware of the specific types of work to be carried out at the workplace

   B. Aware of the design rules, codes, and ethical responsibility of a construction site

   C. Knowledgeable and available all times in a construction site

   D. Jack of all trades

Answer: *A*
In order to be considered competent, a person must be aware of the specific types of work to be carried out at the workplace, as well as of all required companies and legally mandated safety and health practices and procedures.

3.3   Contractors and Safety

Contractors, as the owners of the company, bear the ultimate responsibility for safety and health on their job sites. Contractors are responsible for the following:

- Setting a prosafety tone for the company

- Establishing a complete commitment to safety and health from the top down

- Ensuring that sufficient resources are provided to support a comprehensive, company-wide safety effort

By letting their managers and supervisors know that they expect safe and healthy workplaces, contractors set a tone for the company that ensures high priorities for safety and health. By introducing safety and health into the strategic plan of the company, contractors can establish the necessary commitment on the part of all workers (Fig. 3.2). By providing the resources necessary to support company-wide safety and health programs, contractors ensure that the safety and health program is properly staffed and sufficiently funded.

3.4   Supervisors and Safety

Supervisors are solely responsible for the safety and health of staff and other personnel that they direct or oversee; it's not just a position that solely assigns tasks. Supervisors must ensure a safe and healthful workplace for employees. Employees must be able to

FIGURE 3.2 Understanding the workplace safety.
(Photo courtesy of Aleksey from Pexels.)

report unsafe or unhealthful workplace conditions or hazards to a supervisor without fear of reprisal.

The following is a list of primary responsibilities that supervisors have in the area of occupational safety and health for all employees under their supervision.

Conduct Orientation and Training of Employees: Instruct and train employees so that they can do their work safely (Fig. 3.3). Know what PPE is needed for each task and how it must be properly used, stored, and maintained. When there are mandated safety training courses, ensure that your employees take them and that they are appropriately documented.

Enforce Safe Work Practices: It's the supervisor's responsibility to enforce safe work practices and procedures; failure to do so is an invitation for accidents to occur. Workers must be encouraged to identify unsafe or unhealthful workplace conditions or hazards and absolutely not be disciplined for doing so. The supervisor should ensure that the appropriate PPE and clothing are available, worn when required, and inspected and maintained.

Correct Unsafe Conditions: Supervisors must take immediate steps to correct unsafe or unhealthful workplace conditions or hazards within their authority and ability to do so. When an unsafe or unhealthful workplace condition or hazard cannot be immediately corrected, the supervisor must take temporary precautionary measures. Supervisors

Figure 3.3   Supervisors are directly responsible for the safety and health.
(Photo courtesy of ThisIsEngineering from Pexels.)

must follow up to ensure that corrective measures are completed in a timely manner to address the hazard.

Prevent Lingering Unsafe or Unhealthful Workplace Conditions or Hazards: Many near-miss incidents are caused by unsafe or unhealthful workplace conditions or hazards. It's the supervisor's responsibility to train and periodically remind employees of what to look for and how to correct or report unsafe conditions or hazards. If a hazard is identified, the supervisor must act.

Investigate Workplace Accidents: Supervisors are responsible for conducting accident investigations and for ensuring that all occupationally injured employees report to the medical service immediately.

Promote Quick Return to Work: Employees must be encouraged to return to work as soon as possible. The longer an employee is away from work, the less likely he or she will actually return. When possible, light or limited duties should be identified and considered, to assist in returning the employee to work.

3.5   Office Personnel and Safety

In companies of sufficient size, contractors have a staff of management personnel and other professional personnel. Management positions may include the following:

- Project managers
- Financial managers and accountants

- Marketing representatives
- Purchasing agents
- Cost accountants
- Human resources personnel
- Office managers

Other professional personnel may include:

- Engineers
- Designers and drafting technicians
- Estimators
- Expeditors
- Architects

Managers and other professionals are responsible for setting a good precedent surrounding safety and health and for transforming the contractor's commitment to everyday practice. This is accomplished by the following type of actions:

- Developing job descriptions that make safety and health part of every employee's job
- Developing performance appraisal forms that contain safety and health criteria
- Rewarding safe behavior on the job by making it an important factor in promotions and pay raises
- Developing work procedures that emphasize safety and health
- Recognizing safe work behavior as a part of company incentive programs
- Ensuring that the company has a comprehensive and effective safety and health program
- Keeping up-to-date with the latest Occupational Safety and Health Administration (OSHA) standards and regulations related to construction
- Effectively communicating safety and health information to all employees and subcontractors

3.6   Employee and Safety

The best efforts of contractors, managers, other office personnel, supervisors, etc., may not work well if the employees do not help practicing safety. Employees must:

- Work safely to ensure their safety and health
- Make sure their actions do not cause injury or harm to others
- Follow their employer's instructions on safety and health—ask for assistance if they do not understand the information
- Take care of any protective clothing and equipment in the way they have been instructed and report any concerns about it

- Report any hazards, injuries, or ill health to their supervisor or employer
- Cooperate with their employer when they require something to be done for safety and health at the workplace

The employees' "duty of care" responsibility also applies to contractors, labor hire workers, apprentices, and workers in other labor arrangements.

3.7 Safety-First Culture

Safety-first culture is an organizational culture that positions the importance of safety beliefs, values, and attitudes at the forefront of the task, and these attitudes are reflected by the majority of employees within the company or at the workplace. The importance of workplace safety can't be stressed enough. The following are some reasons:

- Injury
- Death
- Corporate financial loss
- Property damage
- Worker productivity increases
- The service or quality of the product improves
- Corporate reputation/public relations improves

The following 10 steps are followed to establish safety-first culture in construction industry (Goetsch, 2018):

1. Understand the need for a safety-first corporate culture.
2. Assess the current corporate culture as it relates to safety.
3. Plan for a safety-first corporate culture.
4. Expect appropriate safety-related behaviors and attitudes.
5. Model the desired safety-related behaviors and attitudes.
6. Orient personnel to the desired safety-first corporate culture.
7. Mentor personnel in the safety-related behaviors and attitudes.
8. Train personnel in the desired safety-related behaviors and attitudes.
9. Monitor safety-related behaviors and attitudes at all levels.
10. Reinforce and maintain the desired safety-first corporate culture.

3.8 Ethics and Safety

A construction company is judged by the integrity that it demonstrates in conducting its business and how it treats its employees. A key component of a company's reputation is the ethics of the company leaders and the company's employees. This is particularly true with respect to construction safety. Ethics are the moral standards that individuals use to make business and personal decisions. The moral standards that we use to control our decision-making process make up our ethics, helping us decide what is right and what is wrong.

Ethics involves determining what is right in a given situation and then having the courage to do what is right. This means taking action if an unsafe act is observed. Each decision has consequences, often to us as well as to others. Generally, there are three primary ethical directives: loyalty, honesty, and responsibility. Loyalty may be sought by a number of groups and institutions. Honesty is more than telling the truth. It does not include lying, but, most importantly, it includes the correct representation of ourselves, our behavior, and our views. Responsibility means anticipating the potential consequences of our actions and taking measures to prevent harmful occurrences such as accidents.

Construction company employees will face many ethical dilemmas in the execution of a construction project. Ethical behavior is a difficult area because it involves more than simply complying with legal requirements. It includes treating others in a respectful and fair manner. Providing a secure workplace atmosphere is an ethical issue that demonstrates concern for the welfare of anyone working on or passing by the project site. Construction company leaders must create an internal environment that promotes and rewards ethical behavior and must set the example by their own personal actions. This is true in many aspects of company operations and is critical in creating a work safely culture within the company. If a company leader identifies an unsafe activity and refuses to take action to correct the incident, he or she demonstrates a lack of interest in safety. Construction company leaders must, therefore, continue to demonstrate their commitment to safety through their actions and statements.

Example 3.3
The three primary ethical directives are:

A. Loyalty, honesty, and reliability

B. Loyalty, safety, and responsibility

C. Healthy, honesty, and responsibility

D. Loyalty, honesty, and responsibility

Answer: D

3.9   Ethics and Whistle-Blowing

Whistle-blowing is an act of informing an outside authority or a media organ of alleged illegal or unethical actions on the part of an organization or individual.

Whistle-blowing is one of the crucial ethical concerns of any big company. On the one hand, it is rational for managers to stimulate whistle-blowing because it gives a company a chance to become better. Also, whistle-blowing can cause danger because employees reveal various secret aspects of the functioning of the company. A whistle-blower is an individual who provides a precise data about the illegal activities of the company in which he or she works to authorities or specific regulatory agencies. It means that whistle-blower informs the general public about the hidden and improper mechanisms of a company's market performance. For instance, the employees could inform regulatory agencies about the discrimination inside the organization, failure to pay taxes, or the cases of environmental pollution that happened as a result of the company's violation of laws. Accordingly, whistle-blowing is linked to the most diverse spheres of business operation. In such case, almost any employee could become whistle-blower in accordance with his or her ethical values.

It is necessary to note that whistle-blowing is closely connected with the interpretations of morality. For instance, some people consider whistle-blowing immoral because an employee acts as a traitor. This logical error exists due to the subconscious social thinking patterns. It is irrational to state that some whistle-blower is a traitor only because they share some information about their organization. On the other hand, a whistle-blower usually helps to bring justice to the whole district, city, or even country. Consequently, whistle-blowing has to be interpreted as a positive behavioral pattern.

There is a tendency among some people not to speak against unethical or illegal practices, although they hate these activities. This "don't speak" mentality is instilled in children by many parents. Even at adult age, people seem to nurture this attitude. Some other problems of whistle-blowing are listed below.

Retribution.   People who blow the whistle on their own employer may be subjected to retribution such as being fired, transferred to undesirable areas, or reassigned to an undesirable job. They may also be shunned.

Damaged relationship and hostility.   Often, whistle-blowing damages a relationship. After whistle-blowing, some person may be disciplined or punished, which causes bitter relationship between this person and the whistle-blower. Sometimes, the damaged relationship may turn into a form of hostility directed toward the whistle-blower.

Loss of focus.   Whistle-blowing takes time, energy, and attention. Therefore, the whistle-blower may lose focus on their own work and may turn into a looser.

Scapegoating.   Negative consequences often occur as a result of whistle-blowing. This is why it is a common practice to either disregard the problem or only share it with colleagues, and the case closes there. If anyone is injured or fatality occurs, then the associated personnel may be questioned about the unethical practices.

Whistle-blower is protected by OSHA (Fig. 3.4). OSHA deals with the issue of whistle-blowing by provisions of the Energy Reorganization Act (ERA). This provision makes it illegal for an employer to discharge an employee or otherwise discriminate against an employee in terms of compensation, job location, job position, or privileges of employment so similar to other actions. Employees are engaged in protected activity when employees (OSHA, Part No 24 Subpart A):

- Notify their employer of an alleged violation of the ERA
- Refuse to engage in any practice made unlawful by the ERA
- Testify before Congress or at any federal or state proceeding regarding any provision or proposed provision of the ERA
- Commence or cause to be commenced a proceeding under the ERA, or a proceeding for the administration or enforcement of any requirement imposed under the ERA

Figure 3.4 Protection mechanism of whistle-blower by OSHA.
(Photo courtesy of Leo Green: https://leo-green.com/what-is-whistleblowing-essay/.)

- Testify or are about to testify in any such proceeding
- Assist or participate in such proceeding or in any other action to carry out the purposes of the ERA

Employers may not retaliate against an employee for engaging in protected activity by:

- Intimidating
- Threatening
- Restraining
- Coercing
- Blacklisting
- Firing
- Or in any other manner retaliating against the employee

It is illegal for an employer to fire, demote, transfer, or otherwise retaliate against a worker for using their rights under the law. If you believe you have been retaliated in any way, file a whistle-blower complaint within 30 days of the alleged retaliatory action. OSHA administers more than 20 whistle-blower statutes, with varying time limits for filing. The time frame for filing a complaint begins when the adverse action, such as a firing, occurs and is communicated to the employee.

Example 3.4
A whistle-blower is an individual who:

A. Whistles while working in the company if he or she finds something interesting

B. Publishes important update about the company in the media and in the OSHA

C. Provides a precise data about the illegal activities of the company in which he or she works

D. Is hired to monitor the safety policies in the company

Answer: C
A whistle-blower is an individual who provides a precise data about the illegal activities of the company in which he or she works to authorities or specific regulatory agencies.

3.10 Summary

Construction industry involves different types of personnel, such as laborers, owners, etc., of different backgrounds, motives, and attitudes. Construction sites are very dynamic with different types of activities at a time. Therefore, safety is very vital in such an industry. The employer is responsible for most of the safety and health issues in the workplace. He is the first point of starting all safety and health regulations that satisfy the OSHA standard. The other personnel must follow all the safety and health rules set by the company. If PPE is required, the employer must provide these free of cost to the employees. If an employee finds that the company is not following the safety standards up to the level of OSHA standards, he or she may reach out to the OSHA. If he or she files such complaints, OSHA will protect him/her from any adverse actions taken by the employer.

3.11 Multiple-Choice Questions

3.1 Which of the following is the supervisor's primary role in safety?
   A. Conducting daily safety inspections of the workplace
   B. Helping workers file claims for job-related injuries
   C. Alerting top management to OSHA inspections
   D. Setting health and safety standards for the firm

3.2 According to OSHA, how many competent persons should be employed at each construction site?
   A. At least one
   B. At least two
   C. At least three
   D. At least four

3.3 How frequent should jobsite safety inspections be conducted to identify hazards and ensure compliance with job-specific safety rules?
   A. Every day
   B. Every alternative day
   C. Every week
   D. Every month

3.4 Foremen should hold _____ safety meetings to review the safety aspect of the tasks to be carried on that day.

A. Day-to-day
B. Week-to-week
C. Hour-to-hour
D. Month-to-month

3.5 Report to the nearest OSHA office all work-related fatalities within:

A. 8 hours
B. 24 hours
C. 48 hours
D. 72 hours

3.6 Report to the nearest OSHA office all work-related inpatient hospitalizations, all amputations, and all losses of an eye within:

A. 8 hours
B. 24 hours
C. 48 hours
D. 72 hours

3.7 The OSHA citation is to be posted and remain posted until:

A. The violation has been corrected
B. At least three working days
C. At least five working days
D. OSHA directs to remove it

3.8 Discharging a whistle-blower or otherwise discriminating against an employee in terms of compensation, job location, job position, or privileges of employment so similar to other actions is:

A. Unethical
B. Illegal
C. Dependent on the employer
D. Dependent on the community around the area

3.9 If you believe you have been retaliated in any way for whistle-blowing, file a whistle-blower complaint within _____ of the alleged retaliatory action.

A. 10 days
B. 20 days
C. 30 days
D. 40 days

3.10 OSHA administers more than _____ whistle-blower statutes, with varying time limits for filing.

A. 10
B. 15
C. 20
D. 30

3.12 Practice Problems

3.1 List the responsibilities of employers to ensure safety in a construction site.

3.2 List the responsibilities of contractors to ensure safety in a construction site.

3.3 List the primary responsibilities that supervisors have in the area of occupational safety and health for all employees under their supervision.

3.4 List the 10 steps followed to establish safety-first culture in construction industry.

3.5 List the issues involved in whistle-blowing against an unethical or illegal practice.

3.13 Critical Thinking and Discussion Topic

OSHA always forces employers to adopt, maintain, and ensure safety practices that everybody in a company should follow in their working places. If any accident happens, employers are questioned about it. But if the employers train the employees, provide them all PPEs and any other required safety tools, and still accident happens, then employers should not be questioned. The responsibility of the accident falls upon the employees then. Join this discussion and give your opinion.

CHAPTER 4

Accidents in Constructions

(Photo courtesy of 123RF.)

4.1   General

An accident (Fig. 4.1) is an unplanned event that has inconvenient or undesirable consequences, other times being inconsequential. The occurrence of such an event may or may not have unrecognized or unaddressed risks contributing to its cause. The difference between an accident and an incident is slight but worth noting. Accidents denote

Figure 4.1 An accident of a man worker at a construction site.
(Photo courtesy of 123RF.)

a loss-producing, unintended event, and an incident denotes an unintended event that possibly doesn't result in a loss. The severity of accidents can be negligible, marginal, critical, or catastrophic. In order for an accident to happen, two factors should be present: hazard and exposure to the hazard. Hazard is anything that will hurt or harm you through contact. Exposure is how susceptible you are to an injury or illness because of being near to a hazardous substance or material. Accident causation models deal with controlling/eliminating exposure or eliminating, substituting, or controlling the hazard itself. Most accidents are caused by human error. If you learn the cause-and-effect relationship of accident causation and the human element, then you can reduce injuries and illnesses. Although there are additional factors causing accidents, the human aspect is the leading cause of incidents.

4.2 Accident Theories

4.2.1 Single Factor Theory

The single factor theory for accident investigation, the most basic of the accident causation models, assumes that there is one event solely responsible for an accident or incident. It is similar to the "pilot error syndrome" that states that the cause of the accident was due solely to an error caused by the pilot. The single factor theory has little to no value in modern accident causation models.

4.2.2 H.W. Heinrich's Domino Theory

H.W. Heinrich presented the axiom of industrial safety theory in 1931. The first axiom dealt with accident causation, which states, "Invariably, the occurrence of the injury results from complicated factor sequences, the last of which is the accident itself" (Heinrich, 1931). The model that was used to illustrate this point was a lineup of dominoes, which became known as the domino theory. This theory considers three phases of an accident, which are as follows:

1. Precontact phase prior to the event happening
2. Contact phase, which refers to the event as it's happening
3. Postcontact phase, which refers to after the release of unplanned energy, the product downtime, or the injury or illness

According to the domino theory, accidents are caused by unsafe acts and/or mechanical or physical hazards due to unsafe actions by people, or are caused by their ancestry and social environment (Heinrich, 1931). Heinrich states that if one of the dominoes were removed, preferably the middle unsafe act domino, the accident would have never occurred. The later model of the Domino, Bird, and Loftus added managerial influence and managerial error dominoes. In the later model, loss was organizational loss as opposed to just an accident. An example of organizational loss could be property damage, loss of public trust, as well as an illness or injury. A point of contention of the domino theories was that they don't glean the nature of the incidents. As the accident theories developed, more dimensional approaches became common, indicating that the cause of an incident may be multifactual, with events happening in a nonlinear format.

4.2.3 Multiple Factor Theory

The multiple factor theory looks at not just the three phases in the domino theory—precontact, contact, postcontact—but it also analyzes other factors that influence an accident, such as management, machine, media, and man. The role of management would be to create the organizational structures, policy, and procedures. The machinery includes the design, shape, size, and type of energy sources of the equipment. The role of media can be described as an environmental factor. The human factor includes gender, age, attitude, fatigue level, height, weight, etc.

When you analyze all of these factors, it becomes apparent that your model does not include influencers only. Unlike previous theories, management's role is more concrete in the multiple factor theories. Strong management that respects the role of occupational safety will support and lead the safety effort of the organization.

4.2.4 Human Factor Theory

The human factor theory states that the accident is caused by human error. This theory analyzes the factors that lead to the human error, as shown in Fig. 4.2. Such factors are overload, and inappropriate activities and responses. Overload doesn't always mean that the worker is overburdened, but the job itself may have excess stressors such as noise, heat, or unclear instructions. Workers can often feel overloaded when too many tasks are given at a time, or if work instructions do not clearly define the goals. Front-line supervisors are charged with making sure that the workers understand the steps of each task.

There are two types of failures that cause human errors: active failures and latent failures. An active failure is usually caused by the worker or person engaged in an activity. The actions give an immediate consequence, which is the direct cause

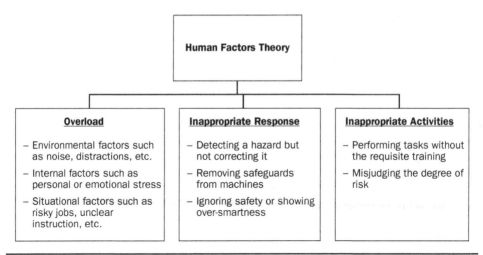

FIGURE 4.2 Details of human factors accident theory.

of the accident. Latent failures lay the foundation of an active failure due to such issues as:

- Ineffective training
- Poor engineering of the equipment or location
- Poor or inadequate supervision
- Ineffective communication
- Unclear roles for the worker

The observation of the human factor theory through a behavioral-based safety (BBS) analysis can help improve the safety system. The BBS is designed to highlight the behavior of the worker that is causing the accident and addressing it systematically. The key to systematic intervention is having a well-planned and thought-out BBS program.

4.2.5 James Reason's Swiss-Cheese Theory

James Reason developed an accident causation model based on the analogy of holes in Swiss cheese (Reason, 1991). In his analysis of an accident, he uses visualization of a complete system with no holes. In a complete system, management is involved, employees are motivated, there is a safety management system in place, and a safety culture exists in the organization.

A breakdown of any one of these components, the blanks, will create a whole similar to the hole in a slice of Swiss cheese. As mortals develop, the system becomes less than ideal and dysfunctional. Eventually, each one of the systems will have their holes line up and allow a release of unplanned energy to create an accident condition. In this model, it does have a linear component, but it also includes multiple factors that influence the accident.

Upper management has a change in leadership with the new ownership group. The ownership group values production more than safety. Therefore, the workers are not given adequate time for safety briefings in the morning. Production becomes the main factor for the organization. Quality and safety may suffer due to the production quotas. Eventually, workers will become fatigued, equipment fails, or safety training is omitted entirely.

During a busy shift, when the worker is trying to meet production quota, he decides not to report a broken guard on a piece of equipment and comes in contact with the unguarded point of operation. The worker sustains a laceration that could have easily been an amputation.

The summary of all the accident theories is presented in Fig. 4.3.

Single Factor Theory	One event is solely responsible for an accident, akin to the "pilot error syndrome."
Domino Theory	Chain of sequential events trigger each other to cause an accident, removing a factor prevents the chain reaction.
Multiple Factor Theory	Analyzes three phases, precontact, contact, postcontact, and influencers like management, machines, and workers.
Human Factor Theory	Analyze factors that lead to human error such as overload and inappropriate activities and responses.
Swiss-Cheese Theory	Breakdown of components of a complete system creates holes that line up creating an accident condition.

Figure 4.3 Summary of the different accident theories.

Example 4.1
The domino theories are based on three phases:

    A. Single factor, multiple factor, and domino

    B. Precontact, Swiss cheese, and single factor

    C. Precontact, contact, and postcontact

    D. Precontact, contact, and Swiss cheese

Answer: C

Example 4.2
There are two types of failures that cause human errors: active and:

    A. Latent

    B. Passive

    C. Inactive

    D. Blatant

Answer: A

4.3   Causes of Accidents

Accidents may result from an unsafe act by a worker or from unsafe working conditions, or both. Research into why construction accidents occur has shown that about 90% of accidents on construction sites are due to unsafe behavior, and about 10% are due to unsafe worksite conditions. Unsafe behavior may result from a worker's state of mind, fatigue, stress, or physical condition. This may involve attempting to do more than he or she is capable of doing, such as picking up a heavy load, engaging in unsafe work activity, or improperly responding to an unsafe situation. Overexertion is a major cause of accidents, because tired workers often are not mentally alert. A major concern in the construction industry today is the aging workforce and the greater susceptibility of older workers to job-related injuries.

Some examples of causes of accidents are:

- A person detects a hazardous condition but does nothing to correct it, and an accident may result. An example may be the use of defective equipment, such as a ladder.
- A person disregards a safety policy or procedure, and an accident may result. For example, a worker not wearing gloves may get a sliver when handling lumber.
- An individual may lack proper training in how to perform a specific construction task safely and may undertake to perform the work in an unsafe manner.
- An individual may misjudge the risks associated with a specific task, and mistakenly choose to perform the task in an unsafe manner.

The following are types of accidents that occur on construction projects in the United States each year:

- A worker is connecting steel structural members on the fourth floor of a commercial building project and falls to the ground.
- A worker is struck in the head by a load being moved by a tower crane.
- A worker is working on a platform that collapses.
- A worker installing a pipe in an open trench is crushed when the sides of the trench collapse.
- A worker installing roofing material slips and falls to the ground.
- An electrical worker installing a circuit breaker is electrocuted.
- A brick mason working on a scaffold falls to the ground.

4.4   Cost of Accidents

Accidents and the corresponding damage that they cause to employees, property, and equipment can have a significant adverse impact on the financial condition of a construction company. It is difficult to measure precisely the cost of accidents, but they have a significant adverse impact on employee well-being, productivity, and morale. Two types of costs are possible:

1. Direct cost
2. Indirect cost

Direct costs of accidents are those costs covered by insurance. Indirect cost is not paid by insurance and thus is unrecoverable. The cost of insurance coverage is a function of the insurance underwriter's assessment of the risk posed and the construction company's claims history. Indirect costs are all other costs not recovered through insurance coverage. The indirect costs associated with an accident often are up to four times the direct costs. The direct costs of accidents include:

- Physical therapy
- Medical expenses
- Repair fees for damaged equipment
- Continuation of pay
- Compensatory damages
- Workers' compensation insurance premium cost
- Liability insurance premium cost
- Equipment liability insurance premium cost
- Legal expenses associated with claim resolution

Workers' compensation insurance is a no-fault insurance that compensates an injured worker (Fig. 4.4) for the cost of medical expenses, provides supplemental income if the worker is unable to work, and provides retraining if the worker cannot perform the duties of his or her job. Insurance premiums are based on the risk presented by the craft of the worker, such as roofer, carpenter, or steelworker, and the claims record of the employer.

The indirect costs of accidents include:

- First-aid expenses
- Damage or destruction of materials

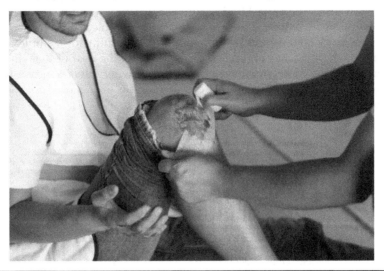

FIGURE 4.4 Construction worker has an accident while working on new house. (Photo courtesy of 123RF.)

- Cleanup and repair cost
- Idle construction machinery cost
- Unproductive labor time
- Construction schedule delays
- Loss of trained manpower
- Work slowdown
- Administrative and legal expenses
- Lowered employee morale
- Third-party lawsuits
- Wages paid to injured workers for absences not covered by workers' compensation
- Lost wages and work stoppage associated with the worker injury
- Overtime due to the accident
- Administrative costs and time spent by safety personnel, clerical workers, and other employees after the injury
- Training for replacement workers
- Lost productivity due to the work unit separation from the injury
- Accommodation of the injured employee within the organization
- Cleanup, preparation, and replacement cost of damaged materials, machinery, and property

The indirect cost of an accident generally greatly exceeds the direct cost incurred. In the event of an accident, all project operations typically cease while an investigation is undertaken. Workers on the site are being paid, but little productive work is being performed, leading to unproductive labor and equipment time, which may adversely affect the project budget and construction schedule. Any damaged materials or equipment will need to be replaced, and the project site may need to be cleaned. Suppose work on a project is stopped because a contractor's employee or a subcontractor's employee has an accident. In that case, the cost of delaying the project is assumed by the contractor and not the project owner. Table 4.1 gives the relationship between direct costs and indirect costs in a ratio that is used to calculate the total accident cost.

Direct Costs	Indirect Cost Ratio
$0–$2999	4.5
$3000–$4,999	1.6
$5000–$9999	1.2
$10,000 or more	1.1

TABLE 4.1   Relationship between Direct Costs and Indirect Costs in an Accident

Example 4.3
To evaluate the total cost of the accident you must combine both _____ and _____ cost.

    A. Incentive, Behavioral

    B. Direct, Indirect

    C. Direct, Incentive

    D. Behavioral, Indirect

Answer: B

Example 4.4
The cost of an accident is paid from company:

    A. Sales

    B. Consultants

    C. Extra budget

    D. Profits

Answer: D

4.5   Incident Investigation Techniques

4.5.1   Root Cause Analysis

A root cause analysis is a detailed procedure used by safety professionals to look beyond the initial knee-jerk reaction to assign blame for an incident. First, the investigator will notice direct causes of any accident, but he or she cannot see indirect or latent causes that become activated to create the accident. A root cause analysis must be conducted to determine all factors and variables responsible for an incident. Most of these types of analysis will require a team of trained professionals alert to accident causation models, as well as good investigators and critical thinkers. There are many types of root cause analysis, but they all seek to do the same thing: address the source cause of any accident.

4.5.2   The 5 Whys

The 5 Whys is an accident investigation technique used by safety professionals in developing the root cause analysis. This technique will help the data entry investigator pass simple questions to date beyond the superficial yes or no answer. These are quantifiable questions that build one upon the next. As each question gets answered, the investigator will uncover more reasons for an accident. Although this technique is called the 5 Whys, it often takes more questionings to get to the root cause. An example of the 5 Whys technique for incident investigation is as follows:

A worker receives a laceration from touching unguarded blade on a handfed ripsaw. Using the 5 Whys technique for questioning, the investigator can begin with these questions:

1. *Why did the worker receive a laceration?—Because, he touched unguarded blade.*
2. *Why did the worker touch the unguarded blade?—Because the guard was not placed on the saw.*
3. *Why was the guard not on the saw?—Because the worker believed he could work faster without the guard.*
4. *Why did the worker believe it is better to work faster and not safer?—Because the worker gets paid on a quota basis.*
5. *Why is management placing a greater importance on production quotas over safety?— Because management receives a bonus for achieving production quotas.*

In this brief example, the use of production quotas has indirectly led the worker to value speed over safety. There may be many other factors related to why the worker received a laceration in the scenario. Such factors can include but are not limited to:

- Poor organizational safety culture
- Untrained worker
- Management and supervision has unfair expectation of production quotas
- Worker fears retribution for slowing down production to address safety concerns
- Upper management has condoned and/or rewarded production goals over safety concerns

When conducting a 5 Whys accident analysis, the investigator will dig deeper to uncover an accident causation. This new revelation may lead the organization to address gaps in their safety management system. The value of operating safely to reduce or eliminate injury and illness must be made apparent to management in order for them to value safety equal with production.

4.6 Impact of an Accident on the Employer

Many accidents are considered to be thought of as expensive when considering lost time events. However, there are many more cost factors related to an accident that can be both direct and indirect cost. To evaluate the total cost of an accident you must combine both of these costs. In order to pay the cost, the organization must use the profits of the company. All profits are derived after the operational costs of the company have been calculated. Accidents affect the organization's profitability, because the cost of the accident must be paid from increased revenue. A company's profit margin is calculated by:

$$\text{Profit Margin} = \text{Total Profits} / \text{Total Sales}.$$

The revenue required for funds to offset an injury is:

Revenue Required = Total Cost of Incident / Profit Margin.

4.7   Common Human Performance Snares

There are several behaviors that can cause a human to perform below expectations. These performance traps or snares will show themselves to be behaviors to be coached in a BBS observation. To overcome these behaviors the coach should be aware of what they are and how to help the worker understand ways to master the behavior. The following are common human performance snares and the ways to overcome them.

4.7.1   Time Constraints

One of the most common human performance snares are workers feeling that they have a time constraint and being forced to cut corners. There are many actual pressures related to jobs such as due dates, daily schedules, personal pressures for performance, and front-line supervision time crunches. Sometimes pressures are legitimate and cannot be adjusted or easily adjusted.

When there is a time constraint due to a pressing engagement such as emergency situation, then the employee will have to make decisions rapidly. During times when rapid decisions are needed, the employee must rely on what is already a habit strength. Habit strength will leave the worker to restoring to their homeostasis for behavior. In the BBS program the workers will be trained to use safe behaviors versus at-risk behaviors which would lead to their habit strength in the time of emergencies.

However, in some cases workers just use time constraints as a way of avoiding the opportunity to use safe behaviors. In these situations, the coaching session should include some tools that can help deal with time constraints. The coach should include certain considerations when coaching to overcome this particular behavior.

- The coach can perform a self-check to see if there was truly a time constraint to perform this duty.
- A peer check of the situation will also reveal if another person is feeling a time constraint for performing this task.
- A prejob briefing would help the worker to see the whole job and visualize how long it will take.
- A careful consideration of the worker's attitude at the time of the job will reveal if they are placing self-pressure on time.
- Create an opportunity to do a three-way communication to ensure that all considerations were taken prior to performing a task.
- Was policy and procedure followed for the employee performing the task or was it bypassed?

4.7.2   Interruptions or Distractions

In some cases, at-risk behavior is caused because the worker is distracted or interrupted during the task. In order to successfully coach this individual, there must be an assessment of what was the distraction and where did it come from. The distraction can come from the worker themselves or an outside source such as a ringing phone.

In coaching distractions or interruptions, first the distraction itself should be removed, or the employee should be moved away from the distraction area. The employee should perform a system check prior to resuming the task to make sure all conditions are still safe for operations.

It may also be a good idea to have the worker seek assistance from a coworker before resuming work in order to assess the situation for any more distractions or interrupting forces. Then all distractions should be removed so that the worker can focus on performing a task with safe behaviors.

4.7.3   Multitasking

As technology has become commonplace, more people are doing what is known as multitasking. The term "multitasking" implies that the person can do more than one task at the exact same time. It is nearly impossible for workers to multitask in the purest sense of the word. Workers can, however, try to switch rapidly from one task to the next in order to multitask.

This can become a very dangerous pattern of behavior that can lead to mistakes on both or more processes. The worker should prioritize a list of tasks he/she needs to accomplish first. Once a task is accomplished then the next task in line should be tackled. If the workers feel that they have too many tasks that are due at the same time then they might feel tempted to multitask. This activity will only slow them down and create substandard outcomes.

4.7.4   Overconfidence

When a BBS observation team finds a worker who is overconfident, they will notice certain behaviors that can lead the workers to be at-risk of hurting and injuring themselves or others. In some instances, the workers feel that they do not need to recheck the process because they performed the work correctly the first time. However, this is not always the case and even the best worker can forget steps in the procedure.

To help workers overcome the feeling of overconfidence (that they are too good to make a mistake), the coaching team should ask them how they would feel if they did not get all steps correct and there was an incident. This may get them to rethink the idea of never making any errors. In addition, the coaching session may include the idea of having them question or challenge their own expectations through a self-check.

The supervisor should routinely reinforce expectations of policies and procedures with this individual. Then they should show the individual, or a work team that may be overconfident, some benchmarks from industry leaders.

4.7.5   Vague Guidance

There are incidents where workers developed at-risk behaviors because they were informed of a job through vague guidance. It is a possibility that the supervisor himself/herself might not have a good understanding of the task at hand, and therefore disseminating vague information to the worker. When this happens, there is no clarity of roles and responsibilities or even procedures.

The BBS steering committee must address this issue through the front-line supervisor. If there are established policies or standard operating procedures for a certain job then it must be reviewed by all parties. Standard operating procedures are there to make sure that each step of the job has been identified.

The worker should be encouraged to ask questions if they're unsure of any guidance given by the supervisor. And the supervisor should have a good understanding of the job and ensure that all workers understand each task that has been assigned to them. In some cases, the supervisor might even require retraining on how to perform any given task that they are responsible for delegating.

4.7.6   Overnight Shift Work

In some instances, there is a human performance letdown for workers that are working the overnight shift. The shift is typically from 11 p.m. at night to 7 a.m. in the morning with some variation of working during the early morning time period. Workers who are on this shift for a continuously long period of time adapt to the schedule. However, new workers will need a break in time to adjust to this change in lifestyle.

During the time that the new late shift operator gets their body adjusted to their work schedule, there are chances of at-risk behavior due to drowsiness or other related factors. It is possible that they are unable to sleep during the day because they are used to sleeping at night and the sudden change in schedule might keep them from getting ample sleep.

Therefore, at home, the overnight shift worker must have systems in place to get good sleep. Some workers are known to use blackout curtains and shut off all electronic devices during the sleep hours. When you are coaching the worker who is showing at-risk behavior during the night shift, consider the factors that lead to fatigue. It must be your common goal to have a worker assimilated to the new schedule when they are new on a job.

4.7.7   Peer Pressure

It is very important to monitor social impact from peers in the workforce. This impact can be both good and bad depending on the individuals involved. When there's a workforce that is very tightknit and the safety culture that values low risk, then many workers will encourage each other to have safe behaviors.

In some cases, however, there may be a "bad apple" among the workers. If this bad apple has some social impact on the workers, then more workers will have at-risk behavior as a result. It is important for the front-line supervisor to be an agent of change and not be the bad influence themselves. When poor behavior is observed among multiple people in one division, then it is easy to assume that this behavior is being pressured or conditioned in that one group.

The assessment of the at-risk behavior may lead to modeling from an agent of influence in that group. It may be the front-line supervisor or someone who's been there for considerable numbers for years.

4.7.8   Change

Some workers are averse to change, so when a change is happening they become more prone to at-risk behaviors. They might become uncertain about what to do and their changed behavior might be a way to get attention. Attention-seeking behavior can lead workers to an injury or illness because their mind is not on the work but on gaining attention.

Change may be inevitable in some organizations, but workers need to feel that they are still in the system. Certainty in the system will enable the workers to feel more comfortable in the idea of change and understand that it is in their best interest. In some cases, it is better to inform workers of the change and give them all scenarios related to the event before any actions are taken.

When dealing with operations such as the process safety management programs, the change analysis must be made prior to any major change. A change analysis is a detailed system that is utilized for the workers to see what domino effects will happen from changing a major element of their system.

This analysis is performed through the entire organization and utilizes systematic steps to analyze all repercussions of the change. For instance, if a company wants to change from using gas chorine to liquid bleach then a change analysis is in order. The release of gas chlorine into the atmosphere can create adverse effects for the whole community and not just the workplace. However, liquid chlorine bleach is not as harmful to a community if released into the atmosphere. The change analysis would incorporate all necessary parts, equipment, training, and regulatory requirements prior to the occurrence of such a major change in the system.

4.7.9 Physical Environment

There may be some performance issues due to the actual layout of the workplace. Things like poor lighting, ventilation problems, or even layout of machines can lead to poor workmen behaviors. Many workplaces have worksite analysis to make sure that there are no environmental factors adversely impacting any part of the job. Workers will be the first ones to see if there are any physical environmental problems that are leading them to at-risk behaviors. They will work around them as best as possible through whatever means available to them. A third-party audit would be a great way to analyze the work environment to ensure that there are no physical issues that would lead the workers to adapt at-risk behaviors. When the assessment is complete, a third-party auditor will have a final report with recommendations. Once recommendations have been read and understood, the organization should start making all the required changes to the physical environment promptly.

4.7.10 Mental Stress

Mental stress is caused by many factors in the work environment, although workers may also be stressed due to some problems at their home. In the past, it was believed that workers could separate their home life from work life, but this is not always the case. Some workers will exhibit at-risk behavior because of home stress.

Mental stress can produce severe outcomes, when they are coupled with at-risk behavior in the workplace. Some jobs are not forgiving when it comes to any form of deviations from safe practices. In some cases, the worker will not only hurt themselves and their coworkers but their behavior might even affect the community or the environment adversely.

Mental stress coupled with fatigue can also be a deadly factor for employees and their coworkers. There can be distractions as well as worker harassment that may be an outcome of mental stress. Some workers get stressed in such a way that they become pressurized with emotions and cause accidents.

Workplaces should have areas where workers can release mental stress to prevent at-risk behaviors. One way to combat mental stress is an employee assistance program which provides the employees with the chance to talk about any stress in their lives to a psychologist or a mental health professional. These mental health professionals are often used to help workers cope with home and work life.

Example 4.5

Mental stress coupled with _____ can also be a deadly factor for employees and their coworkers.

    A. Fatigue

    B. Creep

    C. Health

    D. Limited salary

Answer: *A*

The best answer is fatigue although all others are partially true.

4.8 Accident Prevention

Accident prevention requires a commitment to safety, proper equipment and construction procedures, regular and knowledgeable project site inspection, and good planning. Everyone working on the jobsite must understand the need to comply with all safety procedures and policies and be alert to any unsafe conditions. Accident prevention requires an ongoing program that involves everyone working on a project worksite. Techniques that superintendents should use include:

- Involving everyone in the identification of potential hazards
- Involving everyone in the development of procedures to eliminate or mitigate the hazards identified
- Teaching workers on the proper use of PPE and ensuring that the equipment is properly worn at all times
- Teaching workers safe working practices and ensuring that they do not over-exert themselves
- Teaching workers the importance of good housekeeping practices and ensuring that the jobsite is kept orderly and clean
- Establishing emergency procedures for the project site and ensuring that all individuals working on the site understand what to do in emergency situations

The superintendent must consider whether a hazard can be engineered to a safe state (e.g., using guardrails around floor openings or using trench boxes in open trenches). The next step is to create administrative controls such as policies, procedures, rules, and training sessions that do not present a physical barrier to a hazard but rely on good decision-making. The final step is to enforce compliance with jobsite safety policies.

4.9 Summary

Managing safety and health is a key part of keeping workers free from injury and illness. The company will save money by investing in the safety needs of the employees. An organization that values low risk will have a safer culture that will promote safe behaviors at all levels of the organization. In addition, through accident investigations, the company will get a good understanding of what hazards create the incident and how it can be controlled. To understand accident causation, investigation relies on the idea that most accidents are caused by human error. Therefore, if you learn the cause-and-effect relationship of accident causation and the human element, then you can reduce injuries and illness. In order for an accident to happen, two factors are required: the hazard and exposure to the hazard. An accident theory includes:

- Single Factor Theory
- H.W. Heinrich's Domino Theory
- Multiple Factor Theory
- Human Factor Theory
- James Reason's Swiss-Cheese Theory

Every accident incurs direct and indirect costs. The direct costs are paid by the insurance, but the indirect costs must be paid from the company profit. There are several behaviors that can cause a human to perform below expectations, such as:

- Time constraints
- Interruptions or distractions
- Multitasking
- Overconfidence
- Vague guidance
- Overnight shift work
- Peer pressure
- Change
- Physical environment
- Mental stress

With proper planning, accidents or similar hazards can be minimized in a workplace.

4.10 Multiple-Choice Questions

4.1 Accidents denote:

A. A loss-producing unintended event
B. An unintended event that possibly doesn't result in a loss
C. An incident which can be controlled and diminished
D. An incident—the same concept

4.2 Accidents on construction sites occur primarily due to:

A. Complex jobsite
B. Unsafe practice
C. Over smartness
D. Work pressure and stress

4.3 The direct costs of accidents do not include:

A. Workers' compensation insurance premium cost
B. Liability insurance premium cost
C. Work slowdown
D. Legal expenses associated with claim resolution

4.4 In a construction accident, which cost is higher?

A. Direct cost
B. Indirect cost
C. Both are almost equal in any case
D. Cannot be specified for most cases

4.5 In order for an accident to happen, the factor(s) to be present are most nearly: [select the best two options]

A. The hazard
B. The exposure to the hazard
C. Intention of the victim
D. Use of PPE

4.6 The accident severity factors are delineated by characteristics: _____, marginal, critical, and catastrophic.

A. Hazard
B. Negligible
C. Frequent
D. Exposure

4.7 When conducting a 5 Whys accident analysis, the investigator will dig deeper to uncover an accident _____

A. Severity
B. Risk
C. Causation
D. Reactions

4.8 Examples of a direct cost are:

A. Physical therapy
B. Medical expenses
C. Repair fees for damaged equipment
D. All of the above

4.9 Which cost is not paid by the insurance company and is unrecoverable?

A. Direct
B. Indirect
C. Capital
D. Gross

4.10 A root cause analysis is a detailed procedure used by safety professionals to look beyond the initial knee-jerk reaction to assign blame for an incident.

- True
- False

4.11 Most direct costs of accidents are usually paid by the insurance company.

- True
- False

4.11   Practice Problems

4.1 What is the difference between accident and incident?

4.2 List the two factors that are needed to be present.

4.3 Discuss the Single Factor Theory for accident.

4.4 Discuss the H.W. Heinrich's Domino Theory for accident.

4.5 Discuss the Multiple Factor Theory for accident.

4.6 Discuss the Human Factor Theory for accident.

4.7 What are the causes of incident?

4.8 What are the types of accidents that occur on construction projects in the United States each year?

4.9 Discuss the cost of an incident.

4.10 List some examples of indirect costs of an incident.

4.11 List the "5 Whys" to investigate the causes of an accident.

4.12 List some accident prevention techniques in construction.

4.12   Critical Thinking and Discussion Topic

Construction accident is natural and is controlled by the force-majeure. We cannot change the construction accident or control it. No statistical or psychological theories should work for construction accident. All construction accidents are controlled by the nature or God, whatever you believe in. Join this discussion and give your opinion.

CHAPTER **5**

Construction's Fatal Fours

(Photo courtesy of 123RF.)

5.1 General

According to OSHA, falling, electrocution, hit-by-object, and caught-in/between are the leading causes of worker deaths at construction sites. These Fatal Fours were responsible for more than 58% of the construction deaths in 2014. Out of 4779 worker fatalities in private industry in calendar year 2018, 1008 or 21.1% were in construction—that is, one in five worker deaths last year was in construction. The leading causes of death of private sector workers (excluding highway collisions) in the construction industry were falls, followed by object strikes, electrocution, and caught-in/between. These Fatal Fours were responsible for more than half (58.6%) in 2018. Eliminating the

Fatal Fours would save 591 workers' lives in America every year. Some data provided by OSHA in 2018:

- Falls—338 out of 1008 total deaths in construction (33.5%)
- Struck by object—112 (11.1%)
- Electrocutions—86 (8.5%)
- Caught-in/between—55 (5.5%)

Caught-in/between category includes construction workers killed when caught-in or compressed by equipment or objects, and struck, caught, or crushed in collapsing structure, equipment, or material. Construction sites may be dangerous, but there are ways you can ensure safety at work and reduce accidents in the workplace. Let's take a look at each of the Fatal Fours and see if you can take steps today to avoid such tragedies.

5.2 Falls

5.2.1 Basics

The leading cause of death in construction work in 2018 was a fall, with more than 100,000 injuries and deaths per year due to work-related falls. Falls are a leading cause of fatalities and serious injuries in construction as proper fall protection is not always equipped at sites. A construction with high potential of fall is shown in Fig. 5.1.

It is very important to know the physics of a fall. A body in motion can cover vast distances in a short period of time. Consider this:

- A body in free fall can travel 4 feet in 0.5 seconds.
- A body in free fall can travel 16 feet in 1 second.

Figure 5.1 A construction with high potential of falling.
(Photo courtesy of Immortal Snapshots from Pexels.)

- A body in free fall can travel 64 feet in 2 seconds.
- A body in free fall can travel 144 feet in just 3 seconds.

5.2.2   Fall Prevention Measures

In order to prevent workers from falling, employers must:

- Select fall protection systems appropriate for given situations.
- Use proper construction and installation of safety systems.
- Supervise employees properly.
- Use safe work procedures.
- Train workers in the proper selection, use, and maintenance of fall protection systems.

5.2.3   Areas Required to Have Fall Protection

Depending on the circumstances, the following areas are required to have fall protection:

- Unprotected sides and edges
- Leading edges
- Hoist areas
- Holes
- Formwork and reinforcing steel
- Ramps, runways, and other walkways
- Excavations
- Dangerous equipment
- Overhand bricklaying and related work
- Roofing work on low-slope roofs
- Roofs
- Precast concrete erection
- Residential construction
- Wall openings
- Walking/working surfaces not otherwise addressed

Depending on the circumstances, the following duties are to be performed to avoid fall protection:

- Fall protection is generally required when one or more employees have exposure to falls of 6 feet or greater to the lower level.
- Surfaces must be inspected before the work begins.
- Employees are only permitted to be on surfaces that are strong enough to support them.

Employers are expected to assess the workplace in order to decide if the walking/working surfaces on which employees are to work have the strength and structural integrity to safely support workers. Employees are not allowed to work on these surfaces until it has been determined that the surfaces have the necessary strength and structural integrity to support workers. Once employers have determined that the surface is safe for employees to work on, the employer must select one of the available options for the work operation if a fall hazard is present. For example, if an employee is exposed to falling 6 feet (1.8 meters) or more from an unprotected side or edge, the employer must provide a guardrail system, safety net system, or personal fall arrest system to protect the worker. Similar requirements are prescribed for other fall hazards as follows.

Leading Edge Work.   Every employee who constructs a 6-feet (1.8 meters) leading edge or more above the lower levels shall be properly protected. Guardrail systems, safety net systems, or personal fall-arrest systems may provide adequate protection. Positioning devices shall be secured to anchorage capable of supporting at least twice the potential impact load of an employee's fall or 3000 pounds, whichever is greater.

Hoist Areas.   Each employee in a hoist area shall be protected from falling 6 feet (1.8 meters) or more by guardrail systems, personal fall arrest systems, or other appropriate means. Suppose guardrail systems (or chain gate or guardrail) or portions thereof must be removed to facilitate hoisting operations, as during the landing of materials, and a worker must lean through the access opening or out over the edge of the access opening (to receive or guide equipment and materials, for example). In that case, that employee must be protected by one of the appropriate means.

Formwork and Re-Bar.   During formwork or re-bar assembly, employees shall be protected from falls of 6 feet or more by personal fall arrest systems, safety net systems, or positioning device systems. Positioning devices shall be rigged such that an employee cannot free more than 2 feet (0.9 meters).

Ramps, Runways, and Walkways.   Each employee using ramps, runways, and other walkways shall be protected from falling 6 feet (1.8 meters) or more.

Excavations.   Each employee at the edge of a 6 feet (1.8 meters) or deeper excavation shall be protected from falling by guardrail systems, fences, barricades, or covers when excavations are not easily seen due to plant growth or other visual barriers. Where walkways are provided to permit employees to cross over excavations, guardrails are required on the walkway if it is 6 feet (1.8 meters) or more above the excavation. Covers do not need to be labeled. Covers located in roadways and vehicular aisles shall be capable of supporting without failure, at least two times the maximum axle load of the largest vehicle expected to cross over the cover.

Dangerous Equipment.   Each employee working above dangerous equipment must be protected from falling into or onto the dangerous equipment by guardrail systems or by equipment guards even in those cases where the fall distance is less than 6 feet (1.8 meters).

Overhand Bricklaying.   Except as otherwise provided in the OSHA Fall Protection Standards, every employee who performs overhand bricklaying and related work 6 feet (1.8 meters) or more above the lower level shall be protected from falling by guardrail systems, safety net systems, personal fall arrest systems, or operate in a controlled access zone.

Low-Sloped Roof Work.   Each employee engaged in roofing activities on low-slope roofs, with unprotected sides and edges 6 feet (1.8 meters) or more above lower levels, shall be protected from falling by guardrail systems, safety net systems, and personal fall arrest systems, or a combination of a warning line system and guardrail system, warning line system and safety net system, warning line system and personal fall arrest system, or warning line system and safety monitoring system.

- Safety Monitoring System: a safety system in which a competent person is responsible for recognizing and warning employees of fall hazards.
- Warning Line System: a barrier erected on a roof to warn employees that they are approaching an unprotected roof side or edge, and which designates an area in which roofing work may take place without the use of guardrails, body belts, or safety net systems to protect employees in the area.
- Warning lines must be flagged at not more than 6 feet (1.8 meters) intervals with high-visibility materials.

Steep Roofs.   Each employee on a steep roof with unprotected sides and edges 6 feet (1.8 meters) or more above lower levels shall be protected by guardrail systems with toe boards, safety net systems, personal fall arrest systems, or by other appropriate means. Toe boards shall be capable of withstanding, without failure, a force of at least 50 pounds applied in any downward or outward direction at any point along the toe board.

Precast Concrete.   Each employee who is 6 feet (1.8 meters) or more above lower levels while erecting precast concrete members and related operations, such as grouting of precast concrete members, shall be protected by guardrail systems, safety net systems, or personal fall arrest systems.

Wall Openings.   Each employee working on, at, above, or near wall openings (including those with chutes attached) where the outside bottom edge of the wall opening is 6 feet (1.8 meters) or more above lower levels, and the inside bottom edge of the wall opening is less than 39 inches (1.0 meter) above the walking/working surface must be protected from falling by the use of a guardrail system, a safety net system, or a personal fall arrest system.

Example 5.1
Warning lines must be flagged at not more than _____ feet intervals with high-visibility materials.

    A. 4

    B. 6

    C. 8

    D. 10

    Answer: *B*

Example 5.2
A positioning device system is a fall arrest system.

- True
- False

Answer:
False

Example 5.3
Covers located in roadways and vehicular aisles shall be capable of supporting without failure, at least _____ the maximum axle load of the largest vehicle expected to cross over the cover.

A. Five times

B. Four times

C. Three times

D. Two times

Answer: *D*

Example 5.4
All covers should have any of the following to provide warning of the hazard, except:

A. Color coded

B. Marked with the word "HOLE"

C. Marked with the word "COVER"

D. Unlabeled

Answer: *D*

5.2.4   Protection from Falling Objects

When employees are exposed to falling objects, the employer must have employees wear hardhats and implement one of the following measures:

- Erect toe boards, screens, or guardrail systems to prevent objects from falling from higher levels.
- Erect a canopy structure and keep potential fall objects far enough from the edge so that those objects will not go over the edge if they are accidentally displaced.
- Barricade the area to which objects could fall, prohibit employees from entering the barricaded area, and keep objects that may fall far enough away from the

edge of a higher level so that those objects would not go over the edge if they were accidentally displaced.

Example 5.5

Positioning devices shall be rigged such that an employee cannot free more than 6 feet.

- True
- False

Answer:
False—Positioning devices shall be rigged such that an employee cannot free more than 2 feet (0.9 meter).

5.2.5 Types of Fall Protection—Passive Systems

Passive systems are protective systems that do not involve the actions of employees. An example of a passive system is a catch platform extending around the perimeter of the work area.

Guardrails. Guardrails are one of the most common forms of fall protection. They can be constructed of wood, pipe, structural steel, or wire rope. Flags must be provided on wire rope to increase visibility. Guardrails must have a top rail, a midrail, and posts, and when necessary, a toe board.

- Guardrail systems shall be capable of withstanding, without failure, a force of at least 200 pounds (890 N) applied within 2 inches (5.1 centimeters) of the top edge, in any outward or downward direction, at any point along the top edge.
- Steel or plastic bands must not be used as top rails or midrails.
- Manila, plastic, or synthetic rope being used for top rails or midrails shall be inspected as frequently as necessary to ensure that it continues to meet the mandated strength requirements.

The guardrails design criteria are listed below:

- Top edge height of top rails, or equivalent guardrail system members, shall be 42 inches (1.1 meters) plus or minus 3 inches (8 centimeters) above the walking/working level. When conditions warrant, the top edge height may exceed 45-inch, provided that the guardrail system meets all other criteria in this paragraph.
- Midrails, screens, mesh, intermediate vertical members, or equivalent intermediate structural members shall be installed between the top edge of the guardrail system and the walking/working surface when there is no wall or parapet wall at least 21 inches (53 centimeters) high.
- Midrails, when used, shall be installed at a height midway between the top edge of the guardrail system and the walking/working level.
- Top rails and midrails shall be at least one-quarter inch (0.6 centimeters) nominal diameter or thickness to prevent cuts and lacerations. If wire rope is used for top rails, it shall be flagged at not more than 6-foot intervals with high-visibility material.

- For pipe railings: posts, top rails, and intermediate railings shall be at least one and one-half inches nominal diameter (schedule 40 pipe) with posts spaced not more than 8 feet (2.4 meters) apart on centers.

- For structural steel railings: posts, top rails, and intermediate rails shall be at least 2-inch by 2-inch (5 x 10 centimeters) by 3/8-inch (1.1 centimeters) angles, with posts spaced not more than 8 feet (2.4 meters) apart on centers.

- Screens and mesh, when used, shall extend from the top rail to the walking/ working level and along the entire opening between top rail supports.

- Intermediate members (such as balusters), when used between posts, shall not be more than 19 inches (48 centimeters) apart.

- Other structural members (such as additional midrails and architectural panels) shall be installed such that there are no openings in the guardrail system that are more than 19 inches (0.5 meters) wide.

Safety Net Systems. Safety net systems must comply with the following provisions:

- They must be installed as close as practicable under the walking or working surface on which employees are working, but in no case more than 30 feet below the surface.

- If the net is not vertically more than 5 feet from the working level, the safety net must extend outward from the outermost projection of the work by 8 feet.

- If the net is not vertically more than between 5 feet and 10 feet from the working level, the safety net must extend outward from the outermost projection of the work by 10 feet.

- If the net is vertically more than 10 feet from the working level, the safety net must extend outward from the outermost projection of the work by 13 feet.

- Safety nets must be drop-tested on the jobsite after they have been installed and before use, whenever they have been moved, after a major repair, and at 6-month intervals after installation, if left in one place.

- Drop-tests must consist of a 400-pound bag of sand 28–32 inches in diameter being dropped into the net from the highest working or walking surface, but not from less than 42 inches above that level.

- Safety nets must have enough clearance beneath them to prevent contact with the surface or structures below when a load equal to the drop-test weight is dropped on them.

- Safety nets must be capable of absorbing an impact force that is equal to the drop-test weight.

- Defective nets cannot be used.

- All materials, scraps, equipment, and tools that have fallen in the net must be removed as soon as possible and at least before the next work shift.

- The maximum size of each safety net mesh opening shall not exceed 36 square inches (230 centimeters^2) nor be longer than 6 inches (15 centimeters) on any side, and the opening, measured center-to-center of mesh ropes or webbing, shall not be longer than 6 inches (15 centimeters).

- The safety net must have a border rope with a minimum breaking strength of at least 5000 pounds.

- If the safety nets are connected together, the connection must, therefore, be as strong as the individual nets and not more than 6 inches apart.

5.2.6   Types of Fall Protection—Active Systems

Active fall protection systems require workers to be engaged in ensuring that proper protection is in place. This may include activities such as donning a full-body harness with an attached lanyard and attaching the lanyard to an appropriate anchorage point.

Active systems are designed to operate in free-fall situations. Active systems must be connected to other systems/components or activated to provide protection. Active systems are designed to protect employees from the following:

- Falls

- Forces that can cause injury

The personal fall arrest system (PFAS) is an example of an active system. These are cheap and easy to use. When used according to the manufacturer's instructions, the PFAS can save life if a fall occurs. Generally, a PFAS consists of three major components:

1. A full-body harness.

2. A shock-absorbing lanyard or retractable lifeline

3. Secure anchors

PFAS shall not be attached to a guardrail system or hoists. All components of a fall arrest system must be inspected before each use and after impact. Defective components must be removed from service. Personal fall arrest systems and components subjected to impact loading shall be immediately removed from service and shall not be used again for employee protection until inspected by a competent person, and determined to be undamaged and suitable for reuse. Action must be taken to promptly rescue fallen employees or be assured they can rescue themselves. When stopping a fall, a PFAS must:

- Limit maximum arresting force on an employee to 1800 pounds (8 kilonewton) when used with a body harness.

- Be rigged such that an employee can neither free fall more than 6 feet (1.8 meters), nor contact any lower level.

- Be attached to an anchor that is capable of withstanding 5000 pounds of force or is designed, installed, and used as part of a complete PFAS that maintains a safety factor of at least 2 and is used under the supervision of a qualified person.

- Bring an employee to a complete stop and limit maximum deceleration distance an employee travels to 3.5 feet (1.07 meters).

- Have sufficient strength to withstand twice the potential impact energy of an employee free falling a distance of 6 feet (1.8 meters), or the free fall distance permitted by the system, whichever is less.

5.2.7   Inspecting Fall Protection Equipment

The fall protection equipment must be inspected before each use for:

- Tears, cuts, burns, and abrasions
- Distorted hooks, damaged springs, and nonfunctioning parts
- Manufacturer labels
- Deformed eyelets, D-rings, and other metal parts
- Dirt, grease, oil, corrosives, and acids
- Desirable strength

5.2.8   Lifelines, Safety Belts, and Lanyard

Lifelines, safety belts, and lanyards shall be used only for employee safeguarding. Any lifeline, safety belt, or lanyard actually subjected to in-service loading, as distinguished from static-load testing, shall be immediately removed from service and shall not be used again for employee protection until inspected and determined by a competent person to be undamaged and suitable for reuse.

Vertical lifelines shall have a minimum breaking strength of 5000 pounds (22.2 kilonewton). Self-retracting lifelines and lanyards that automatically limit free fall distance to 2 feet (0.61 meters) or less shall sustain a minimum tensile load of 3000 pounds (13.3 kilonewton) applied to the device with the lifeline or lanyard in the fully extended position. All safety belt and lanyard connectors shall be made of drop forged, pressed, or formed steel, or equivalent materials. Each connector shall have a corrosion-resistant finish, and its surface shall be smooth and free of sharp edges.

5.2.9   Safety Monitoring Systems

The employer must designate a competent person to monitor the safety of other employees, and the employer has the duty to ensure that the safety monitor complies with the following requirements:

- He/she must be competent to recognize fall hazards.
- He/she must warn the employee when it appears that the employee is unaware of a fall hazard or is acting in an unsafe manner.
- He/she must be on the same walking/working surface and within visual sighting distance of employee being monitored.
- He/she must be close enough to communicate orally with the employee.
- He/she must not have other responsibilities which could take attention from monitoring function.

Each employee working in a controlled-access zone must be directed to comply with all instructions from the monitor. It is recommended that you have a written plan for using the safety monitoring system to address:

- Identification of the monitor
- Roles of employees in monitoring system
- Training for using the monitoring system

Example 5.6
Each employee working in a controlled access zone must be directed to comply with all instructions by the_____.

    A. Any designated co-worker.

    B. Safety monitor

    C. Employer

    D. Employer and the safety monitor

Answer: B

5.2.10   Fall Protection Plan

The option of a fall protection plan is available only to employees engaged in lead edge work, precast concrete construction work, or residential construction work who can demonstrate that it is unfeasible or creates a greater risk of using conventional fall protection equipment.

If used, the plan should be strictly enforced.

- A fall protection plan must be prepared by a qualified person and developed specifically for each site.
- The fall protection plan must be maintained up to date.
- Any changes to the plan must be approved by a qualified person.
- A copy of the plan with all approved changes must be maintained at the site.
- The fall protection plan shall document the reasons why the use of conventional fall protection systems (guardrail systems, personal fall arrest systems, or safety nets systems) is infeasible or why their use would create a greater hazard.

A fall protection plan must consist of the following elements:

- Statement of policy
- Fall protection systems to be used
- Implementation of plan
- Enforcement
- Accident investigation
- Changes to the plan

The sequence of fall protection plan should be as follows:

Step 1. Develop a written fall protection plan and train employees in fall protection.

Step 2. Assess potential fall hazards prior to each project.

Step 3. Use administrative controls to eliminate hazards by changing or rescheduling tasks.

Step 4. Inspect fall protection equipment to ensure its condition and appropriateness.

Site Conditions	Selection
There is a hoisting area without guardrails	Personal fall arrest system
Workers are working 4 feet above paving equipment	Equipment guards
An 8-foot deep hole is hidden by overgrown grasses	Fences
Roofers are working on low-slope roofs with unprotected sides	Warning line and safety monitoring systems
Workers may be hit by falling tools and materials	Hardhats and barricades
Falling tools and materials	Toe boards
Work on a roof with an unprotected edge	Warning line system
Workers working on a 12-foot high wall	Positioning device system
Workers on a roof with large holes	Covers

TABLE 5.1   Selection of Fall Protection System based on Site Conditions

Step 5. Before starting work, remind employees of specific fall hazards that are present on the jobsite.

Step 6. Once work has begun, observe workers to ensure that fall protection is being used properly.

Some examples of fall protection system based on site conditions are summarized in Table 5.1.

5.3   Struck-by-Object

5.3.1   Basics

The second leading cause was struck-by-object, which includes objects that are falling (suspended load coming loose), flying (thrown tools or debris), swinging (load swaying), or rolling (vehicle or heavy equipment in motion). In these scenarios, employees are sometimes caught off guard and don't have enough time to respond and move out of the way. According to OSHA, being struck by objects is a leading cause of construction-related deaths. Only falls rank higher and is the number one cause of death in the construction industry. OSHA estimates that 75% of struck-by fatalities involve heavy equipment like trucks or cranes. The number of workers who died as a result of being struck by a vehicle was 7 years higher in 1998.

Safety and health programs must include ways to limit or eliminate the many ways struck-by accidents can occur because one of the major causes of construction-related deaths is from being struck by objects.

Typically, struck-by accidents are associated with:

- Vehicles
- Falling or flying objects
- Masonry walls
- Nailing

Example 5.7
OSHA estimates that _____ of struck-by fatalities involve heavy equipment like trucks or cranes.

   A. 10%

   B. 50%

   C. 75%

   D. 100%

Answer: C

5.3.2   Vehicles

Suppose vehicular safety practices are not followed at a worksite. In that case, workers are at risk of being pinned (caught) in between construction vehicles and walls or stationary surfaces, struck by swinging equipment, crushed beneath overturned vehicles, or many other similar accidents. When working near a public roadway, workers are additionally exposed to being struck by trucks, cars, or other vehicles. Improper operation of heavy vehicles poses a life-threatening danger to construction workers. Always follow safe practices to minimize injuries and save lives. Important engineering controls include:

- Always install, use, and maintain vehicle back-up alarms.
- Station flaggers behind vehicles that have obstructed rear views.
- Keep nonessential workers away from areas of vehicle use.
- Keep workers away from temporary overhead activities.
- Place barriers and warning signs around hazardous operations and public roadways.

The use of seat belts during construction equipment or other motor vehicles must be made mandatory to reduce the effects of a crash. Research shows that the use of a seat belt reduces the risk of a fatal injury by 45% to front-seat occupants of a car and 60% by light truck occupants. Workers must wear seat belts in all vehicles that are equipped with seat belts. In the event of an accident, workers can be struck by the frame of the cab. Rollover accidents can pierce tools or material into the operator. There are many ways to protect workers from being struck by objects and equipment. Two important general rules to follow are:

- Never put yourself between moving or fixed objects.
- Always wear bright, highly visible clothing when working near equipment and vehicles.

Additional safe practices are:

- Use manufacturer-approved safety restraints unless the vehicle is not designed for them.

- Never allow workers to drive equipment in reverse without an alarm or flagger.
- Enforce a limited access zone before dumping or lowering buckets.
- Properly turn off and block all equipment, including accessories.
- Set parking brakes and use chock wheels if parked on an incline.
- Install cab shields on hauling vehicles to protect against struck-by and rollover injuries.
- Never exceed the vehicle's rated lift capacity or carry unauthorized personnel.
- Use signs, barricades, and flaggers to protect workers near roadways.
- Use proper lighting and reflective clothing/vests at night.

Example 5.8
Always wearing bright, highly visible clothing when working near equipment is a method to avoid a _____ accident.

    A. Struck-by

    B. Burned

    C. Shocked

    D. None of Above

Answer: A

5.3.3 Falling or Flying Objects

Workers are at risk from falling objects when they are required to work around cranes, scaffolds, overhead electrical line work, etc. There is a danger from flying objects when using power tools, or during activities like pushing, pulling, or prying, that can cause objects to become airborne. Flying/falling objects can also roll off rooftops, or be accidentally kicked off walkways, scaffold platforms, etc., if they are not properly constrained. Depending on the situation, injuries from being struck by flying or falling objects range from minor ones like bruises to severe ones like concussions, blindness, and death.

Loose debris left on the roof could easily be blown away by a gust of wind and carried to the ground below where the worker could be standing. When working in this kind of an environment, accidents are inevitable. Workers must be trained to be careful and to remain constantly on the lookout for such conditions, to secure all materials in an appropriate manner. Employers must educate their employees on how to prevent accidents and exposures. Employers have a responsibility under OSHA standards to educate and train their employees to recognize and avoid unsafe conditions that may lead to struck-by injuries. Workers can be struck by falling or flying objects or by materials that slide, collapse, or otherwise fall on them. To protect workers from these types of injuries, OSHA requires that employers:

- Require workers to use hardhats/helmets when appropriate.
- Train employees to stack materials to prevent sliding, falling, or collapsing.

- Install protective devices onsite, such as toe boards on elevated platforms and walkways.

- Install debris nets beneath overhead work.

Safety nets must be installed as close as practicable under the walking/working surface on which employees are working, but in no case more than 30 feet (9.1 meters) below such levels. When nets are used on bridges, the potential fall area from the walking/working surface to the net must be unobstructed.

Example 5.9

Consider that you are a part of a three-man crew who has a job of digging a trench. The materials for the sewer line are already here and the installers are waiting to get started. Suddenly you found that the backhoe you are supposed to use is old and not well maintained. You hop on and pass the starter button but it did not work. The other two crew are saying that the boss is furious and let us get this show on the road. What should you do?

   A. Tell your friends to put on bright and highly visible clothing

   B. Go complain to your boss that you can't get anything done with old equipment

   C. Tag the backshoe as defective and have it removed from the worksite

   D. Take a screw driver and start fixing the issue.

Answer: C
Any equipment that doesn't meet the OSHA requirements must be identified and cannot be used at all.

5.3.4   Masonry Walls

Because of the tremendous weight of a masonry wall or slab, it can cause permanent injury or death if one collapses on a worker. Proper safeguards should be used, and all jackets and equipment used to support and position such walls and slabs should be reliably maintained and safeguarded.

Only essential workers should be allowed near this type of operation. To enforce this, set up a limited access zone around operations. Additionally, be sure to:

- Have concrete structures checked by qualified persons before placing loads.

- Adequately shore or brace structures until they are permanently supported.

- Secure unrolled wire mesh so it cannot recoil.

- Never load a lifting device beyond its intended capacity.

Example 5.10

There are several types of struck-by-hazards. Mark by an 'X' to the category it belongs to as listed in Table 5.2.

	Flying	Falling	Swinging	Rolling
Sawdust blown by compressed air	X			
A hammer dropped from a scaffold		X		
Being hit by a truck while unloading freight				X
A nail ejected from a nail gun	X			
A collapsing stack of water pipe				X
Debris blown off a roof		X		
A crane load blown by wind			X	
Bricks tumbling from a masonry wall		X		
Being hit by the boom of crane			X	
Being run over by a tractor trailer				X

TABLE 5.2   Matching the Types of Struck-by-hazards with Its Category

5.3.5   Nailing

One of the most used power tools in the construction field is the nail gun. It is also responsible for an estimated 37,000 emergency room visits. A study of apprentice carpenters by OSHA states that:

- Two out of five were injured using a nail gun during their 4 years of training.
- One out of five was injured twice.
- One out of ten were injured three or more times.

Most of the injuries obtained from a nail gun accident involve hand and finger injuries that involve structural damage to tendons, joints, nerves, and bones. Some serious injuries related to being struck-by a nail from a nail gun (nailer) are:

- Paralysis
- Blindness
- Brain damage
- Bone fractures
- Death

There are various types of specialized nailers such as for framing, roofing, and flooring. The framing nail guns are powerful pieces of equipment that fire larger nails. Framers are therefore even more at risk from the mishandling and misuse of nailers.

The following tips will ensure proper handling of a nail gun:

1. Use the full sequential trigger nail gun for the safest trigger mechanism. This type of trigger reduces the risk of unintentional nail discharge or double fires. New workers should be restricted to using the full sequential trigger nail guns only until they are fully oriented with other trigger types.

2. All workers that use nail guns must be trained on how to use the tool and its safety features. Hands-on training is always the best form of training, so the

worker can see how to use the equipment first hand. OSHA recommends the following training topics:

   a. How nail guns work and how triggers differ.

   b. Main causes of injuries, especially differences among types of triggers.

   c. Instructions provided in manufacturer tool manuals and where the manual is kept.

   d. Hands-on training with the actual nailers to be used on the job. This gives each employee an opportunity to handle the nailer and to get feedback on topics such as:

- How to load the nail gun
- How to operate the air compressor
- How to fire the nail gun
- How to hold lumber during placement work
- How to recognize and approach ricochet-prone work surfaces

3. Establish nail gun work procedures for workers that will include:

   a. Conduct mandatory reviews of the tool operations and maintenance manual.

   b. Have operations and maintenance manuals onsite for review.

   c. Check tools and power source for proper operations and require broken or malfunctioning equipment to be taken out of service immediately.

   d. Check lumber surfaces to ensure that there are not knots, nails, hangers, or anything that can impede the nail from going through the material.

   e. Keep hands at least 12 inches away from the point of impact of the nailer.

   f. Disconnect the compressed air when servicing, traveling, or clearing a nail jam from the equipment.

   g. Analyze the dangers of nail gun work and mitigate as many hazards as possible prior to working in the area.

4. Provide personal protective equipment (PPE) such as hard hats, high impact eye protection, and hearing protection.

5. Encourage reporting and discussion of injuries and near misses to help workers learn how to identify hazards. Once the hazards have been identified, the prompt correction of the problem is needed.

6. Provide first aid and medical treatment for workers at the job location. Get workers medical care as quickly as possible to limit the impact of the accident.

5.4   Electrocutions

5.4.1   Basics

The third leading cause of death in construction work in 2018 was electrocution. In addition, electrical hazards cause more than 300 deaths and 4000 injuries in the entire workplace every year. Electricity causes 12% of young worker deaths in the workplace.

OSHA's electrical standards address electrical workplace hazards, equipment, work practices, safety practices, and more. Employees working on, near, or around electricity may be exposed to dangers such as electric shock, electrocution, burns, fires, and explosions. The objective of the standards is to reduce potential danger by specifying safety design characteristics when installing and using electrical equipment and systems.

Electricity is the flow of electrons from a voltage source through a conductor and back to the source. An electrical shock is obtained as the electrical current travels through the body. You will get an electrical shock if parts of your body complete an electrical circuit by:

- Touching an exposed energized circuit with one part of your body and a grounded point with another part of your body
- Contacting two different energized conductors at the same time

Low voltage does not mean that there is a low hazard of electrical shock. The severity of the shock depends on:

- The path of current through the body
- The amount of current flowing through the body (Ampere)
- The duration of the shocking current through the body

Example 5.11
Electricity causes _____ of young worker deaths in the workplace.

   A. 6%

   B. 12%

   C. 18%

   D. 24%

Answer: *B*

5.4.2   Electrical Injuries

The following are considered to be direct electrical injuries:

- Electrocution (death due to electrical shock)
- Electrical shock and related symptoms resulting from the shock (e.g., tissue damage, neurological disorders, muscle contractions which can cause falls and injuries, etc.)
- Burns
- Arc flash/blast (usually resulting in burns, concussion injuries, etc.)

An arc flash occurs when a flashover of electric current leaves the intended path and travels through the air from one conductor to another or to ground.

The following are considered to be indirect electrical injuries:

- Falls
- Back injuries
- Cuts to the hands

Electric shocks may also cause indirect injuries. Workers on ladders and in elevated areas that encounter shocks can slip, resulting in severe injury or death. Electrical shocks, fires, or falls result from many conditions, including the following hazards:

- Exposed electrical parts
- Overhead power lines
- Inadequate wiring
- Defective insulation
- Improper grounding
- Overloaded circuits
- Wet conditions
- Damaged tools and equipment
- Improper personal protective equipment (PPE)

The following are some of the dangers associated with electricity:

- More than five workers are electrocuted every week.
- It takes very little current flow to cause harm to a person who comes in direct contact with an electrical circuit.
- There is a significant risk of fires due to electrical malfunctions.

Example 5.12

An _____ flash occurs when a flashover of electric current leaves the intended path and travels through the air from one conductor to another or to ground.

A. Ice

B. Fire

C. Ace

D. Arc

Answer: *D*

5.4.3 Controlling Electrical Accidents

Electrical accidents are caused by many factors, including these:

- Unsafe equipment and/or installation
- Unsafe workplace environments
- Unsafe work practices

The following tasks can be followed to control the electrical hazards.

Exposed Electrical Parts. Live parts of electric equipment operating at 50 volts or more must be guarded against accidental contact by cabinets or other forms of enclosures or by any of the following means:

- By location in a room, vault, or similar enclosure that is accessible only to qualified persons.
- By partitions or screens so arranged that only qualified persons will have access to the space within reach of the live parts. Any openings in such partitions or screens shall be so sized and located that persons are not likely to come into accidental contact with the live parts or to bring conducting objects into contact with them.
- By location on a balcony, gallery, or platform so elevated and arranged as to exclude unqualified persons.
- By elevation of at least 8 feet or more above the floor or other working surface and so installed as to exclude unqualified persons.

Conductors Entering Boxes, Cabinets, or Fittings. Conductors entering boxes, cabinets, or fittings must be protected from abrasion. Openings through which conductors enter must be effectively closed. Unused openings in cabinets, boxes, and fittings also must be effectively closed.

Covers and Canopies. All pull boxes, junction boxes, and fittings shall be provided with covers. If metal covers are used, they shall be grounded. In energized installations, each outlet box shall have a cover, faceplate, or fixture canopy. Covers of outlet boxes having holes through which flexible cord pendants pass shall be provided with bushings designed for the purpose or shall have smooth, well-rounded surfaces on which the cords may bear.

5.4.4 Safety Tips

Electricity is a serious workplace hazard, and sadly many of these injuries and fatalities could be easily avoided by taking the following precautions:

- Provide safety training and the proper personal protective equipment (PPE) for your employees.
- Emphasize the importance of always wearing PPE even if workers think it slows them down or observe them, so they are not at risk.

- For arc flash protection, in particular, use PPE that has an arc rating equal to or greater than the calculated incident energy.

- Assume that all overhead wires are energized at lethal voltages. Never assume that a wire is safe to touch even if it is down or appears to be insulated.

- Never touch a fallen overhead power line. Call the electric utility company to report fallen electrical lines.

- Stay at least 10 feet (3 meters) away from overhead wires during cleanup and other activities. Many lines require a much more significant safe working distance. If working at heights or handling long objects, survey the area before starting work for the presence of overhead wires.

- If an overhead wire falls across your vehicle while you are driving, stay inside the vehicle and continue to drive away from the line. If the engine stalls, do not leave your vehicle. Warn people not to touch the vehicle or the wire. Call or ask someone to call the local electric utility company and emergency services.

- Never operate electrical equipment while you are standing in water.

- Never perform repairs to electrical cords or equipment unless qualified and authorized.

- Have a qualified electrician inspect electrical equipment that has gotten wet before energizing it.

- If working in damp locations, inspect electric cords and equipment to ensure that they are in good condition and free of defects, and use a ground-fault circuit interrupter (GFCI).

- Always use caution when working near electricity.

More discussion on this topic is provided in Chapter 18.

Example 5.13

How does electricity work?

   A. Electrons flow from a voltage source through a conductor and back to the source.

   B. Electrons flow from a voltage source through an insulator and back to the source.

   C. A current's circuit path through a conductor is broken.

   D. Heat ignites a voltage source which sends electrons through an electrical current.

Answer: *A*

Electrons flow from a voltage source through a conductor and back to the source.

Example 5.14
How do you get an electrical shock?

  A. Touching a grounded part with a part of your body

  B. Contacting one energized conductor

  C. Touching an exposed energized circuit with rubber gloves

  D. Parts of your body completing an electrical circuit

Answer: D
Parts of your body completing an electrical circuit. Then, electrical current passes through the body.

Example 5.15
Which of the following statements about electrical shocks is correct?

  A. You cannot get an electrical shock if you are standing on the ground.

  B. Low voltage means there is a low hazard of electrical shock.

  C. The severity of the shock depends on the path of the current through the body.

  D. You cannot get an electrical shock if you contact two different energized conductors at the same time.

Answer: C
The severity of the shock depends on the path of the current through the body.

5.5   Caught-in-Between

5.5.1   Basics

The fourth leading cause of deaths in construction workplaces in 2018 was caught in between. These are accidents in which the body part of a worker is trapped, crushed, or squeezed between two or more objects and occurs as a result of material collapse; parts of the body pulled into unguarded machinery and equipment rollover.

When workers do not pay attention to their body position in relation to a caught-in-between hazard, they place themselves between an immovable and a moveable object. Sometimes the worksite has limited access to and travel between staging materials. In such a condition, employees will find themselves squeezed between filled forklifts and pallets of materials. If the forklift operator loses his focus or has an event in which they lose faculties, then the workers walking between the two loads can be crushed. A clear walking path must be established, with the workers being trained to keep those paths free from heavy equipment.

5.5.2   Caught-in-Between Hazards

The recognized hazards related to caught-in-between accidents come from:

- Cranes and heavy equipment
- Tools and equipment
- Material handling
- Masonry and stone work
- Vehicles
- Trenching and excavations

Cranes and Heavy Equipment.   Cranes and heavy equipment can cause a variety of injuries to the workers in a dangerous location. The worker must never place their body between the tracks and the superstructure of the crane. Though it is the closest area to communicate with the driver, this is an extremely dangerous practice. The crane operator and the personnel must have alternative means of communication to avoid this practice. Workers that position themselves between a fixed object, such as a wall, and have heavy machinery running behind or next to them are often in the line of fire. If a backhoe operator is starting an excavation next to the wall where the person is standing, then the possibility of an accident becomes greater.

When a crane is actively moving a load from one area to the next, it produces a swing radius for the rotating part of the equipment with the load. The individual within the swing radius of the crane can be struck by the load or caught-in between the material and the ground if the load drops. Heavy debris can fall from a swinging bucket. A crane can accidently break something loose and send it flying. If hoists break during use, their loads can tumble down and strike workers.

It is necessary for the worker to always keep a safe distance from the equipment so that the equipment does not crush their feet or legs. Some workers are caught up in the job and forget the boundaries that must be maintained from equipment, vehicles, and themselves. It is common for individuals to get their feet in the path of a skip loader or backhoe and get their feet crushed by the equipment. When operating cranes and hoists during construction, always:

- Secure tools and building materials to keep them from falling or being pushed over.
- Barricade areas underneath operation and post warning signs.
- When using hoists for scaffold work, use toe boards, screens, or guardrails to keep materials and tools from falling.
- Use debris nets and other appropriate safeguards to intercept falling objects.

Tools and Equipment.   Most tools and equipment that are not used per the manufacturer's recommendation will lead to some misuse or even a hazardous condition. Guarding of portable power tools and bench tools often has guards taken off for reasons that are behaviorally driven. The worker may feel that they can see the work better, so they remove the guard on the equipment creating a new hazard.

When a guard is missing, it becomes easier for loose clothing, gloves, or jewelry to get caught up in the rotating parts of the equipment. Once that occurs, the speed of most equipment will snag the loose item and pull it into the machinery. If this happens, the part of the body that the loose item is attached to is more difficult to pull out. This would lead to a crushing injury for the worker. It's best practice never to place your hands or body close to moving parts.

The construction manager should have a daily inspection of the integrity of the equipment that is on the construction site. Any person using the equipment must also keep a log of when it has been checked and if there are issues such as broken parts or missing safety functions. This will ensure a regular check of the equipment and help in avoiding any future caught-in-between accidents.

Material Handling.   Workers must use extreme caution when handling material from one location to another. It is common to see workers "stabilizing" the load by having their hands on the material as they travel by cranes or rough terrain forklifts. This practice can lead to being crushed by the load if there is a shift due to road conditions, driver error, or poor rigging.

Workers who must guide a load should not use their own hands on the load but must use an authorized tagline or guideline. The stacking and storing of material is important because the worker that is walking next to the load will be more susceptible to getting trapped under the load if there is a shift in the balance of weight. A clear walking path for pedestrians needs to ensure that if any material that is being stored tips, it will not land on a worker. Storage of materials must be carried out in a way that will aid in the stability of the product. They must be stacked or interlocked in a way as to not create a falling object hazard. The height of the material also matters as to the stability of the cargo. If the product is too high, then tipping one side or the next will be easier. This can lead to someone getting trapped underneath the load.

Workers must be ever mindful as to not place themselves in a way that will pin them against an immovable structure. This will come from hazard recognition tools and training. A system of near-miss reporting can bring to light any conditions that may be hazardous and cause a debilitating injury.

Masonry and Stone Work.   The hazards associated with handling concrete slabs include being caught in-between slabs if they fall or shift onto a worker. Some caught-in-between hazards have been documented while transporting granite and marble slabs. During loading, transport, and unloading of these slabs, the loads have been known to shift and tip over. Workers can either be trapped between slabs or they can be struck by such shifting or falling slabs.

Jacking equipment must be capable of supporting at least two and one-half times the load being lifted during jacking operations, and the equipment must not be overloaded. Lifting inserts mounted or otherwise connected to tilt-up wall panels must be capable of supporting at least two times the maximum expected load applied or transmitted to them. Lifting inserts for other precast members, excluding tilt-up members, must be capable of supporting four times the load. Lift hardware members must be capable of supporting five times the actual expected load applied to the lifting hardware.

Erected shoring equipment must be inspected immediately before, during, and after concrete placement. All base plates, shore heads, extension devices, and adjustment screws must be in close contact and, where necessary, secured with the form and foundation. Shoring equipment that is found to be damaged or weakened after erection must be reinforced immediately. Only essential workers should be allowed near masonry construction. Have concrete structures checked by qualified persons before placing loads. Adequately shore or brace structures until they are permanently supported. Secure unrolled wire mesh so it cannot recoil. Never load a lifting device beyond its intended capacity.

Vehicles.   We have already addressed the need to provide vehicles with backup alarms or to have flaggers when drivers do not have a clear view to the rear. It is bad enough if a worker is struck by a vehicle, but if he or she is also pinned or caught in between another stationary surface, there is a high likelihood that life or limb will be lost.

Blind spots on construction vehicles must always be checked for. When a vehicle is large and has an enclosed cap, it can make blind areas around the equipment which are hard to see. This can be hazardous for ground workers and pedestrians, specifically on roadway work zones.

Trenches and Excavations.   If a trench collapses on a worker, he or she may be caught in between the rubble. In addition to the collapse hazard, at times, a backhoe may be used to lower material like a precast pipe section into a trench with a worker present. In this case, he or she may be adequately protected by remaining in a trench box while the backhoe is operated. If a trench worker was to stand directly between the hoisting path and the trench box wall, he or she would be vulnerable to both the struck-by and caught-in-between hazards. However, if a long trench box (or several adjoining ones) was provided and the worker was far enough away from the backhoe and hoisting path to eliminate a struck-by or caught-in-between hazard, then he or she could safely remain in the trench box.

Example 5.16
The worker must never place their body between the tracks and the superstructure of the crane. Though it is the closest area to communicate with the driver, this is an extremely dangerous practice.

- True
- False

Answer:
True—Cranes and heavy equipment can cause a variety of injuries for the worker in a dangerous location. The worker must never place their body between the tracks and the superstructure of the crane. Though it is the closest area to communicate with the driver, this is an extremely dangerous practice. The crane operator and the personnel must have alternative means of communication to avoid this practice.

5.5.3   Preventing Caught-in-Between Hazards

General Safety Measures.   Engineering controls like shoring, fall protection systems, and properly stacking building materials can help prevent caught-in-between hazards. Some strongly recommended safety practices are:

- Never allow workers to enter an unprotected trench (or excavation) that is 5 feet or deeper unless an adequate protective system is in place; in many cases, trenches less than 5 feet deep may also require such a system.

- Ensure the trench (or excavation) is adequately protected by sloping, shoring, benching, or trench shield systems.

- Always properly stack building materials so they are clear of work areas and so they do not suddenly shift or slide onto a worker.

Trenches. Trenches 5 feet or deeper must be protected using any of the following protective systems. In many cases, even trenches that are less than 5 feet deep must be secured. Protective systems are used to ensure that trenches do not collapse onto workers. All trench protective systems must be designed or verified by a competent person and/or an engineer. These systems include:

- Sloping

- Shoring

- Benching

- Trench shield systems

Fall Protection. While guardrails are a critical engineering control used to protect workers from falling, they can pose a caught-in-between hazard under certain circumstances.

Human Performance Snares. The following are common human performance snares, and care should be taken to overcome them.

- Time constraints to finish the job

- Interruptions or distractions during working

- Multitasking at the jobsite

- Overconfidence of the employee

- Vague guidance by the supervisors

- Too much overnight shift work

- Peer pressure in the workplace

- Frequent changes in the tasks

- Poor physical environment such as poor lighting, ventilation problems, or even layout of machines

- Mental stress

5.6 Summary

When employees might be exposed to falling objects, the employer must have employees wear hardhats and erect toe boards, screens, or guardrail systems to prevent objects from falling from higher levels. This means that employers must either erect a canopy structure or ensure that potential fall objects are far enough from the edge so that those objects will not go over the edge, if they are accidentally displaced, or barricade the area to which objects could fall, thereby prohibiting employees from entering the barricaded area and keeping objects that may fall far enough away from the edge of a higher level so that those objects will not go over the edge if they were accidentally displaced.

Steel or plastic bands must not be used as top rails or midrails. Midrails, when used, shall be installed at a height midway between the top edge of the guardrail system and the walking/working level. Each employee engaged in roofing activities on low-slope roofs, with unprotected sides and edges 6 feet or more above lower levels, shall be protected from falling by guardrail systems, safety net systems, and personal fall arrest systems, or a combination of a warning line system and guardrail system, warning line system and safety net system, warning line system and personal fall arrest system, or warning line system and safety monitoring system. During formwork or rebar assembly, employees shall be protected from falls of 6 feet or more by personal fall arrest systems, safety net systems, or positioning device systems. Passive systems are protective systems that do not involve the actions of employees. A positioning device system is a body belt or body harness system rigged to allow an employee to be supported on an elevated vertical surface, such as a wall, and work with both hands free while leaning. A warning line system is an awareness device erected on a roof to warn employees that they are approaching an unprotected roof side or edge, and which designates an area in which roofing work may take place without the use of guardrail, body belt, or safety net systems to protect employees in the area. A fall protection plan must be prepared by a qualified person and developed specifically for each site. The fall protection plan must be maintained up to date. Any changes to the plan must be approved by a qualified person. A copy of the plan with all approved changes must be maintained at the site. The fall protection plan shall document the reasons why the use of conventional fall protection systems (guardrail systems, personal fall arrest systems, or safety nets systems) is infeasible or why their use would create a greater hazard.

Burns often occur on the hands, although other parts of the body may be affected. In the case of arc flash, additional internal injuries may occur with the burns as a result of the concussive force produced by the explosion from the arc flash. When an arc occurs, current that is available from the source of electrical energy passes from one conductor to another at the point of the arc fault. In an arc flash incident, a large amount of concentrated radiant energy explodes outward from electrical equipment, creating pressure waves that can damage a person's hearing, a high-intensity flash that can damage eyesight, and a superheated ball of gas that can severely burn a worker's body as well as melt metal.

There are many ways to protect workers from being struck by objects and equipment. Two important general rules to follow are:

1. Never put anyone between moving or fixed objects.
2. Always wear bright, highly visible clothing when working near equipment and vehicles.

Many struck-by accidents are associated with vehicles, falling or flying objects, and masonry walls. For example, workers are at risk from falling objects when they are required to work in the vicinity of cranes, scaffolds, and overhead electrical lines. There is also danger from flying objects when using power tools or during activities like pushing, pulling, or prying that can cause objects to become airborne. Flying/falling objects can also roll off rooftops or be accidentally kicked off walkways or scaffold platforms if not properly constrained. Depending on the situation, injuries from being struck by flying or falling objects range from minor ones like bruises to severe ones like blindness or death. Because improper operation of heavy vehicles and equipment poses a life-threatening danger to construction workers, always follow safe practices to minimize injuries and save lives.

Operational plans must always allow for adequate work areas in which to move suspended loads. While guardrails are a critical engineering control used to protect workers from falling, they can pose a caught-in-between hazard under certain circumstances. Guardrail requirements can actually create a hazard at the leading edge of installed floors or roof sections by creating a possibility of employees being caught in between guardrails and suspended loads. Because workers can also be caught in between a collapsed trench that is not properly braced, or warehoused construction materials which were not correctly stacked to prevent sliding, engineering and workplace controls like shoring, fall protection systems, and properly stacking building materials can help in preventing caught-in-between hazard. Sometimes the workers fall into a common human performance trap that leads to at-risk behaviors which can put them in the way of hazards. It is important to recognize such behaviors and address them through coaching efforts.

5.7   Multiple-Choice Questions

5.1 You must be provided with fall protection if you are:

    A. Exposed to fall hazards
    B. Seeing fall hazards.
    C. Thinking about hazards
    D. Taking about fall hazards

5.2 The purpose of fall protection is to:

    A. Elevate tools above the ground
    B. Eliminate danger
    C. Prevent you from falling
    D. All of the above

5.3 You must be provided with fall protection at:

    A. Two feet and above
    B. Four feet and above
    C. Exactly 6 feet
    D. Below 10 feet

5.4 Covers are used to prevent you from falling into:

    A. Stairs
    B. Hotel
    C. Tools
    D. Aisles

5.5 Designated areas can only be used on top of:

    A. Low-slope roofs
    B. Equipment
    C. Trailers
    D. Platforms

5.6 A guardrail is a physical barrier that consists of:

    A.   Top rail

    B.   A midrail

    C.   Posts

    D.   All of the above

5.7 Types of personal fall protection include:

    A.   Personal fall arrest

    B.   Travel restraint

    C.   Positioning devices

    D.   All of the above

5.8 Personal fall protection must be inspected before use for:

    A.   Mildew

    B.   Wear

    C.   Damage

    D.   All of the above

5.9 A ladder safety system prevents falls from:

    A.   Portable ladders

    B.   Mobile ladder stands

    C.   Damage

    D.   All of the above

5.10 If provided fall protection, you must be trained on how to:

    A.   Inspect it

    B.   Use it

    C.   Maintained it

    D.   All of the above

5.11 Which of the following is not a part of the fatal fours?

    A.   Electrical exposure

    B.   Struck-by objects

    C.   Caught-in-between objects

    D.   Eye infection

5.12 Positioning devices shall be secured to anchorage capable of supporting at least twice the potential impact load of an employee's fall or _____ pounds, whichever is greater.

    A.   500 pounds

    B.   160 pounds

    C.   3000 pounds

    D.   5000 pounds

5.13 Toe boards shall be capable of withstanding, without failure, a force of at least _____ applied in any downward or outward direction at any point along the toe board.

    A.   25 pounds

    B.   50 pounds

    C.   100 pounds

    D.   200 pounds

5.14 It is ok to use extension cords that have loosen instillation and have exposed wires.

- True
- False

5.15 If vehicular safety practices are not followed at a worksite, workers are at risk of being struck-by swing equipment.

- True
- False

5.16 _____ debris left on a roof could easily be blown and strike a worker.

A. Secure
B. Chained
C. Loose
D. All of the above

5.8   Practice Problems

5.1 List the fatal fours of construction industry. Why are they fatal fours?

5.2 List some fall prevention measures.

5.3 List some areas required to have fall protection.

5.4 List some duties that are to be performed to avoid fall protection.

5.5 List some OSHA regulations for low-sloped roof work.

5.6 List some OSHA regulations for protection from falling objects.

5.7 Discuss the types of passive fall protection.

5.8 Discuss the guardrails design criteria.

5.9 Discuss the types of active fall protection.

5.10 What are the different ways struck-by accidents happen?

5.11 List some direct and indirect electrical injuries.

5.12 List some safety tips for electrical hazards.

5.13 List some recognized hazards related to caught-in-between accidents.

5.14 List some preventing measured to avoid caught-in-between hazards.

5.9   Critical Thinking and Discussion Topic

Construction fatal fours sound very basic type of construction accidents. Any grown-up human being has the common sense of avoiding these fatal fours. With careful safety practices, these fatal fours can be avoided with ease. There is no need to provide formal training to employees for these fatal fours. Join this discussion and give your opinion.

CHAPTER 6

Worker Compensation and Injured Worker Management

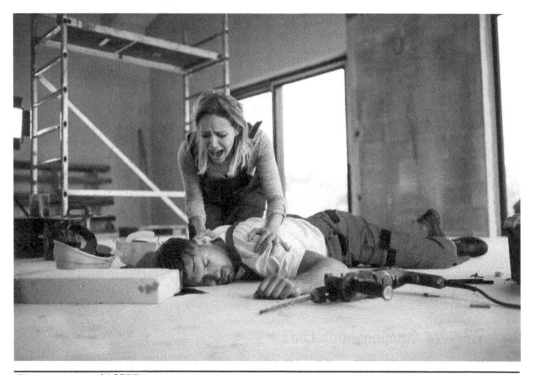

(Photo courtesy of 123RF.)

6.1   General

Construction work is dangerous, and sites present risks beyond other workplaces due to the complexities of the work environment, including operation and movement of heavy equipment, uneven terrain, and exposure to the elements. Despite the danger, construction workers have the right to work in a position that is relatively safe and protected from unnecessary risks.

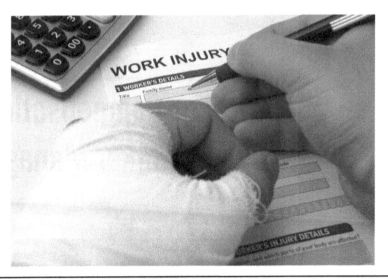

Figure 6.1   Filling up a work injury claim form.
(Photo courtesy of 123RF.)

The concept of workers' compensation (Fig. 6.1) has been developed in order to remove the need for workers to go to court and obtain compensation for accidents suffered in the workplace. Proving that an accident was the result of employer negligence often was costly and time-consuming. Today, all states have enacted workers' compensation laws requiring employers to provide their employees with no-fault insurance. These laws require employees to give up the right to sue their employers for compensation resulting from injury or illness sustained in the workplace and employers to provide compensation irrespective of whether or not the employee's negligence contributed to the injury. The reasoning behind the development of workers' compensation was the fairness of employees and the reduction of costs to employers as a result of injuries at the workplace. It is a no-fault approach to resolving workplace accidents by providing needed medical care and replacement income as well as rehabilitating injured workers.

6.2   Workers' Compensation Laws

All 50 states, the District of Columbia, Guam, and Puerto Rico, have enacted statutes establishing workers' compensation programs. A separate program covers federal civilian workers. Workers' compensation schemes vary across the states in terms of who is allowed to provide insurance that compensates for injuries and illnesses and the level of benefits. Generally, state laws require employers to obtain insurance or to prove that they have the financial capacity to carry their own risk and self-insurance. Some states have state funds that provide workers' compensation insurance. These state funds may be monopolistic or may be competitive. Private insurance coverage can be purchased in states that do not have monopolistic state funds. The premiums paid by employers are based on the occupational classifications of their workers and their record of injury frequency and benefit payments (experience based).

Private insurance providers are the primary sources of workers' compensation insurance. They are authorized to sell insurance in all but five states that have monopolistic state funds—Ohio, North Dakota, Washington, West Virginia, and Wyoming. In these five states, employers must purchase workers' compensation insurance from a state fund, unless they are allowed to self-insure.

States have different policies on how to pay for permanent disabilities. Some pay benefits lifetime or retirement benefits while others limit benefits to a specified period or to a specified dollar amount. Many states have developed a schedule to compensate for permanent impairments based on body parts, for example, loss of hearing. An example partial award schedule for the State of Washington is shown in Table 6.1. For conditions not addressed in the schedules, benefits generally are determined based on the amount of impairment or the inability of the worker to compete for a position with comparable earnings.

The objective of workers' compensation is to minimize the need for legal action to resolve issues related to accidents at work. Workers' compensation laws vary from state to state, but, in general, they all contain the following provisions:

- Income Replacement. Injured employees will lose their income if they are unable to work. Thus, workers' compensation provides replacement income until the injured worker is able to return to work.

- Rehabilitation. Medical costs for an injured worker are covered until he or she recovers from injuries sustained during the accident. In the event that the employee has a permanent disability and is unable to return to work in his or her original work classification, vocational training shall be provided to enable the injured worker to return to work in a different work classification.

- Accident Prevention. Employer's cost of workers' compensation shall be determined in such a way as to provide an incentive for investment in the prevention of accidents.

- Cost Allocation. Accident costs should be allocated based on accident severity and based on the injury history and hazards presented by the classification of the work. The hazards encountered by a roofer, for example, are greater than those a plumber experienced.

- Workers' compensation provides benefits to workers who are injured on the job or who contract a work-related illness. Benefits include medical treatment for work-related injuries or illness. Temporary total disability benefits are paid while the worker recuperates away from work. If the worker is unable to return to work in his or her classification of work, rehabilitation training in another classification of work shall be provided. If the condition of the worker has lasting consequences after the worker heals, permanent disability payments may be paid. In the case of a fatality, the worker's dependents receive survivor benefits.

- Workers' compensation laws were enacted by the federal government and the various states to:
  - Provide income and medical benefits to injured workers or provide income to injured workers' dependents, regardless of fault.

LEG	
Leg above the knee joint with short thigh stump (3" or less below the tuberosity of ischium)	123,291.12
Leg at or above the knee joint with functional stump	110,961.99
Leg below knee joint	98,633.04
Leg at ankle (syme)	86,303.85
FOOT	
Foot at mid-metatarsals	43,151.97
TOE	
Great toe with resection of metatarsal bone	25,891.14
Great toe at metatarsophalangeal joint	15,534.60
Great toe at interphalangeal joint	8,219.46
2nd lesser toe with resection of metatarsal bone	9,452.28
3rd lesser toe with resection of metatarsal bone	9,452.28
4th lesser toe with resection of metatarsal bone	9,452.28
5th lesser toe with resection of metatarsal bone	9,452.28
2nd lesser toe at metatarsophalangeal joint	4,602.84
3rd lesser toe at metatarsophalangeal joint	4,602.84
4th lesser toe at metatarsophalangeal joint	4,602.84
5th lesser toe at metatarsophalangeal joint	4,602.84
2nd lesser toe at proximal interphalangeal joint	3,411.09
3rd lesser toe at proximal interphalangeal joint	3,411.09
4th lesser toe at proximal interphalangeal joint	3,411.09
5th lesser toe at proximal interphalangeal joint	3,411.09
ARM	
Arm at or above the deltoid insertion or by disarticulation of the shoulder	123,291.12
Arm at any point below the deltoid insertion to below the elbow joint at the insertion of the biceps tendon	117,126.51
Arm at any point from below the elbow joint distal to the insertion of the biceps tendon to and including mid-metacarpal amputation of the hand	110,961.99
FINGER	
All fingers except the thumb at the metacarpophalangeal joints	66,577.08
Thumb at metacarpophalangeal joint or with resection of carpometacarpal bone	44,384.85
Thumb at interphalangeal joint	22,192.38
Index finger at metacarpophalangeal joint or with resection of metacarpal bone	27,740.55
Index finger at proximal interphalangeal joint	22,192.38
Index finger at distal interphalangeal joint	12,205.80
Middle finger at metacarpophalangeal joint or with resection of metacarpal bone	22,192.38

Middle finger at proximal interphalangeal joint	17,753.91
Middle finger at distal interphalangeal joint	9,986.67
Ring finger at metacarpophalangeal joint or with resection of metacarpal bone	11,096.22
Ring finger at proximal interphalangeal joint	8,877.03
Ring finger at distal interphalangeal joint	5,547.99
Little finger at metacarpophalangeal joint or with resection of metacarpal bone	5,547.99
Little finger at proximal interphalangeal joint	4,438.53
Little finger at distal interphalangeal joint	2,219.25
MISCELLANEOUS	
Loss of one eye by enucleation	49,316.34
Loss of central visual acuity in one eye	41,097.00
Complete loss of hearing in both ears	98,633.04
Complete loss of hearing in one ear	16,438.71
Compensation for unspecified disabilities of 100% as compared to total bodily impairment	205,485.09

Source: Washington State Department of Labor and Industries, https://lni.wa.gov/claims/for-workers/claim-benefits/permanent-partial-disability. Accessed March 27, 2020.

TABLE 6.1   Permanent Partial Disability Awards Schedule for Dates of Injury for FY 2019

- Eliminate the need for legal proceedings to resolve claims associated with workplace injuries.
- Promote employer interest in providing safe working environments and establishing safe working processes and procedures.
- Prevent accidents by encouraging implementation of good workplace safety and health procedures.

6.3   Workers' Compensation Insurance

Workers' compensation insurance is a no-fault insurance that is mandated by state law, in that the employer cannot deny a claim by an insured employee on the basis that the employee was negligent in causing the injury. Similarly, the employee cannot sue the employer on the basis that the employer's negligence has contributed to the injury or disease of the worker. While an employee is unable to sue his or her employer, an employee of a subcontractor or supplier may be able to sue the general contractor, who is not his or her employer, if the general contractor allowed unsafe conditions to exist on the job site. These are known as third-party lawsuits.

The dollar value of the construction firm's liability is usually established by statute. The benefits offered are usually medical benefits, loss of earnings, and retraining if the employee is unable to perform the duties of the current job. As stated in the previous section, workers' compensation insurance is offered by monopolistic state agencies in some states, while in other states, it is available from insurance companies or carriers. Construction firms that meet state requirements may choose to be self-insuring. The essence of a workers' compensation insurance contract is that the insurer agrees, for a price (premium), to assume the liability imposed on the insured construction company by the workers' compensation statute of the state named in the policy.

Only injuries incurred while undertaking the work prescribed by the employee's job description, assigned by the supervisor or normally performed in the course of performing the work classification, shall be covered. Falling off a ladder at the employee's home would not be covered because it did not occur as a result of his or her employment.

Example 6.1
Which of the following is false for workers' compensation insurance?

A. It is a no-fault insurance.

B. The employer cannot deny a claim by an insured employee on the basis that the employee was negligent in causing the injury.

C. The employee cannot sue the employer on the basis that the employer's negligence has contributed to the injury or disease of the worker.

D. It is mandated by many states in the United States.

Answer: D
Workers' compensation insurance is mandated by all states.

6.4 Types of Disabilities

A work-related injury or illness can range in severity from relatively mild to devastating. In many cases, workers who suffer from work-related injuries may be going back to work in a matter of weeks, while others may face life-long consequences from injury. Workers who are disabled due to work-related injuries and will never recover will be eligible for unique workers' compensation benefits. All workers who are injured or become ill in the course of performing their job duties have their medical expenses, including all relevant treatments and prescriptions, covered by their workers' compensation benefits. In addition, where their injury or illness prevents them from working, they may be granted temporary disability benefits to cover lost wages, which is usually a percentage of their average weekly earnings. If it becomes clear that their injuries are too severe for them to recover or resume gainful employment, they may be eligible for permanent disability.

Injuries that are compensable typically fall into one of the following four categories:

- Temporary Partial Disability (TPD). Injured worker is capable of light or part-time duties and is expected to recover fully. For example, a broken toe that can allow for office work but not manual labor.

- Temporary Total Disability (TTD). Injured worker is incapable of any work for a period of time but is expected to recover fully with no permanent disability. This is the most common type of injury. An example would be a broken bone that requires surgery.

- Permanent Partial Disability (PPD). Injured worker is not expected to recover fully but will be able to work again. An example would be the loss of a finger.

- Permanent Total Disability (PTD). Injured worker is not expected to recover and is unable to work in any job classification.

Injured workers must be referred to a physician in order to determine the extent of injuries, prescribe appropriate medical treatment, determine temporary work restrictions, and set a timeline for recovery.

The benefits paid for each type of injury vary from state to state. Medical costs for each type of injury would be covered. Workers with temporary partial disabilities are typically given tasks that they are capable of performing. Workers with a temporary total disability are also paid two-thirds of their wages during the time of disability. Workers sustaining a permanent partial disability may receive benefits based on a standard schedule adopted by the state or, if not a scheduled disability, based on the amount of disability. For example, a worker with a disability of 25% can receive 25% of his or her wages. Schedule disabilities typically result from the loss of a body part, such as an arm, eye, or finger. If the worker is permanently disabled, typical benefits are two-thirds of their wages for life, until retirement, or for a set period. In the case of a worker's fatality at the workplace, the spouse and minor children shall be given the death benefit. In addition to the benefits discussed above, workers' compensation insurance may also fund medical and/or vocational rehabilitation. Medical rehabilitation means providing whatever treatment is needed to restore, to the extent possible, any lost ability to function normally. Vocational rehabilitation means providing education and training needed to enable the injured worker to find employment in another work classification. Under the workers' compensation statutes of the various states, the employer's liability is limited to a specific benefit level when an employee is injured or killed on the job. The benefits are automatic and cannot be appealed by the employer. Since workers' compensation insurance is a type of no-fault insurance, a construction company cannot be charged with additional compensation for injury or death by its employees or their heirs.

6.5 Cost Reduction Strategies

Because compensation premiums for workers are a significant business expense, construction companies need to take strategies to minimize their costs. The following strategies are suggested as methods for reducing workers' compensation costs:

- Have a good accident prevention program. The most effective strategy for reducing the compensation costs for workers is to minimize the potential for accidents occurring. A site-specific safety plan to identify all project hazards and strategies to mitigate those hazards provides a framework for preventing accidents. Company leaders must emphasize safety and ensure that field supervisors do not tolerate unsafe behavior.

- Stay in touch with all employees who were injured. Workers who are injured should know that their employer is concerned about their well-being. They should be encouraged to return to work, even in a limited manner, as soon as they are medically able.

- Have a good return-to-work program and use it. The sooner an injured worker returns to work, the lower the claim will be. Ensure that the worker's physician agrees that he or she can return to work and determine what work limitations the physician stipulates.

- Carry out comprehensive investigations of any accidents. Eliminating accidents is the key to reducing workers' compensation costs. Any accident that occurs must be investigated thoroughly to determine its cause. Measures must then be implemented to avoid such an accident happening again.

6.6   Remedy for Injured Workers

If a construction worker gets injured at work, they are eligible for certain benefits. Each state has its own version of workers' compensation. This is a form of insurance that employers are required to carry and cover expenses related to workplace injuries. Compensation benefits for the workers provide compensation for lost wages and medical expenses. There may also be additional benefits to vocational education. Also, compensation for permanent disabilities caused by a workplace injury is usually covered. Death benefits may also be available for a deceased worker's surviving family members.

6.7   Compensable Injuries

Almost any injury to the workplace that is serious enough to cause missed work is covered by workers' compensation. Common construction-related injuries and their causes include:

- Broken bones from falls or impact with heavy equipment
- Traumatic brain injuries from being struck in the head by falling objects
- Burns from welding sparks, fires, explosions, or electrocutions
- Eye injuries from welding or flying debris

Chronic exposure to certain hazards can result in injury, which can also be covered and include:

- Heat stroke from working in the sun on hot days
- Hand-arm vibration syndrome from operating jack hammers
- Hearing loss from working around loud equipment
- Repetitive motion injuries from performing the same tasks repeatedly
- Back strain from lifting heavy objects

Workers' compensation can also cover workplace illnesses that typically result from repeated chronic exposure, such as:

- Respiratory disease from inhaling solvent, paint, and other construction-related substances
- Hearing loss from exposure to loud noise

- Other diseases, such as mesothelioma caused by exposure to asbestos
- Nerve damage from exposure to lead or lead products

Example 6.2
Which of the following injuries is not covered by workers' compensation?

    A. Broken bones from falls or impact with heavy equipment

    B. Traumatic brain injuries from being struck in the head by falling objects

    C. Burns from welding sparks, fires, explosions, or electrocutions

    D. Falling off a ladder at the employee's home

Answer: D
Almost any injury at the workplace (not at home) that is serious enough to cause missed work is covered by workers' compensation.

6.8 How to Obtain Workers' Compensation Benefits

You must have been injured from your work or become ill as a result of your work to qualify for workers' compensation, and you must report the injury to your employer. Workers' compensation is a no-fault insurance program. There is no need to prove that the employer is negligent to qualify. However, you are required to follow the reporting and filing requirements of your state. States differ in the way in which they manage the compensation programs of their workers. The timeframe for filing claims is relatively short. You will likely be required to submit to a physical examination by a physician chosen by the program as opposed to your usual doctor.

Example 6.3
To qualify for workers' compensation you must:
[select all that apply]

    A. Have been injured from your work

    B. Become ill as a result of your work

    C. Report the injury to your employer

    D. Be healthy before joining the work

Answer: A, B, C

6.9 Death Benefits

Workers' compensation insurance provides compensation to victims of work-related injuries. When a worker dies as a result of his or her work-related injury or illness, death benefits may be available to members of his or her family who depended on the victim for financial support, including spouses, children, and relatives who lived with the deceased employee. The laws on these benefits vary by state.

6.10   Injured Worker Management

Workers injured should be returned to the work force as soon as possible. If a worker is injured in a project, the first step is to bring the injured worker to the nearest medical clinic to have a doctor inspect the worker. The doctor will determine the extent of the injuries and the type of job the injured worker can do. Based on the doctor's instructions, the construction company should devise a return-to-work strategy for the worker. This may involve recuperation, physical therapy, shorter work hours, and/or alternative work assignments. Keeping the worker connected with the workplace is both good medicine for the worker and good business for the company. Research has shown that effective return-to-work strategies encourage faster recovery and avoid a spiral downward into disability.

Providing return-to-work options benefits the injured worker and reduces the financial impact on the company's workers' compensation premiums. In addition to reducing claim costs and insurance premiums, an effective return-to-work strategy:

- Encourages communication between the employer and the injured employee—a key factor in the employee's recovery
- Allows an experienced employee to continue working for the company
- Keeps the loss of productivity to a minimum
- Reduces the cost of training replacement employees
- Keeps the injured employee active and speeds medical recovery
- May reduce the risk of reinjury
- Provides a sense of job security to the injured employee
- Allows the injured employee to maintain contact with coworkers
- Shows that the company values the injured employee and his or her contributions to the company

The employer may offer a transitional job to an injured employee to enable him or her to return to work when restrictions preclude performing the job held when the injury occurred. This reconnects the injured employee with the construction company and illustrates the employer's concern for the employee's welfare. There are three types of transitional jobs that may be offered:

1. Modified work involves adjusting or altering the way in which a job is normally performed in order to accommodate the severe disabilities of the employee.
2. Part-time work involves working less than a normal work schedule because the doctor has not released the injured worker for full-time work.
3. Alternative work is a different job within the company that meets the physical restrictions the doctor prescribed. For example, an injured carpenter may be employed in the estimating department until his or her physician authorizes return to field work.

Individuals who sustained an injury that results in permanent disability need to be given vocational training in a work classification for which they are medically qualified. This enables the injured employee to return to the work force in a different job classification. An injured worker's physician needs to verify that he or she is physically qualified for the new work classification before being allowed to start the vocational training.

Example 6.4

Individuals who sustained an injury that results in permanent disability need to be given:

A. Vocational training in a work classification for which they are medically qualified

B. Total home rest for the rest of his/her life

C. A handsome money so that he/she can spend during the rest of his/her life

D. Some time to return to workplace

Answer: A

6.11 Summary

Workers' compensation is a form of insurance providing wage replacement and medical benefits to employees injured in the course of employment in exchange for mandatory relinquishment of the employee's right to sue his or her employer for the tort of negligence. While plans differ among jurisdictions, provision can be made for weekly payments in place of wages (functioning in this case as a form of disability insurance), compensation for economic loss (past and future), reimbursement or payment of medical and like expenses (functioning in this case as a form of health insurance), and benefits payable to the dependents of workers killed during employment. The system of collective liability was created to prevent that, and thus to ensure security of compensation to the workers.

6.12 Multiple-Choice Questions

6.1 The concept of workers' compensation was developed to eliminate the need for workers to:

A. Go to court to obtain compensation for injuries incurred in the workplace.

B. Go to OSHA to obtain compensation for injuries incurred in the workplace.

C. Go to employer to obtain compensation for injuries incurred in the workplace.

D. Go to their worker association to obtain compensation for injuries incurred in the workplace.

6.2 Construction work is dangerous and sites pose risks beyond other workplaces due to the:

A. Static nature of the work environment

B. Dynamic nature of the work environment

C. Noise nature of the work environment

D. Crowded nature of the work environment

    6.3 All of the following actions are effective ways in which employers reduce workers' compensation costs, except:

        A. Enforcing safety policies
        B. Communicating safety policies
        C. Providing classroom training
        D. Using appropriate safety equipment

    6.4 Which insurance does carry the primary sources of workers' compensation insurance?

        A. Private insurance
        B. Company
        C. OSHA
        D. Workers association

6.13 Practice Problems

    6.1 Discuss the workers' compensation laws.

    6.2 Discuss the workers' compensation insurance policy.

    6.3 Discuss the four categories of disabilities.

    6.4 Discuss the cost reduction strategies for workers' compensation.

    6.5 Discuss the remedial measures for injured workers.

    6.6 List some common construction-related injuries and their causes.

    6.7 List some workplace illnesses that typically result from repeated chronic exposure.

    6.8 Discuss the procedure to obtain workers' compensation benefits.

6.14 Critical Thinking and Discussion Topic

Adopting, maintaining, and ensuring safety in a construction job cost a lot of money, effort, and time. However, these activities may decrease the occurrence of accidents. On the other hand, if safety is not forced or ensured, the company cost will decrease. However, there might be more chances of having accidents and the accidents cost money. If you consider these two scenarios, both have cost saving strategies. What is your opinion on these two strategies?

CHAPTER 7

Hazards Analysis

(Photo courtesy of Anamul Rezwan from Pexels.)

7.1 General

To understand the behavior-based safety program, it is necessary to understand what a hazard is first. A hazard is anything that can hurt or injure someone. While you know that a workplace can have different hazards, it does not mean that there will be an injury. Hazards must be exposed in order for there to be an injury or an incident. Hazard identification is primary to understanding how an accident can happen. Later in this program, we will review accident causation from a few different models.

The hazard itself may be the safety issue for workers, or the exposure to that hazard may be the issue for the employer. For example, if there is a cave-in hazard for trenching and excavation, the cave-in is possible even without the presence of

FIGURE 7.1   Complexity of a construction site.
(Photo courtesy of Pixabay from Pexels.)

workers. The hazard becomes an accident or incident when workers are present, meaning exposure.

Construction sites can be hazardous working environments (Fig. 7.1). Safety managers and project managers are tasked with identifying and eliminating hazards to ensure everyone's safety on the jobsite. Conducting a job hazard analysis for each project job is an effective way to mitigate potential hazards and protect workers.

A job hazard analysis is a technique that focuses on job tasks as a way to identify hazards before they occur. It focuses on the relationship between the worker, the task, the tools, and the working environment. Ideally, you will take steps after you identify uncontrolled hazards to eliminate or reduce them to an acceptable level of risk. The risk can be determined as the combined effect of the probability of occurring and its severity of hazards.

Example 7.1
The formula for Risk is:

    A. Risk = Probability x Severity

    B. Risk = Reward x Severity

    C. Risk = Severity / Probability

    D. Risk = Probability / Severity

Answer: *A*

7.2   Types of Hazards

7.2.1   Chemical Hazards

A hazardous chemical is any substance classified as:

- A physical hazard
- A health hazard
- A simple asphyxiant
- Combustible dust
- Pyrophoric gas
- Another hazard not otherwise classified

7.2.2   Physical Hazards

Physical hazards are chemicals that are classified as posing one or more of the following hazardous effects:

- Explosive
- Flammable gases, aerosols, liquids, or solids
- Oxidizer (liquid, solid, or gas)
- Self-reactive
- Pyrophoric (liquid or solid)
- Self-heating
- Organic peroxide
- Corrosive to metal
- Gas under pressure
- Emits flammable gas when in contact with water

In addition, a pyrophoric gas is a chemical in a gaseous state that will ignite spontaneously in air at a temperature of 130°F (54.4°C) or below. You should also be aware that some dust can be combustible in certain situations. Substances that can create combustible dust when they are processed could include:

- Grains, sugar, and other types of food and agricultural products
- Charcoal, soot, and similar materials
- Chemicals, such as sulfur
- Metals, such as magnesium or aluminum
- Plastics and resins

A spark or flame can cause a combustible dust fire or explosion if, for example, a buildup of the dust is disturbed so that the dust is suspended in the air in a confined area.

7.2.3 Health Hazards

Health hazards are chemicals that are classified as posing one or more of the following hazardous effects:

- Acute toxicity from any route of exposure (e.g., ingestion, inhalation, etc.)
- Skin corrosion or irritation
- Serious eye damage or eye irritation
- Respiratory or skin sensitization (e.g., an allergic reaction)
- Aspiration hazard
- Carcinogenicity
- Reproductive toxicity
- Germ cell mutagenicity
- Specific target organ toxicity from a single or repeated exposure (e.g., a chemical that can damage the liver)

In addition, a simple asphyxiant is a substance or mixture that displaces oxygen in the air to create an oxygen-deficient atmosphere that can lead to unconsciousness and death. Some of these health hazards can occur rapidly, following a brief exposure (and acute effect). Health hazards can also cause long-term effects that usually follow repeated long-term exposure (a chronic effect).

7.3 Importance of Job Hazard Analysis

Job hazard analysis is a tool that is used across organizations to help workers visualize the work that needs to be done and break it down into steps. It is different from the standard operating procedure because each step will have the hazard identified and the control identified. The job hazard analysis is commonly used for training workers as to how to perform a job in the safest manner possible. A standard operating procedure will only show the person how to do the job by breaking down its components. The components of any job are usually communicated through instructions by people who have done it in the past. These can be front-line supervisors or the manufacturers of the equipment.

Many workers are injured and killed every day at the workplace in the United States. Safety and health can add value to your business, work, and life. You can help prevent workplace injuries and illnesses by looking at your workplace operations, establishing proper job procedures, and ensuring that all employees are trained properly. One of the best ways to identify and establish proper work procedures is to conduct a work hazard analysis. A job hazard analysis is one component of the larger commitment of a safety and health management system.

Supervisors can use the findings of a job hazard analysis to eliminate and prevent hazards in their workplaces. This is likely to result in fewer injuries and illnesses for workers, safer, more efficient work methods, reduced workers' compensation costs, and increased worker productivity. The analysis also can be a valuable tool for training new employees in the steps required to perform their jobs safely. For a job hazard analysis to be effective, management must demonstrate its commitment to safety and health and ensure that any uncontrolled hazards identified are corrected. Otherwise, management will lose credibility, and employees may hesitate to go to management when dangerous conditions threaten them.

7.4 Appropriate Jobs for Hazard Analysis

A job hazard analysis can be conducted on many jobs in your workplace. Priority should go to the following types of jobs:

- Jobs with the highest injury or illness rates
- Jobs with the potential to cause severe or disabling injuries or illness, even if there is no history of previous accidents
- Jobs in which one simple human error could lead to a severe accident or injury
- Jobs that are new to your operation or have undergone changes in processes and procedures
- Jobs complex enough to require written instructions

7.5 Getting Started for Hazard Analysis

A job hazard analysis is a process used to identify and eliminate or mitigate potential hazards. The process involves the assignment of tasks, the identification of potential hazards for each task, the risk assessment to determine the likelihood and severity of each hazard, and the development of preventive measures to eliminate each hazard.

The first step is to determine which jobs you want to create a job hazard analysis for and prioritize which ones you will tackle first. You might want to start with jobs with a high accident rate at your company or jobs with the highest potential to cause serious injury or death. Jobs with low accident rates or potential for serious injuries should go to the bottom of your list. New jobs and jobs that are rarely performed should be given priority because your employees are unlikely to have as much experience in performing these tasks. They might not be as familiar with the necessary safety measures compared to the jobs they perform on a regular basis.

You will eventually create a library of job hazard analyses on all the jobs that your employees do. Periodically, these will need to be updated and altered when the environment or site conditions change or if the sequence of steps to perform a job is modified or changed.

a. Involve your employees. It is very important to get your employees involved in the hazard analysis process. They have a unique understanding of their job, and this knowledge is invaluable in identifying hazards. Involving employees will help minimize oversights, ensure a quality analysis, and get workers to "buy in" to the solutions because they will share ownership in their safety and health program.

b. Review your accident history. Review with your employees your worksite's history of accidents and occupational illnesses that needed treatment, losses that required repair or replacement, and any "near misses"—events in which an accident or loss did not occur, but could have. These events are indicators that the existing (if any) hazard controls may not be sufficient and deserve more scrutiny.

c. Conduct a preliminary job review. Discuss with your employees the hazards they know exist in their current work and surroundings. Brainstorm with them for ideas to eliminate or control those hazards. If there are any hazards that pose an immediate risk to the life or health of an employee, take immediate action to protect the worker. Any problems which can be easily corrected should be corrected as soon as possible. Do not wait to complete your job hazard analysis. This will demonstrate your commitment to safety and health and enable you to focus on the hazards and jobs that need more study because of their complexity. For those hazards determined to present unacceptable risks, evaluate types of hazard controls.

d. List, rank, and set priorities for hazardous jobs. List jobs with hazards that present unacceptable risks, based on those most likely to occur and with the most severe consequences. These jobs should be your first priority for analysis.

e. Outline the steps or tasks. Nearly every job can be broken down into job tasks or steps. When beginning a job hazard analysis, watch the employee perform the job and list each step as the worker takes it. Be sure to record enough information to describe each job action without getting overly detailed. Avoid breaking down steps in such detail that it becomes unnecessarily long or so wide that it does not include basic steps. You may find it valuable to get input from other workers who have performed the same job. Later, review the employee's job steps to make sure you haven't omitted anything. Point out that you are evaluating the job itself, not the employee's job performance. Include the employee in all phases of the analysis—from reviewing the job steps and procedures to discussing uncontrolled hazards and recommended solutions. Sometimes, it may be helpful to photograph or videotape a worker performing a job when conducting a job hazard analysis. These visual records can be useful references when performing a more detailed analysis of the work.

7.6   Breaking the Jobs Down for Hazard Analysis

Once you've decided which jobs to focus on first, it's time for each job to be broken down into individual sequential tasks. The best way to do this is to have a supervisor or safety manager observe an experienced worker performing the job. The observer should take detailed notes of each step, focusing on what is being done rather than how the worker is performing each task.

Taking a video recording of the worker performing the job can be helpful in ensuring you haven't missed a step in the process or gotten any of the steps out of the order they were performed. The video can also be used to review the steps and verify their accuracy, as well as to provide visual assistance to your team to identify potential hazards.

When sequencing out steps in a job, be sure to start each step with an action verb like "lift" or "cut" in order to focus on what is being done. Try to keep the number of steps or tasks in each job to 10 or fewer. If 10 or more steps are needed for a particular job, it might be useful to separate the job into two separate jobs.

Avoid being too general, because you might be leaving out a crucial step in the process. On the other hand, being too detailed may cause you to combine individual tasks or overcomplicate the steps needed to complete the job.

7.7 Identifying Potential Hazards

Now that you have the task broken down into individual tasks, it's time to identify all the potential hazards for each task. Get workers to know the work involved in this step, because they already know some of the potential hazards. Remember, you're trying to identify potential hazards, not just the hazards you've seen or that have been eliminated when you saw a worker performing a job.

For example, let's say you observed a worker cutting concrete as part of conducting a job hazard analysis. During the observation, the worker used a concrete saw equipped with a blade guard and a dust collector. The worker was wearing safety goggles and earplugs to protect his eyes and ears.

Just because the worker implemented these safety measures does not mean that there were no potential hazards such as flying debris or exposure to silica dust and loud noise. Potential hazards have been eliminated because the worker has followed proper safety procedures. You can't rely on every worker to do this, which is why conducting a job hazard analysis is so important.

OSHA suggests asking the following questions during your job hazard analysis to identify potential hazards:

- What can go wrong?
- What are the consequences?
- How could it arise?
- What are other contributing factors?
- How likely is it that the hazard will occur?

In addition to these questions, it is important to take into account potential hazards from working at heights, hazards from the use of specific tools and equipment, and changes in weather conditions. Falls, electrocutions, objects being struck and caught in or between are some of the deadliest hazards at construction sites.

7.8 Preliminary Hazard Analysis

Preliminary hazard analysis (PHA) is an initial high-level screening exercise that can be used to identify, describe, and rank major hazards during conceptual stage of a facility design. This technique can also be used to identify possible consequences and likelihood of occurrence and provide recommendations for hazard mitigation. PHA is performed to identify areas of the system, which will have an effect on safety by evaluating the major hazards associated with the system. It provides an initial assessment of the identified hazards. A sample of hazard analysis template is provided in Table 7.1.

PHA typically involves:

- Determining hazards that might exist and possible effects.
- Determining a clear set of guidelines and objectives to be used during a design.
- Assigning responsibility for hazard control (management and technical).
- Allocating time and resources to deal with hazards.

Job Title	Job Location	Analyst	Date
Task #	Task Description		
Hazard Type	Hazard Description		
Consequence:	Hazard Controls		
Rational or Comment:			

TABLE 7.1   Sample Job Hazard Analysis Form by OSHA

"Brainstorming" techniques are used during which the design or operation of the system is discussed on the basis of the experience of the people involved in the brainstorming activity. Checklists are commonly used to assist in identifying hazards. The results of the PHA are often presented in tabular form, which would typically include information such as but not limited to:

- A brief description of the system and its domain
- A brief description of any subsystems identified at this phase and the boundaries between them
- A list of identified hazards applicable to the system, including a description and unique reference
- A list of identified accidents applicable to the system including a description, a unique reference, and a description of the associated hazards and accident sequences
- The accident risk classification
- Preliminary probability targets for each accident
- Preliminary predicted probabilities for each accident sequence
- Preliminary probability targets for each hazard

- A description of the system functions and safety features
- A description of human error which could create or contribute to accidents

The advantages of using the PHA method include:

- It identifies the potential for major hazards at a very early stage of project development.
- It provides basis for design decisions.
- It helps to ensure plant to plant and plant to environment compatibility.
- It facilitates a full hazard analysis later.

Example 7.2
When conducting a required assessment of the various hazards that may be present, an employer would consider all the following except:

A. Source of motion of objects

B. Source of workers' mental stress

C. Source of high and low temperatures

D. Source of electric hazards

Answer: C

7.9   Correcting or Preventing Hazards

After reviewing your list of hazards with the employee, consider which control methods will eliminate or reduce them. The most effective controls are engineering controls that physically change the machine or work environment to prevent employees from being exposed to hazards. The more reliable or less likely, the control of hazards can be circumvented, the better. Administrative controls may be appropriate if this is not feasible. This may involve changing the way employees do their jobs. Discuss your recommendations with all employees who carry out the work and consider their responses carefully. If you are planning to introduce new or modified job procedures, be sure that they understand what they need to do and the reasons for the changes.

7.10   Detailed Hazard Analysis

Hazard analysis is the process of identifying hazards that may arise from a system or its environment, documenting their unwanted consequences and analyzing their potential causes. In most cases, project hazard analysis (PHA) is good enough to identify and prevent hazards. In cases where there is a potential for serious injury, multiple injuries, catastrophic illness, etc., a detailed hazard analysis shall be performed. The most widely used detailed hazard analysis methods are as follows:

- Failure mode and effects analysis
- Hazard and operability review

- Technic of operations review
- Human error analysis
- Fault tree analysis
- Risk assessment

7.10.1   Failure Mode and Effects Analysis

Failure mode and effects analysis (FMEA) is a formal step-by-step analytical method that is a spin-off of reliability analysis—a method used to analyze complex engineering systems. FMEA steps proceed as follows:

1. Critically examine the system in question.
2. Divide the system into its various components.
3. Examine each individual component and record all of the various ways in which the components may fail. Rate each potential failure according to the degree of hazard posed on a scale from 0 to 4:

   0 = no hazard

   1 = slight

   2 = moderate

   3 = extreme

   4 = severe
4. Examine all potential failures for each individual component of the system, and decide what effect the failures could have.

The process or system is broken down into seven components: die backer, die, billet, dummy block, stem pressing, container liner, and container fillet. Types of failures that may occur are identified as corrosion, cracking, shaking, bending, and surface wear. Of the various components, only the dummy block poses an extreme hazard and a corresponding hazard for workers.

An FMEA shall produce an extensive analysis of a particular process or system. However, the FMEAs have their own limitations. First of all, the element of human error is missing. This is a major weakness because human error is more likely to be at the heart of a workplace accident than a system or a process failure. This weakness can be overcome by coupling human error analysis with an FMEA. Second, the FMEAs focus on the components of the system, as if the components were operating in a vacuum. The mechanism of interface between components or between systems is not taken into account. It is at these interface points that problems often arise.

7.10.2   Hazard and Operability Review

Hazard and operability review (HAZOP) is a method of analysis that allows problems to be identified even before a body of experience has been developed for a given process or system. It is particularly useful for operations involving chemicals or toxic materials. Although originally intended for use with new processes, it does not need to be limited to new operations. HAZOP works equally well with old processes and systems.

HAZOP involves forming a team of experienced, knowledgeable people from various backgrounds related to the process or system and having team members brainstorm potential hazards. The person responsible for safety and health should chair the team and serve as a facilitator. The chair's role is to elicit and record team members' ideas, make sure that one member does not dominate or intimidate other members, encourage maximum participation from all members, and assist members in combining ideas, when appropriate, coform better ones. A variety of approaches can be used with HAZOP.

The most widely used approach is based on the following guide words: no, less, more, part of, as well as reverse, and other than.

HAZOP proceeds in a step-by-step manner which is summarized as follows:

1. Select the process or system to be analyzed.
2. Form the team of experts.
3. Explain the HAZOP process to all team members.
4. Establish goals and time frames.
5. Conduct brainstorming sessions.
6. Summarize all input.

7.10.3   Human Error Analysis

Human error analysis (HEA) is used to predict human error and not to review what has occurred. Although records of past accidents can be studied in order to identify trends that can, in turn, be used to predict accidents, this should be done as part of an accident investigation. HEA should be used to identify hazards prior to the occurrence of accidents. Two approaches to HEA can be effective:

1. Observing employees at work and noting hazards (the task analysis approach)
2. Actually performing job tasks to get a firsthand feel for hazards

Regardless of how HEA is conducted, it is a good idea to perform in conjunction with FMEA and HAZOP. This will enhance the effectiveness of all three processes.

7.10.4   Technic of Operations Review

Technic of Operational Review (TOR) is a method of analysis that allows supervisors and employees to work together to analyze accidents, failures, and incidents at work. It answers the question, "Why did the system allow this accident to happen?" TOR's aim is to enable construction professionals and safety professionals to identify the root causes of an operational failure. TOR is initiated by a specific incident that occurs at a specific time and place involving specific persons.

A weakness of TOR is that it is designed as an after-the-fact process. It is triggered by an accident or incident. A strength of TOR is its involvement of line personnel in the analysis.

1. Establish the TOR team. The team should be made up of workers who were present at the time of the accident or incident, the supervisor, and the safety and health director. The safety and health professional should chair the team and act as a facilitator.

2. Conduct a roundtable discussion to establish a common knowledge base among the members of the team. Five team members may have five different versions of the accident or incident at the beginning of the discussion. There should be a consensus at the end.

3. Identify one major systematic factor that led to or played a significant role in causing the accident or incident. This one TOR statement, about which there must be a consensus, serves as the starting point for further analysis.

4. Use the Consensus Group to respond to a sequence of yes/no options. Through this process, the team identifies a number of factors that contributed to the accident or incident.

5. Evaluate the identified factors carefully to ensure that there is a consensus on each team. Prioritize the contributing factors, starting with the most serious ones.

6. Develop corrective or preventive strategies for each factor. Include corrective or preventive strategies in the final report that are forwarded through the normal channels for appropriate action.

7.10.5   Fault Tree Analysis

Fault tree analysis (FTA) can be used to predict and prevent accidents or, after the event, as an investigative tool. FTA is an analytical method that uses a graphical model to visually display the analysis process. A fault tree is built using special symbols, some of which are derived from the Boolean algebra. As a result, the resulting model resembles a logic diagram or flow chart. Fault trees can be made quantitative by assigning probability figures to the various events below the top box. However, this is rarely done because reliable probability figures are seldom available. A fault tree is developed using the following steps:

1. Decide on the accident or incident that should be placed at the top of the tree.

2. Identify the widest possible level of failure or fault event that could contribute to the top event. Assign the appropriate symbols to you.

3. Move down through successively more specific levels until basic events are identified.

7.10.6   Risk Assessment

After identifying the potential hazards for each task of the job, the next step is to conduct a risk assessment of each hazard. The risk assessment of the job hazard analysis will identify the probability of the occurrence of the hazard and the severity of the possible injury to the accident.

Knowing which potential hazards pose the highest risk will help to determine which preventive measures should be put in place to mitigate those hazards. Higher risk hazards should take precedence over hazards with a low chance of occurring.

Conducting a risk assessment during your job hazard analysis will also come in handy for safety training purposes. More emphasis should be placed on hazards that are more likely to occur and cause serious injury or death.

Example 7.3

What are the recommended actions for identifying all workplace hazards? [Select all that apply]

A. Conduct comprehensive baseline and periodic surveys for safety and health.

B. Perform routine job hazard analyses supported by checklists and job analysis forms.

C. Analyze injury and illness trends over time to identify patterns with common causes with an eye toward prevention.

D. Provide for investigation of accidents and "near miss" incidents to identify causes and means for prevention.

E. Provide a reliable system for swift punishment of employees who point out hazards.

F. Conduct site safety and health inspections no more than once a year.

G. Do not investigate or analyze "near miss" incidents.

H. Require monthly medical checkups of all employees.

Answers: *A, B, C,* and *D*

7.11   Hierarchy of Hazard Prevention and Reduction

Now it's time to determine what measures need to be implemented to mitigate the hazards. Having the hazard eliminated whenever possible should be your top priority. If the hazard cannot be removed, try to contain the hazard using engineering controls such as machine guards or cordoning off workspaces. Other hazard mitigation methods include revising the work procedures by modifying or adding steps or changing the task sequence. OSHA's standards and regulations for construction are a good starting point when determining what preventive measures to implement.

Reducing hazard exposure should be your last line of defense, as it is generally considered to be the least effective hazard mitigation method. This could include limiting the amount of time a worker is permitted to perform a specific task or using personal protective equipment (PPE) to reduce injury potential in the event of an accident.

The use of job hazard analyses is a proven method to help identify and eliminate hazards at the construction site. They may be used to conduct safety training and as topics to be covered when conducting safety meetings or toolbox talks. Job hazard analysis is also a useful tool in accident investigations when mishaps occur.

Each job hazard analysis should be reviewed periodically. Updates should be made when conditions or work procedures change in order to ensure that the job hazard analysis is accurate and effective in identifying and eliminating hazards.

Information obtained from a work hazard analysis is useless unless the hazard control measures recommended in the analysis are incorporated into the tasks. Managers should recognize that not all hazard controls are the same thing. Some are more effective

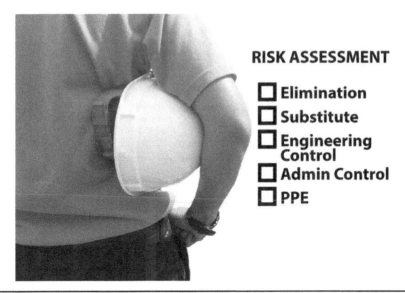

FIGURE 7.2   Hierarchy of hazard control.
(Photo courtesy of 123RF.)

than others in reducing risk. There is a hierarchy of hazard control beginning with (Fig. 7.2):

1. **Elimination** is the primary way to protect the worker from any hazard. If workers do not have any exposure to that hazard because it is not present, then they will not be injured or harmed.

2. **Substitution** is a secondary way to protect the worker from injury or illness. If there is work with the gas chlorine and the workplace decides to go to liquid bleach, that substitution would protect the workers from exposure to a hazardous gas.

3. **Risk transfer:** Another type of hazard control would be to transfer exposure using a contractor. If the contractor is used to perform the work that your employees would have done, the exposure will go to the contractor and not to your employees. Many companies decide to use this method when hiring specialists, such as mold remediation companies, asbestos removal companies, or private fire rescue companies.

4. **Engineering controls.** Engineering controls include the following:

   • Elimination/minimization of the hazard—Designing the facility, equipment, or process to remove the hazard, or substituting processes, equipment, materials, or other factors to lessen the hazard

   • Enclosure of the hazard using enclosed cabs, enclosures for noisy equipment, or other means

   • Isolation of the hazard with interlocks, machine guards, blast shields, welding curtains, or other means

- Removal or redirection of the hazard such as with local and exhaust ventilation

5. **Administrative controls.** Administrative controls include the following:
   - Written operating procedures, work permits, and safe work practices
   - Exposure time limitations (used most commonly to control temperature extremes and ergonomic hazards)
   - Monitoring the use of highly hazardous materials
   - Alarms, signs, and warnings
   - Buddy system
   - Training

6. **Personal protective equipment.** Personal protective equipment, such as respirators, hearing protection, protective clothing, safety glasses, and hardhats, is acceptable as a control method in the following circumstances:
   - When engineering controls are not feasible or do not totally eliminate the hazard
   - While engineering controls are being developed
   - When safe work practices do not provide sufficient additional protection
   - During emergencies when engineering controls may not be feasible

Use of one hazard control method over another higher in the control precedence may be appropriate for providing interim protection until the hazard is abated permanently. In reality, if the hazard cannot be eliminated entirely, the adopted control measures will likely be a combination of all three items instituted simultaneously.

7.12   Components of Hazard Prevention and Reduction

The components of safety programs for injury and illness include:

- Management leadership
- Worker participation
- Hazard identification and assessment
- Hazard prevention and control
- Education and training
- Program evaluation and improvement
- Communication and coordination for most employees, contractors, and staffing agencies

7.12.1   Management Leadership

Leaders are all considered managers in some form or fashion. A committed management unit provides clearly defined objectives and goals for organizational safety behavior. They finance the safety activities through purchases and resource allocations. Every level management values safety practices and accomplishments as much as regulatory compliance and water quality. Steps to implement leadership commitment to safety are:

- Writing or personally singing a clearly defined safety policy that acknowledges safety and health as important as productivity, water quality, regulatory compliance, and customer service.
- Communicating the policy and value to all levels of the generation.
- Visually set examples of safety and health.
- Allocate resources for safety and health.
- Hold all level of the organization accountable for safety performance.

7.12.2 Worker Participation

Worker's participation is very important in controlling the hazard. They can help the employers by identifying the hazards areas and following the rules by the company. Worker participation refers to any process in the company that allows workers to exert influence over their work or their working conditions. Worker participation is obligatory in various processes in the company due to European legislation. In practice, it can be seen as a powerful instrument in safety and health management and is strongly recommended by OSH experts as well as by the European Commission to generally involve the workers and their representatives.

7.12.3 Hazard Identification and Assessment

A hazard is any condition or action that can cause an organizational loss. An organizational loss can come in the form of an injury, illness, damaged equipment, or even workers' turnover. When a loss occurs, the organization must determine the root cause of the loss and not just the symptoms leading to the loss event. The assessment process must be structured in detail and must deliver actionable measures to address the root cause. Hazard identification and assessment can be accomplished by:

- Analyzing past, present, and predictive date from reports, instrumentation, and maintenance logs, even worker injury and illness records
- Inspecting worksite for safety hazards
- Investigating each accident until the root cause is completely disclosed
- Identifying hazards that may arise outside of normal operating conditions including emergencies, start-up, or shut-down operations
- Characterizing the true composition of hazards, giving a priority value to them, and identifying appropriate hazard controls.

7.12.4 Hazard Prevention and Control

The prevention and control of hazards protect the worker from injury and illness, but also gives employees a clean sign that the company cares about their well-being. Eliminating of hazards is the best way to avoid an organizational loss. However, this may not be possible in all situations. Therefore, hazard control is appropriate for some hazards that are still present when workers are performing their daily tasks. Although some worksites are practicing substitution of highly hazardous chemicals such as gas chlorine with liquid chlorine, it is mostly because they are trying to avoid the Risk Management Program, regulated by EPA, and not primarily for worker safety. The hierarchy of hazard controls after elimination and substitution are:

1. Engineering (physical barrier or device such as a machine guard)
2. Administrative (work rule such as work rotation)
3. PPE (protection worn by workers as a barrier to hazards such as a hardhat)

Tips for implementing hazard prevention and controls are as follows:

- Identify what controls are available for each type of hazard.
- Select the proper controls by doing a detailed hazard assessment.
- Develop, maintain, and update a hazard control plan.
- Select controls that are applicable for all aspects of the organization and conditions.
- Implement the selected hazard controls with a priority on elimination and substitution of hazards.

7.12.5   Education and Training

Education and training can be thought of as a tool that binds each step together to keep the efforts cohesive. Some companies have relied on safety training from organizations or even video tapes with outdated material. The role of education and training must be a factor in developing both management and workers to meet the overall safety culture. Employers must educate employees on how to prevent accidents and exposures. General workers should have safety awareness training with regular operations or maintenance training. However, if they work in a specialized area that exposes them to unique hazards, then training must be applicable to the hazard. Effective training can be done peer-to-peer, formal classrooms, online, or at the worksite. Some action items that OSHA suggests are as follows:

- Provide program awareness training
- Train employers, managers, supervisors on their individual safety roles
- Train workers on their specific role in the safety program
- Train workers on the hazard identification and controls

Example 7.4
.....................must educate employees on how to prevent accidents and exposures.

    A. Unions
    B. Employers
    C. OSHA
    D. Compliance Officer

Answer: *B*

7.12.6   Program Evaluation and Improvement

Every program in an organization must be vetted and improved in order to stay viable and productive; safety programs are no different. Program evaluation must be done at given intervals by a competent group. If there are deficiencies found in a program, then the corrections must be made in a systematic way by high-risk issues being fixed then lower risk areas lastly.

Probability. Risk can be calculated as the product of probability and severity of the hazard. The probability of a loss event occurring can be broken down into five categories:

1. Improbable
2. Unlikely
3. Probable
4. Likely
5. Frequent

Severity. Security speaks of the consequences of a loss event when it does occur.

1. Minor
2. Marginal
3. Serious
4. Catastrophic

7.12.7   Communication and Coordination for Host Employers, Contractors, and Staffing Agencies

The host company must take responsibility for all workers including contact and staffing agency workers. Many public-sector businesses are not under the jurisdiction of federal OSHA or even a state OSHA, but the contract companies are under an occupational safety agency that will regulate and cite them for violations. However, local government officials have a moral obligation to make sure that workers of all types who do business with them are protected from hazards. To keep the workers safe, the company should:

- Communicate with all outside contractors the importance of worker safety.
- Coordinate with supervisors, owners, and workers throughout the project to make sure the worksite is safe.
- Hold all workers and agencies accountable for operating a safe worksite.
- Verify that the bids and contracts specify that safe work practices are must for working with the company.

A safety culture will protect the workers from injury and illness because the company places a value on the lives of the workers. This is a deposit into the "good will" bank of the worker and will be rewarded with loyalty. A deep commitment to a safety culture will lead to worker relation and organizational benefits far beyond regulatory compliance.

Example 7.5
New policy and procedures are an example of what type of hazard control?

    A. Engineering

    B. PPE

    C. Administrative

    D. All of the above

    Answer: C

7.13 Summary

Hazard analysis is defined as the process of collecting and interpreting information on hazards and conditions leading to their presence to decide which are significant for safety. Hazard analysis is the first step in risk management. You need to know where the hazards are, how big the likely damage is, and what the chances are of the damage materializing. Construction sites can be hazardous working environments. Conducting a job hazard analysis for each project job is an effective way to mitigate potential hazards and protect workers. A job hazard analysis is a technique that focuses on job tasks as a way to identify hazards before they occur. It focuses on the relationship between the worker, the task, the tools, and the working environment. Preliminary hazard analysis (PHA) is an initial high-level screening exercise that can be used to identify, describe, and rank major hazards during conceptual stage of a facility design. A detailed analysis may be required then considering the complexity of the project. The results of hazard analysis are used in the hazard control actions. There are broadly five hierarchy levels of hazard control, namely:

- Elimination
- Substitution
- Engineering control
- Administration control
- PPE

After following the levels of controls, a safe and smooth project can be expected. An effective program includes provisions for system at identification, evaluation, and prevention or control of hazards and goes beyond specific requirements of the law to address all hazards. It has been found that effective management of worker safety and health programs reduces the extent and severity of work-related injuries and illnesses, improves employee morale and productivity, and reduces workers' compensation costs.

7.14 Multiple-Choice Questions

7.1 Substances that can create combustible dust when they are processed include:

    A. Agricultural grains

    B. Charcoal

    C. Soot

    D. All of the above

7.2   An asphyxiant as a substance or mixture that creates an:

A.   Oxygen-deficient atmosphere
B.   Oxygen-enhanced atmosphere
C.   Oxygen-balanced atmosphere
D.   Oxygen-unstable atmosphere

7.3   The results of the preliminary hazard analysis (PHA) are often presented in:

A.   Tabular form
B.   Graphical form
C.   Numerical form
D.   Comparison form

7.4   "Brainstorming" techniques are used during the:

A.   Design or operation of the system
B.   Maintenance of the system
C.   Job hazards analysis of the system
D.   Closing the operation of the system

7.5   The advantages of using the PHA method include:

A.   It identifies the potential for major hazards at a very early stage of project development.
B.   It provides basis for design decisions.
C.   It helps to ensure plant-to-plant and plant-to-environment compatibility.
D.   All of the above

7.6   Failure mode and effects analysis (FMEA) is a method used to analyze:

A.   Complex financial system
B.   Complex engineering system
C.   Complex corporate system
D.   Complex operation system

7.7   HAZOP means:

A.   Hazard and opportunity review
B.   Hazard and obligation review
C.   Hazard and operability review
D.   Highlights and operability review

7.8   Administration controls include the following:

A.   Elimination/minimization of the hazard
B.   Enclosure of the hazard using enclosed cabs
C.   Isolation of the hazard with interlocks, machine guards, etc.
D.   Monitoring the use of highly hazardous materials

7.9   An effective hazards identification process will meet all these criteria, except:

A.   Comprehensive
B.   Periodic
C.   Secret
D.   Routine

7.10 Management commitment and employee involvement in a safety program is demonstrated by all these actions, except:

    A. Annual review and revision of the program
    B. Employee involvement in safety meeting and inspections
    C. Union oversight alone, if the workers are unionized
    D. Management observance of safety rules

7.11 Exemplary workplaces share all these common characteristics, except:

    A. Re-evaluate effectiveness of existing programs
    B. Regular inspections for hazards
    C. Strict adherence to original safety plan
    D. Training of all employees about hazards

7.12 Comprehensive worksite analysis should involve all these hazards, except:

    A. Existing
    B. Potential
    C. Anticipated
    D. Corrected

7.15   Practice Problems

7.1 List some hazardous chemicals.
7.2 List some physical hazards.
7.3 List some substances that can create combustible dust when they are processed.
7.4 List some health hazards.
7.5 Why is job hazard analysis important?
7.6 What are the appropriate jobs for hazard analysis?
7.7 How is hazard analysis getting started?
7.8 List the questions that can be asked during the job hazard analysis?
7.9 What are the tasks for the preliminary hazard analysis?
7.10 What are the steps for the detailed hazard analysis?
7.11 What is HAZOP? What are the steps of it?
7.12 What is TOR? What are the tasks of it?
7.13 What is FTA? What are the steps of it?
7.14 How can hazard be prevented or reduced?

7.16   Critical Thinking and Discussion Topic

It might look that construction hazard analysis is impractical as we exactly do not know what types of hazards actually exist in workplace until the construction work starts. Hazard analysis is also very complicated as so many uncertainties are involved in it. Many small constructions do not execute any formal hazard analysis and they end up as projects with no fatality. Some construction professionals may show this logic. Some others may comment that if we try to see different unseen potential hazards, it will help us in eliminating a lot of hazards. As the construction becomes big and complex, the chances of rising hazards are very high and uncertain. A formal hazard analysis will help finish the project on time and smoothly without fatality. Join this discussion and give your opinion.

(Photo courtesy of Pixabay from Pexels.)

CHAPTER 8

Promoting Safety and Preventing Violence

(Photo courtesy of 123RF.)

8.1   General

The safety and health of your employees is essential to your business. It's your job as an employer to help keep your workers as safe and healthy as possible. Don't skimp on these costs, as an injured worker could end up costing you more in the long run. Depending on your industry, you will need to incorporate different safety elements into the day-to-day routine of your workers. The needs of the chemical laboratory, for example, are different from those of the general office. Carefully plan programs and work toward keeping things up to date.

8.2   Safety Rules and Regulations

In the whole body of the text, safety rules and regulations of different components of construction are discussed. From a legal point of view, an employer must:

- Have rules ensuring a safe and healthy workplace.
- Ensure all employees know the safety rules.
- Ensure that safety rules are enforced objectively and consistently.

Some guidelines for developing safety rules and regulations are discussed below:

- Minimize the number of rules to the extent practicable.
- Write rules in clear and simple language; a sample is shown in Fig. 8.1. Be brief and to the point to avoid ambiguous and overly technical language.
- Write only the rules necessary to ensure a safe and healthy workplace.
- Involve employees in the development of rules that apply to their specific areas of operations.
- Develop the rules that are to be enforced.
- Be practical and apply common sense in developing rules.

8.3   Promoting Safety

Some important summary of promoting safety are listed below:

1. **Post signs that draw attention to safety issues.** The types of signs will depend on your business. For example, if employees work with heavy equipment, all

Figure 8.1   A sample of safety warning in a site.
(Photo courtesy of 123RF.)

machines should have a label warning of possible injuries. You may also want to place signs on doors, reminding employees to close doors behind them in order to avoid unwanted guests, signs on wet floors, or signs indicating areas of the building that require special safety clothing or items such as hats or goggles.

2. **Provide the proper safety equipment.** Safety materials aren't just hard hats and plastic goggles, though those are certainly necessary in some industries. Safety equipment for office workers includes ergonomic desks, keyboards, and chairs.

3. **Train employees in safety.** Hold a safety training workshop at least once a year to draw attention to potential safety concerns in your office and to remind employees to be safe. Topics could include everything from awareness of not letting strangers into the building to stacking boxes safely. If your business is in an area that is prone to natural disasters, such as earthquakes or hurricanes, you should also inform your employees about what to do in the event of an emergency.

4. **Encourage employees to report things that seem unsafe.** An employee may, for example, notice a cord that someone could trip over or a machine that has been acting up. Once you've heard these reports, act on them.

5. **Provide fitness equipment for the use of employees.** A full gym may not be feasible, but having a few pieces of equipment, such as a treadmill or resistance bands, can help workers get fit on their breaks.

6. **Choose a health plan that focuses on preventive care.** This may include regular checkups, cancer screenings, health and wellness education, and treatments such as acupuncture.

7. **Encourage your workers to adopt healthy behaviors.** For example, you may want to start a smoking cessation or a weight-loss group. A company-wide healthy challenge could be a great motivator.

8. **Bring in health workers to give flu shots.** Flu outbreaks can seriously impede productivity levels. Immunizations at work are convenient and will benefit people who may not take the time to get shots on their own.

9. **Reduce employee stress.** Physical injuries are not the only safety threat to employees; work stress poses a serious threat to the well-being of employees. When employees file stress-related claims, human resource managers need to investigate several factors. Conversely, employees need to prove that work is the primary cause of overwhelming anxiety. Personal concerns contribute to the stress of work. They must also determine whether the circumstance in question is the true cause of a psychologically harmful condition. Managers need to find out if pre-existing conditions or personal issues have caused anxiety on an employee. Human resource managers also consider whether employees are receiving professional treatment for their work-related stress. Managers must weigh these factors in order to determine whether claims are legitimate and decide whether or not to grant compensation.

10. **Develop a safety-first culture.** Developing a safety-first culture is the best approach to adopt, manage, and enforce safety among all personnel in

construction. If it is developed, everybody will be careful from his own position to be safe and keep others safe.

11. **Create incentive plan.** Similarly to bonus program for finishing job early, an incentive plan can be awarded to employees to finish a job without any injuries.

8.4   Factors of Safety Violence

Workplace violence takes many forms and has many causes. Angered former employees, customers who feel wrong, stressed-out employees, or a conflict between coworkers can elevate the office to the point of a violent altercation. Personal life does spill over into the workplace, and sometimes those personal issues present themselves at the office with dangerous consequences. Companies that recognize the potential for violence at work are in the best position to prevent it.

Lack of Pre-employment Screening.   Companies that do not perform thorough background screenings on potential employees run the risk of hiring someone who might be prone to violence, or who has a violent past. Although many companies use psychological tests during the recruitment process to try to get rid of potentially violent candidates, the test is not comprehensive and should not be substituted for a thorough background check.

Stress.   As companies attempt to dictate, employees do not leave their personal issues at the door. Workplace stress and personal stress can cause an employee to snap and lash with whoever the perceived enemy is. Overworking employees can create a hostile working environment, and if employees are also struggling with personal issues, the combination can have disastrous, if not deadly, results.

Lack of Employee Assistance Program.   An employee assistance program (EAP) can diffuse a situation with a potentially violent employee before the employee has a chance to act. Most employees become violent as a last resort—they feel that no one addresses or recognizes their needs. With the EAP program in place, employees have an impartial party with whom they can discuss stressors in their life and possibly receive the assistance or treatment they need before things escalate into violence.

Denial.   Companies that take the approach that nothing bad can happen, such as a disgruntled employee returning to harm his former boss, actually fuel the fire of violence in the workplace. Ignoring the potential for violence in an organization will make it impossible for an organization to put in place proper safety and conflict resolution measures, leaving the organization vulnerable to such attacks by employees and customers, and lack the tools to diffuse the situation.

Disgruntled Customers and Former Employees.   A person who feels that a company has wronged him in some way can lash out at the company. It could be a customer who feels he has been cheated, or it could be an employee who has been fired, laid off, or wronged by a coworker. An angry client or employee who returns and opens fire at the office is the most talked about type of workplace violence, but in actuality, it is a very small percentage of the total cases.

8.5   Workplace Violence

8.5.1   Basics

Workplace violence is any act or threat of physical violence, harassment, intimidation, or other threatening disruptive behavior occurring at the workplace (Fig. 8.2). It ranges from threats, verbal abuse, physical assault, and even homicide. It can affect and involve employees, clients, customers, and visitors. Acts of violence and other injuries are currently the third leading cause of fatal occupational injuries in the United States. According to the Bureau of Labor Statistics Census of Fatal Occupational Injuries (CFOI), of the 5147 fatal occupational injuries that occurred in the United States in 2017, 458 were intentional injuries to another person. However, it is self-evident; violence in the workplace is a major concern for employers and employees across the country.

The number of violent events, some involving disgruntled employees, shows that violence in the workplace is an increasingly problematic issue that employers must learn to minimize and work effectively to prevent. This is important in the construction industry where there is interaction with the public, open worksites, high turnover, and increased responsibility for working with hazardous products and equipment. The recent rise in the number of violent events, some involving disgruntled employees, demonstrates that violence in the workplace is an increasingly problematic issue that employers must learn to effectively minimize and prevent. This is especially important in the construction industry, where there is interaction with the public, open jobsites, high turnover, and the added responsibility of working with dangerous products and equipment.

Employers must recognize and understand the need to take constant and active steps to address both internal and external potential threats to workplace violence and to know that they cannot afford to take the "wait and see" approach to onsite safety. While it is not possible to predict who may be at risk of workplace violence, it is possible to anticipate and head off some incidents. Consider adopting some of the following measures to protect employees against internal and external threats of violence.

FIGURE 8.2   A supervisor may be in an occurrence of verbal abusing. (Photo courtesy of Moose Photos from Pexels.)

Example 8.1

It is practically possible to:

A. Predict who may be at risk of workplace violence

B. Anticipate and head off some incidents

C. Remove workplace violence completely

D. Keep track of all aspects of workplace violence

Answer: *B*

8.5.2   Sources of Violence

Occasionally, violence occurs within an organization. When disputes among workers, supervisors, and managers remain unresolved, arguments, threats, harassment, vandalism, arson, assault, or other violent acts may be the result.

The greatest risk of work-related murder comes from violence by third parties, such as robbers and muggers. Risks have been identified with the workplaces involved in dealing with the public; the exchange of money or valuables; the delivery of passengers, goods, or services; working on a mobile worksite (taxicab or police cruiser); working with unstable or volatile people in health care or social services settings; working late at night or early in the morning; working alone or in small numbers; or working in high crime areas.

8.5.3   Recognize Potentially Violent Situations

Learn to recognize situations that could result in violence. A coworker or client will often express troubled feelings before they become angry or violent. Sometimes, listening and concern is all that is needed in the early stages of trouble. If you sense the problem is getting worse, or if someone is threatening you, take it seriously and report the incident to your supervisor.

Be aware of the places where the assailant could hide, and be empty of someone who is loitering or not in the workplace.

8.5.4   Risk of Workplace Violence

Many American workers report that they have been victims of workplace violence every year. Unfortunately, many more cases have not been reported. Research has identified factors that may increase the risk of violence for some workers at certain workplaces. These factors include exchanging money with the public and working with volatile, unstable people. Working alone or in isolated areas can also contribute to the potential for violence. Providing services and care, and working where alcohol is served, may also affect the likelihood of violence. In addition, time of day and location of work, such as working late at night or in areas with high crime rates, are also risk factors that should be considered when addressing issues of workplace violence. Those at higher risk include workers who exchange money with the public, delivery drivers, healthcare professionals, public service workers, customer service agents, law enforcement personnel, and those who work alone or in small groups.

8.5.5 Response to Onsite Threats of Violence

Because construction worksites are open to all types of employees, subcontractors, etc., monitoring external threats can be more difficult. Employers still need to remain diligent in preparing and updating their emergency action plans or onsite active emergency plans. Having plans in place can help to anticipate and address some incidents.

At a minimum, employees should be trained on the emergency action plans and know the:

- Means of reporting emergencies or an active shooter
- Evacuation procedures and emergency escape route assignments
- Person who is responsible for calling the police
- Onsite security operations (if there are any)
- Person or persons who account for all employees after an emergency evacuation has been completed

If practical, the investment in security cameras and/or monitoring can also play a major deterrent in internal and external acts of violence. They can also provide an immediate review of the situation at the jobsite if needed. While there are no guaranteed signs that an employee or visitor is going to engage in violent acts, there are signs of unacceptable behavior that the workforce can be trained to identify and address.

Develop specific training based on work setting, location, and safety layout, as well as training to ensure that employees have general situational awareness. Consider professional instruction by an active shooter expert who can provide onsite, simulation-based training.

8.5.6 Handling a Violent Situation

If you become involved in a violent situation:

- Report threats or suspicious activity to your supervisors of the police. Keep your distance from the situation. Stay safe, and let the authorities handle it.
- If you are confronted, talk to the person. Stay calm, maintain eye contact, stall for time, and cooperate. If you are threatened by a weapon, freeze in place. Never try to grab a weapon.
- Attempts to escape from a confrontation should be the closest secure area where you can quickly contact others and get help.
- Telephone threats, like any threat, must be taken seriously. Write down as much detail as you can about the caller—accent, pitch of voice, background noise, etc. Immediately report the call to your supervisor or the police.

8.5.7 After Violence Occurs

Any violent act upsets people. Think ahead to have a plan on how to react after a violent incident. Some items that your organization will want to be prepared for are:

- Getting medical attention for anyone who was hurt during the incident
- Reporting the incident to supervisors and/or the police
- Securing the area, so evidence is not disturbed

- Identifying witnesses and interviewing them to get detailed notes on what happened
- Analyzing what happened and making plans to prevent it from being repeated
- Conducting employee assistance counseling or debriefing sessions to help employees reduce stress and fear so they can better understand and handle the situation

8.5.8 Policy Reviews and Preventing Internal Threats of Violence

In addition to responding to incidents, employers can also work to prevent issues internally. Most of the workplace violence events involve current employees who have developed an issue with a supervisor or coworker. Internal threats of violence in the workplace require separate considerations from external threats. Analyzing complaints and preventing problems require a focus on policies, hiring, monitoring, and effective training.

8.5.8.1 Prehire Background Screenings

During the hiring process, the use of effective screening can help avoid individuals with red-flag behavior. Some steps that construction employers can take include:

- Companies are allowed to carry out background checks for applicants, including criminal ones. Note, however, that the use of background checks by employers is often governed by state and local laws, so consult legal counsel on best practices for the implementation of this screening tool.
- Social media. Despite valid privacy concerns, there are currently no federal laws that prevent employers from reviewing the public social postings of prospective employees. An individual's social media postings can reveal whether warning signs like fascination with violence, for instance, are present.

8.5.8.2 Policies on Bullying and Unprofessional Behavior

Construction sites and offices can be high-stress environments (Fig. 8.3). Employers should have policies and training in place to look for signs of aggression or bullying

FIGURE 8.3 A woman looking disappointed.
(Photo courtesy of Engin Akyurt from Pexels.)

and outright ban such trigger behaviors. In training employees on how to identify and report inappropriate behavior, the team should also have an effective procedure for handling employees that have been reported, and demonstrate that:

- Management takes all threats seriously
- When the reports are of internal conflicts, harassment, or bullying, management works quickly to understand the root of the conflict and defuse it
- Conflict resolution points are identified
- Management takes steps to provide appropriate discipline in order to clearly demonstrate that any such violent behavior will not be tolerated

8.5.8.3 *Paying Attention to Outgoing Issues*

Finally, having a process in place to ensure safety upon an employee's termination is also key to maintaining a safe environment. Prior to an employee's termination, employers should notify their security team. Let's hope the manager can avoid escalation, but security may need to escort the employee to an isolated area where they can meet with management.

Employers should also have a process in place to monitor the behavior of terminated employees from the time they are informed of the news until they leave the workplace. Things to monitor include:

- Did they make threats?
- Do they have a history of bullying or unprofessional behavior?
- What is the response if they do?
- If there are any threats, call the police immediately. Don't wait for the irate employee to return.

Workplace violence continues to occur at an alarming rate, and yet many employers have not addressed this concern in their safety training programs. No perfect plan is currently available, but begin by taking proactive steps to avoid these situations and minimize the risk to the workplace.

Because several legal issues may arise with an employer's decision to both implement effective pre-employment screening tools, and construct workplace violence plans, consult with legal counsel as procedures are developed.

8.6   Reduction of Workplace Violence

In most workplaces, where risk factors can be identified, the risk of assault may be prevented or minimized if appropriate precautions are taken by employers. One of the best protections employers can offer their workers is to establish a policy of zero tolerance toward violence in the workplace (Fig. 8.4). This policy should cover all workers, patients, clients, visitors, contractors, and anyone else who may come in contact with company personnel.

By assessing their worksites, employers can identify methods for reducing the likelihood of incidents occurring. OSHA believes that a well-written and implemented workplace violence prevention program combined with engineering controls, administrative inspections, and training can reduce the incidence of workplace violence in both the private and federal workplaces.

FIGURE 8.4 A supervisor warning against workplace violence.
(Photo courtesy of Lukas from Pexels.)

This can be a separate program for workplace violence prevention or can be incorporated into a safety and health program, employee handbook, or standard operating procedures manual. It is essential to ensure that all workers are aware of the policy and understand that all claims of workplace violence are promptly investigated and remedied. In addition, OSHA encourages employers to develop additional methods as necessary to protect employees in high-risk industries.

8.7   Prevention of Workplace Violence

Although most contractors work hard to promote job safety, the fatal injury rate in the construction industry—which employs almost 6.5 million people—still exceeds that of any other U.S. industry. The Occupational Safety and Health Act (OSHA) has a whole section of regulations for contractors only. OSHA regulations help contractors mitigate jobsite hazards such as falling, electrocution, and chemical exposure. Outside of these known jobsite risks, there is a less familiar, but possibly just as dangerous, threat of workplace violence. Workplace violence may involve any act of violence against an employee by any individual. Employers in all industries may face citations from OSHA for failing to prevent it properly. Yet OSHA does not have a single standard that specifically addresses violence at work.

Although OSHA does not control workplace violence onsite, "General Duty Clause" allows employers to use "feasible means" to avoid known threats of violence. The General Duty Clause requires employers to provide "employment and a place of employment free of recognized hazards that cause or are likely to cause death or serious physical harm." The elements of a General Duty Clause violation are:

1. A hazard in the workplace.

2. The employer or the employer's industry recognizes the hazard.

3. The hazard is likely to cause death or serious physical harm.

4. There is a feasible means of eliminating or materially reducing the hazard.

In the case of workplace violence, which may give rise to the OSHA citation, the main factor is that the hazard must be known to the employer or to the employer industry. In the field of construction, these citations are much less frequent. However, the contractor may breach the General Duty Clause by ignoring or failing to acknowledge the obvious threats or signs that a person might commit an act of violence against other employees.

While the foregoing focuses on OSHA citations, a contractor can also face civil liability in a lawsuit filed against an employer by an injured employee. However, while the availability of workers' compensation can prohibit many such lawsuits, contractors should not blindly rely on workers' compensation insurance as a shield. Unless the contractor has known and ignored the obvious threat, the employee will be able to circumvent the usual bar and recover directly against the contractor.

Despite the lack of specific regulation beyond the General Duty Clause, OSHA has voluntary guidelines to prevent and mitigate workplace violence. The guidelines provide a helpful outline of a preventative program:

- Identify and authorize individuals within the company to implement antiviolence programs.
- Assess what positions or tasks are most likely to lead to violent incidents.
- Create measures to control the risk.
- Train employees to identify potential violence and handle violent incidents.
- Evaluate the effectiveness of the company's program.
- Make sure that the hiring process thoroughly vets potential employee backgrounds.

Contractors should seek to prevent violence on the part of employees and third parties, just like any other employer. Most contractors have numerous projects taking place at the same time, and workers may face different levels of risk based on the location of those projects. Contractors should, therefore, tailor preventive measures to reflect the location and nature of the projects. For example, if a project is in a neighborhood with a high crime rate, the contractor should devote more resources to safety training and onsite management for the prevention and mitigation of harm.

A critical element, and good starting point, is general awareness of potential harm in the first place. This starts with the hiring process and carries through to evaluating the general safety of workers on particular projects and raising awareness of threats daily.

8.8 Workplace Security

8.8.1 Basics

In the past, security personnel have been concerned with things like the theft of equipment or computer hackers breaking through the firewall of the computer system and planting computer viruses. While these types of crimes are still a concern, employers are facing new security risks these days. Security breaches can result in threats, abuse, injury, sickness, property loss, or business shutdowns. In order to keep employees, clients, and visitors safe, employers and employees must be alert for various types of security risks:

- Penetration of secure areas
- Assaults
- Theft of hazardous materials or other property
- Chemical and biological terrorism
- Misuse of classified materials

8.8.2 Start with Parking Lot and Garage Security

When entering an employee's parking lot or garage, remember to think about your own personal safety and security. While there is no one criteria for determining "safe" parking spots, try to park:

- Where you can see for a distance around your vehicle
- Where you can be seen (a highly visible location)
- In busy areas of the lot, close to the building entrance or public walkways, if possible

The organization will often issue parking stickers. Parking spaces are usually designated for employees- or visitors-only use. Such processes help to distinguish employees and individuals who may not be part of the operation. If you encounter suspicious activity in a parking lot or garage or find a car that is out of place or appears to be abandoned, report to security personnel promptly.

Example 8.2
To improve your personal security in a parking lot:

 A. Try to hide your car in a dark area.
 B. Park in spaces reserved for handicapped drivers.
 C. Park away from the building entrance to get some exercise.
 D. Park where you can see for a distance around your vehicle.

Answer: D

8.8.3 Going to and From Your Vehicle

Before you get out of your locked automobile, scan the area around the vehicle. Are there any suspicious people in the immediate area? If so, move to another parking spot. When you exit your vehicle:

- Quickly move toward the building entrance, scanning the area in front of you for suspicious people and behavior.
- Walk in well-lit areas.
- Stay alert, and if a suspicious person does approach, change directions and head for a safe area.
- Use a security alarm if you are threatened.

When you leave the building and return to your vehicle:

- Ask for a security escort if this service is available.
- Scan the area around your vehicle as you approach.
- Approach your vehicle with care; have your keys ready.
- If something appears out of the ordinary, or if you see a suspicious person approaching, stop, and go back into the building or other safe area. Contact security or call the police.

8.8.4   Maintain Secure Entrances

One way to protect a building is to make it difficult to get close to it. This is often accomplished through a fence or wall. Fences should be hard to climb or penetrate, and they could be complemented by security cameras to make them even more effective. Many workplaces now operate controlled entrances. These are usually security checkpoints that are either manned or unmanned. That means that there may be a:

- Security person or receptionist at the gate or door to screen the people entering the grounds or building.
- Device that automatically scans employee identification (ID) cards or badges. Without a valid ID, the door or gate will not open.

Security cameras are often used to provide surveillance around the entrance to the building. The mere sight of a security camera will often dissuade criminals. These cameras must, of course, be monitored by a receptionist or a security person.

Example 8.3
Security features for a building's entrance area include all of the following, except:

   A. Fences

   B. Cameras

   C. A receptionist to screen people who enter

   D. Poor lighting and shrubs that create hiding spots

Answer: D

8.8.5   Security Measures inside the Facility

Just because you've made it safe in the workplace, you're not supposed to let your guard down. If your organization has relatively few employees, you may know who works there. But, your workplace may be just too large and spread out for employees to be able to recognize everyone who is allowed to be inside. Or, your business may be working with the public or other clients who come and go throughout the day. Watch for people or other employees:

- Attempting to access something that is normally denied them, such as: classified materials, chemicals, hazardous materials, or admission to prohibited or secure areas
- Behaving in a strange manner

- Making unusual request or demands
- Carrying a weapon (unless they are authorized to do so)

Report suspicious activities to the facility's security office.

8.8.6 Work at Working Safely

Some fear is normal and helps people remain vigilant, but there are a number of precautions that both employers and employees can take to help ensure a secure workplace where risks are reduced.

- Take personal safety and security precautions in parking lots and garages.
- Follow all secure entry policies and procedures.
- Report suspicious activities in the workplace.
- Always wear or carry your ID in the workplace.
- Report lost or stolen employee ID badges.
- Make sure visitors check-in and out when they are allowed in your facility.
- Limit public access to your workplace.
- Know what actions to take if you receive a suspicious package.

8.9 Summary

Workplace violence is any act or threat of physical violence, harassment, intimidation, or other threatening disruptive behavior that occurs at the worksite. It is the policy of the company and the responsibility of its managers and all of its employees to maintain a workplace free from threats and acts of violence. The company will work to provide a safe workplace for employees and for visitors to the workplace. With violence becoming a growing problem in society, the importance of taking measures to prevent workplace violence has become increasingly urgent to businesses that want to protect the safety of their employees. Some guidelines to prevent and mitigate workplace violence are summarized below:

- Identify and authorize individuals within the company to implement antiviolence programs.
- Assess what positions or tasks are most likely to lead to violent incidents.
- Create measures to control the risk.
- Train employees to identify potential violence and handle violent incidents.
- Evaluate the effectiveness of the company's program.
- Make sure that the hiring process thoroughly vets potential employee backgrounds.

8.10 Multiple-Choice Questions

8.1 Verbal threats:

    A. Will always lead to physical assault
    B. Are a form of workplace violence
    C. Are not a form of workplace violence
    D. Will never lead to physical assault

8.2 Third-parties, such as robbers and muggers:

    A. Are the source of the greatest risk of work-related homicide
    B. Cause more fatalities than vehicle crashes
    C. Are only a risk in a warehouse setting
    D. Are only a risk in areas with high crime rates

8.3 To help improve security, many workplaces:

    A. Require a visitor to obtain a pass
    B. Have key-card access systems
    C. Have workers wear identification badges
    D. All of the above

8.4 All of the following methods deter criminals, except:

    A. Having good visibility to eliminate hiding spots
    B. Allowing customers or clients to enter through any doorway
    C. Installing video cameras
    D. Posting signs to show that little cash is on-hand

8.5 If you witness suspicious activity that could become violent, follow all of the following, except:

    A. Report it to supervisors or the police.
    B. Confront the person to find out what's going on.
    C. Let the authorities handle the situation.
    D. Keep your distance.

8.6 To help you stay in touch if you work alone or in a small isolated group:

    A. Use a phone to check in with coworkers.
    B. Ask for an escort in parking lots.
    C. Test the burglar alarm before your shift.
    D. None of the above.

8.7 To avoid the risk of becoming a victim of violence if you travel on the job:

    A. Assume the vehicle won't break down.
    B. Stay in the shadows to remain less visible when walking at night.
    C. Go to the closest safe area to report a threat to the police.
    D. Keep your vehicle unlocked so you can enter it quickly.

8.8 Risks for violence have been found in workplace that:

    A. Handle money
    B. Have a mobile worksite
    C. Work late at night or early in the morning
    D. All of the above

8.9   After a violent incident, get medical attention for anyone who was injured, and:

    A.   Report the incident to supervisors and/or the police.

    B.   Pick up all of the evidence you can find.

    C.   Encourage people to leave the area so they stay safe.

    D.   All of the above.

8.10   After a violent incident:

    A.   Talk about it to help reduce stress and fear.

    B.   Stop following your facility's security guidelines until they're improved.

    C.   Take matters into your own hands if it happens again.

    D.   Become angry if someone threatens you.

8.11   Practice Problems

8.1   List some guidelines for developing safety rules and regulations.

8.2   List some guidelines for promoting safety.

8.3   List some factors of safety violence.

8.4   Define workplace violence. What are the sources of workplace violence?

8.5   What are the ways to recognize potentially violent situations?

8.6   What is the risk of workplace violence?

8.7   What should be your response for onsite threats of violence?

8.8   What are the ways to handle a violent situation?

8.9   What should you do after workplace violence occur?

8.10   List some ways to prevent internal threats of violence.

8.11   List some ways to reduce the workplace violence.

8.12   List some ways to prevent the workplace violence.

8.13   List some ways to adopt workplace security.

8.12   Critical Thinking and Discussion Topic

If a company hires a safety manager then the manager is responsible for any safety-related issues as safety is not teamwork. Some other construction professionals may say that the hired safety manager is the team leader for promoting safety and preventing violation as safety is a teamwork. All employees must cooperate with the safety manager to ensure safety. This way, there might be different opinions about safety. Join this discussion and give your opinion.

CHAPTER 9

OSHA Construction
Safety Programs

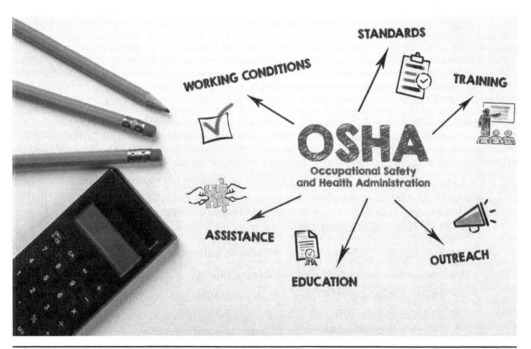

(Photo courtesy of 123RF.)

9.1 History of the OSH Act

The Occupational Safety and Health Administration (OSHA) was set up in 1971. Since then, OSHA and our state partners, together with the efforts of employers, health and safety professionals, trade unions, and advocates, have had a dramatic impact on safety at work. Fatality and injury rates have dropped rapidly. Although accurate statistics were not kept at the time, it is estimated that around 14,000 workers were killed at work in 1970. In 2009, that number dropped to approximately 4340. At the same time, U.S. employment has almost doubled and now includes more than 130 million workers in more than 7.2 million workplaces. Since the passage of the OSH Act, the rate of reported serious injury and illness in the workplace has decreased from 11 per 100 workers in 1972 to 3.6 per 100 workers in 2009. OSHA safety and health standards, including trenching, machine guarding, asbestos, benzene, lead, and blood-borne pathogens, have prevented countless work-related injuries, illnesses, and deaths. This timeline highlights key milestones in occupational safety and health history since the creation of OSHA.

Before 1970, there were no comprehensive and uniform provisions governing the protection of workers against occupational safety and hazards. The number of work-related accidents was as high as 14,000 deaths and 2 million injuries a year. The productivity lost by disabled workers was 10 times higher than that caused by strikes. Although some states had occupational safety measures, the enforcement action was ineffective. There were also issues relating to varying requirements across different states and higher operating costs in states that had their own safety acts.

On December 29, 1970, President Richard Nixon enacted the OSH Act into law to assure, as far as possible, safe and healthy working conditions for every working man and woman in the nation and to preserve the human resources of our nation. In 1971, on the basis of the OSH Act, Congress established OSHA under the Department of Labor and the National Institute for Occupational Safety and Health (NIOSH) under the Department of Health and Human Services to establish safety standards at the federal level. As a result of the OSH Act, the number of work-related accidents dropped to about 4000 deaths in 2010, even though U.S. employment nearly doubled.

OSHA's assigned missions are to:

- Encourage employers and employees to reduce workplace hazards.
- Implement new safety and health programs.
- Improve existing safety and health programs.
- Encourage research that leads to innovative ways of dealing with workplace safety and health problems.
- Establish the rights of employers regarding the improvement of workplace safety and health.
- Monitor job-related illnesses and injuries through a system of reporting and record keeping.
- Establish training programs to increase the number of safety and health professionals and to improve their competence continually.
- Establish mandatory workplace safety and health standards and enforce those standards.
- Provide for the development and approval of state-level workplace safety and health programs.
- Monitor, analyze, and evaluate state-level safety and health programs.

9.2   OSHA Standards and the General Duty Clause

OSHA standards were initially drawn from three main sources: consensus standards, proprietary standards, and laws already in effect. Specifically, OSHA incorporated consensus standards from the American National Standards Institute (ANSI) and the National Fire Protection Association (NFPA). As a result, a significant portion of the original OSHA standard verbiage read as "specification language" (e.g., "should") and has been changed to "process regulation language" (e.g., "shall") over the years.

The OSHA standards are compiled and published in Title 29 (also referred to as "29 CFR") of the Code of Federal Regulations. Figure 9.1 shows an example of an OSHA standard on tool safety for the construction industry. Although vertical standards, such as 29 CFR 1926, specifically designate OSHA standards for construction, horizontal standards, such as 29 CFR 1910, also apply to the general industry, because the standards, as the name suggests, are not industry-specific and thus have a broad coverage. All OSHA standards are freely available online at http://www.osha.gov, and its "regulations" page provides quick access to all the standards. Using the search function on OSHA's main page is another option to locate needed standards but might take multiple tries unless one is sophisticated in forming effective search queries or remembers the specific standard coding numbers.

Part 1926 of the OSHA regulations are entitled "Occupational Safety and Health Standards for Construction." The major subparts of the 1926 standards are:

- Subpart A: General
- Subpart B: General Interpretations
- Subpart C: General Safety and Health Provisions
- Subpart D: Occupational Health and Environmental Control
- Subpart E: Personal Protective and Life Saving Equipment
- Subpart F: Fire Protection and Prevention
- Subpart G: Sign, Signals, and Barricades
- Subpart H: Materials Handling, Storage, Use and Disposal
- Subpart I: Tools—Hand and Power
- Subpart J: Welding and Cutting
- Subpart K: Electrical
- Subpart L: Scaffolds
- Subpart M: Fall Protection

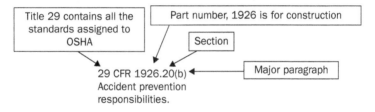

FIGURE 9.1   Number coding of the OSHA standards.

- Subpart N: Cranes, Derricks, Hoists, Elevators, and Conveyors
- Subpart O: Motor Vehicles, Mechanized Equipment, Marine
- Subpart P: Excavations
- Subpart Q: Concrete and Masonry Construction
- Subpart R: Steel Erection
- Subpart S: Underground Construction, Caissons, Cofferdams, and Compressed Air
- Subpart T: Demolition Requirements
- Subpart U: Blasting and the Use of Explosives
- Subpart V: Power Transmission and Distributions
- Subpart W: Rollover Protective Structures, Overhead Protection
- Subpart X: Stairways and Ladders
- Subpart Y: Commercial Diving Operations
- Subpart Z: Toxic and Hazardous Substances
- Subpart CC: Cranes and Derricks in Construction

OSHA standards oversee a number of areas of concern. Even if there is no specific requirement for a particular hazardous situation, the General Duty Clause in the OSH Act (5)(a)(1) should still be followed. The clause states: "Each employer shall furnish to each of his employees' employment and a place of employment which are free from recognized hazards that are causing or are likely to cause death or serious physical harm to his or her employees." The General Duty Clause is often cited by the compliance officers when no specific OSHA standard applies to an observed hazard. However, this does not mean that an OSHA compliance officer can use the General Duty Clause to cite any condition that he or she believes is unsafe.

Overall, OSHA places more responsibility on employers than on employees. While employers are responsible for providing a workplace free of hazards likely to cause death or injury, they are not responsible for ensuring that employees will always comply with OSHA requirements. Employers, however, must do what can reasonably be expected to ensure that employees work under conditions that comply with OSHA requirements. In addition, although employees must conform and obey safety rules and regulations, OSHA does not cite or punish employees for violations of standards. On the other hand, if an employer is not aware of any hazardous condition, the employer could still be liable due to ignorance, even though it was an employee who was responsible for violating safety standards. All the OSHA regulations, expert advisors, directives, publications, and other tools are available at the OSHA's website.

There are four elements that must be present to prove a violation under the General Duty Clause:

1. There must be a hazard to which employees are exposed.
2. The hazard must be recognized.
3. The hazard was causing or was likely to cause death or serious injury.
4. The hazard can be corrected in a feasible manner.

Example 9.1

Where can anyone go to access OSHA regulations, expert advisors, directives, publications, and other tools?

    A. Government Publications section of a public library

    B. Employee resource room at major companies

    C. One of 10 OSHA regional centers

    D. OSHA's website

Answer: D

9.3 OSHA Jurisdiction and State Programs

The Occupational Safety and Health (OSH) Act covers the majority of employers in the private sector and their employees, as well as some state and local government employers and workers in the 50 states and certain territories and jurisdictions under federal authority. State Plans are OSHA-approved workplace safety and health programs operated by individual states or U.S. territories as shown in Fig. 9.2. Currently, there are 22 State Plans covering both the private sector and state and local government workers, and there are six State Plans (Connecticut, Illinois, Maine, New Jersey, New York, and Virgin Islands) covering only state and local government workers. State plans are monitored by OSHA and must be at least as effective as OSHA in the protection of workers and in the prevention of work-related injuries, illnesses, and deaths.

Certain types of organizations or agencies may be exempt from OSHA requirements because they are considered self-employed or are subject to other federal statutes or standards. Examples of such organizations are LLC/partnership companies (self-employed) and coal mining companies (already covered by the Federal Mining Safety and Health Act). Except for these organizations, the OSHA standards still apply to other business establishments, even if they are as small as having only one employee. Employers of 10 or fewer employees could, however, be exempt from the OSHA record-keeping requirement.

OSH Act does not cover the following as well:

- Self-employed
- Immediate family members of farm employers that do not employ outside employees
- Workers who are protected by another Federal agency (e.g., the Mine Safety and Health Administration, FAA, Coast Guard)

9.4 OSHA Recording, Reporting, and Posting Requirements

OSH Act requires that "the Secretary shall issue regulations requiring employers to maintain accurate records of, and to make periodic reports on, work-related deaths, injuries and illnesses other than minor injuries requiring only first aid treatment and which do not involve medical treatment, loss of consciousness, restriction of work or

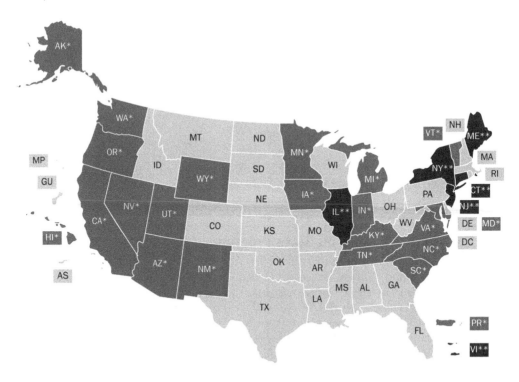

This state's OSHA-approved State Plan covers private and state/local government workplaces.

This state's OSHA-approved State Plan covers state/local government workers only.

This state (with no asterisk *) is a federal OSHA state.

FIGURE 9.2 States and territories with own occupational safety and health programs. (Source: https://www.osha.gov/stateplans/)

motion, or transfer to another job." Recording or reporting a work-related accident or illness does not mean that an employer or employee was at fault or has violated an OSHA standard. Instead, the record-keeping and reporting requirement is used to identify the causes and prevent future occupational illnesses and injuries. For this reason, the Bureau of Labor Statistics (BLS) uses these records to develop nationwide occupational illnesses and injuries, while OSHA uses the records to inform standard development and resource allocation.

Employers with 11 or more employees, including temporary employees who are supervised daily by the employers, must maintain records with occupational injuries and accidents. In particular, OSHA recommends that employers use OSHA Forms 300, "Log of Work-Related Injuries and Illnesses"; 300A, "Summary of Work-Related Injuries and Illnesses"; and 301, "Injury and Illness Incident Report" for record-keeping purposes. The number of entries in OSHA Form 300 (i.e., the log) indicates the total number of recordable accidents and illnesses which can be converted into the overall recordable case rate (i.e., a type of incidence rate, for comparison of safety performance).

Incidence rate is a statistical to assess the safety performance and to compare the safety performance of various employer groups. In business terms, it is generally used for job prequalification, field safety performance monitoring and benchmarking, and determination of workers' compensation premiums.

Record-keeping forms are conducted on a calendar-year basis and must be retained at the place of business for 5 years, and available for review by members of the federal or state OSHA. Employers with 10 or fewer workers can be excluded from the OSHA record-keeping provision, but they do need to report fatalities and multiple cases of hospitalization.

As per 29 CFR 1904.7, the general record-keeping criterion states that an injury or illness must be recorded if it is work-related and results in one or more of the following situations:

- Death
- Loss of consciousness
- Days away from work
- Restricted work activity
- Job transfer
- Medical treatment beyond first aid
- Diagnosed by a physician or other licensed health care professional
- Cancer-related illnesses
- Chronic irreversible diseases
- Fractured or cracked bone
- Punctured eardrum
- Other special conditions

Situations such as when an employee is infected (e.g., through cuts) with blood or other potentially infectious materials of another person or has experienced a standard threshold shift (STS) in hearing may qualify as "other special conditions" for record-keeping purposes.

An injury or illness is considered to be work-related if an event or exposure in the work environment caused or contributed to the condition or aggravated a pre-existing condition significantly. The work environment is not limited solely to job locations. Injuries or illnesses that occur when workers move can still be considered work-related when employees participate in work activities (e.g., traveling from one job site to another) at the time of the injuries/illnesses. Where an injury or illness involves (1) the death of any worker as a result of a work-related incident or (2) in-patient hospitalization of three or more workers as a result of a work-related incident, employers must report the injury or illness to OSHA within 8 hours of learning. Some state plans have more strict guidelines for disclosing injury/illness related to the job. For example, in Washington State, an employer is required to report any accident at work that results in any employee's death or hospitalization. Where an employee is discriminated against or disciplined for disclosing work-related deaths, accidents, illnesses, or complaints, the employee may file a complaint with OSHA by calling, mailing, faxing, online, or in person.

Example 9.2

The OSH Act _____ employment retaliation against an employee who complains to an employer regarding a workplace safety issue or condition, files a complaint related to workplace safety or health conditions, initiates a proceeding, contests an abatement date, or testifies under the Act.

    A. Encourages

    B. Allows

    C. Prohibits

    D. None of the above

Answer: C

Example 9.3

At the end of how many months should an employer review the OSHA Form 300 for completeness, accuracy, and correction of any deficiencies?

    A. Three

    B. Six

    C. Twelve

    D. Twenty-four

Answer: C

9.5   OSHA Inspections and Citations

OSHA enforces standards through inspections. There are only about 2400 inspectors for over 7 million workplaces. At this rate, it would take about 100 years for OSHA to inspect every workplace once. Therefore, OSHA targets the most dangerous workplaces: industries with fatalities and serious injuries (e.g., grain handling in Colorado) and construction (i.e., falls). The OSH Act of 1970 also authorized OSHA to conduct workplace inspections and investigations to determine whether employers meet the OSHA standards, as shown in Fig. 9.3. Inspections are mostly carried out without prior notice, except under special circumstances. If an employer refuses to admit an OSHA compliance officer or attempts to interfere with the inspection, the compliance officer may obtain a warrant in accordance with the OSH Act to inspect the employer. Normally, inspections are scheduled based on the following priorities:

- Imminent Danger. It is any condition where there is a reasonable belief that a danger exists that can be expected to cause death or serious injury immediately or before the danger can be eliminated through normal enforcement procedures. This may include health hazards such as dangerous fumes or dust.

- Catastrophes and Fatal Accidents. Accidents that result in a fatality or hospitalization of three or more workers.

Figure 9.3   OSHA conducts workplace inspections and investigations.
(Photo courtesy of OSHA.)

- Complaints and Referrals. Complaints from employees or others of alleged safety regulation violations or unsafe working conditions.

- Programmed High-Hazard Inspections. Inspections aimed at specific high-hazard workplaces are based on past death, injury, or illness incidence rates.

- Follow-Up Inspections. Inspections to verify whether previously cited violations have been corrected.

The imminent danger is a hazardous situation in which fatalities or serious injuries are likely to occur immediately before the hazard can be eliminated. An example is a worker in an unprotected trench excavation more than 5 feet (1.52 m) deep. It thus receives a higher priority for inspection than those places where disasters or fatal accidents have occurred. Employees may file a confidential complaint with OSHA requesting inspections of the workplace. Programmed inspections aim at high-hazard industries. For new construction projects, programmed inspections take place after major construction activities commence, and when projects are 30 to 60% complete.

The inspection process begins with the compliance officer displaying his or her credentials and making an opening conference. The officer then walks through and inspects the workplace, accompanied by selected employees and representatives of the employer. In addition to identifying unsafe or unhealthful working conditions, during a walk-through inspection, the officer will also review the safety and health records, review the Accident Prevention and Hazard Communication Programs, and determine whether the employer satisfies the OSHA record-keeping requirements.

The inspection process ends with a closing conference to discuss all of the unsafe/unhealthy conditions and violations identified. The compliance officer will notify the employer of the right to appeal, but will not discuss any proposed penalties during inspection.

When violations are identified by a compliance officer, OSHA may issue citations stating penalties, violated standards, and correction deadlines. It is not an option to issue warnings in lieu of citations. In general, a violation means noncompliance with OSHA standards, and consequently, one or more employees are exposed to hazards. On multiemployer workplaces, such as most of the projects in construction, citations shall normally be issued only to employers whose employees are exposed to hazards. However, the primary controlling employer (such as the general contractor with responsibility for correcting the hazards) may be cited instead.

There are five types of violations that may be cited:

1. Other-Than-Serious ($0–$1000). A violation that has a direct relationship to job safety and health, but probably would not cause death or serious physical harm. For example, an employer fails to comply with OSHA's information posting requirement.

2. Serious ($1500–$7000). A violation where there is a substantial probability that death or serious physical harm could result. For example, employees walking/working on surfaces more than 6 feet above lower levels are not protected from falling through holes (including skylights).

3. Willful ($5000–$70,000). A violation that the employer intentionally and knowingly commits.

4. Repeat (Up to $70,000). A violation of any standard, regulation, rule, or order where, upon reinspection, a substantially similar violation is found and the original citation has become a final order.

5. Failure to Abate ($7000 per day). Failure to correct a prior violation may bring a civil penalty of up to $7000 for each day.

For other-than-serious and serious violations, OSHA may adjust a penalty based on the employer's demonstrated good faith, previous inspection history, size of business, and gravity of violation. As mentioned previously, the employer must post copies of the citations at or near the location of alleged violations for 3 days, or until the violations are abated, whichever is longer.

Example 9.4

The _____ protects workers who complain to their employer, OSHA, or other government agencies about unsafe working conditions.

   A. OSH Act

   B. EPA Act

   C. Clean Air Act

   D. Clean Work Act

Answer: *A*

9.6　OSHA Services and Programs

9.6.1　Consultation Services

Consultation assistance is available upon request to employers who want help in establishing and maintaining a safe and healthful workplace. Funded equally by OSHA and each state, the service is provided at no cost to the employer. Each state, whether a federal plan state, or a state plan state, operates the consultation program for the state. Primarily developed for smaller employers with more hazardous operations, the consultation service is delivered by state governments employing professional safety and health consultants. Comprehensive assistance includes a hazard survey of the worksite and appraisal of all aspects of the employer's existing safety and health management system. In addition, the service offers assistance to employers in developing and implementing an effective safety and health management system. No penalties are proposed or citations issued for hazards identified by the consultant. The employer's only obligation is to correct all serious hazards identified by the consultant within the agreed-upon correction timeframe. OSHA provides consultation assistance to the employer with the assurance that his or her name, firm, and any information about the workplace will not be reported to OSHA enforcement staff.

Example 9.5
The consultation service provided by OSHA involve:
[Select all that apply]

A. A hazard survey of the factory

B. An appraisal of all aspects of existing safety and health management system

C. Penalties are proposed for violations

D. Citations issues for hazards identified by the consultant

E. The requirement that the factory will fix all serious hazards that the consultant identifies

F. If serious hazards are identified, the consultant will report to the OSHA enforcement staff

G. Help implementing an effective safety and health management system

H. Funding for engineering control installation and upgrades in the factory

Answers:　*A, B, E, and G*

9.6.2　Safety and Health Achievement Recognition Program (SHARP)

Under the consultation program, certain exemplary employers may request participation in OSHA's Safety and Health Achievement Recognition Program (SHARP). Eligibility for participation in SHARP includes, but is not limited to, receiving a full-service, comprehensive consultation visit correcting all identified hazards, and developing an effective safety and health program management system. Employers accepted into SHARP may receive an exemption from programmed inspections (not complaint or accident investigation inspections) for a period of 2 years initially, or up to 3 years upon renewal.

Example 9.6
OSHA's Strategic Partnership Program is intended to improve workplace safety by:

A. Building cooperative relationship among groups of employers and employees

B. Developing one-on-one relationship with single companies

C. Supporting small business on their safety training efforts

D. Enabling employees to report safety violations anonymously

Answer: A

9.6.3 Voluntary Protection Programs (VPPs)

The Voluntary Protection Program and onsite consultation services, when coupled with an effective enforcement program, expand worker protection to help meet the goals of the OSH Act. The three levels of VPP—Star, Merit, and Demonstration—are designed to recognize outstanding achievements by companies that have developed and implemented effective safety and health management systems. The VPPs motivate others to achieve excellent safety and health results in the same outstanding way as they establish a cooperative relationship between employers, employees, and OSHA.

9.6.4 Strategic Partnership Program

OSHA's Strategic Partnership Program, the newest member of OSHA's cooperative programs, helps encourage, assist, and recognize the efforts of partners to eliminate serious workplace hazards and achieve a high level of worker safety and health. Whereas OSHA's Consultation Program and VPP entail one-on-one relationships between OSHA and individual worksites, most strategic partnerships seek to have a broader impact by building cooperative relationships with groups of employers and employees. These partnerships are voluntary, cooperative relationships between OSHA, employers, employee representatives, and others such as trade unions, trade and professional associations, universities, and other government agencies.

9.6.5 Training and Education

OSHA's area offices offer a variety of information, services, such as compliance assistance, technical advice, publications, audiovisual aids, and speakers for special engagements. OSHA's Training Institute in Arlington Heights, IL, provides basic and advanced courses in safety and health for federal and state compliance officers, state consultants, federal agency personnel, and private sector employers, employees, and their representatives. OSHA awards grants to nonprofit organizations through its Susan Harwood Training Grant Program in order to provide safety and health training and education to employers and workers in the workplace.

The grants focus on programs that will educate workers and employers in small businesses (fewer than 250 employees) about new OSHA standards or about high-risk activities or hazards. Grants are awarded for 1 year, and may be renewed for an additional 12- to 24-month period, depending on whether or not the grantee has performed satisfactorily. Emergency drills and training should occur at least every 12 months.

9.7   Summary

OSHA awards grants to nonprofit organizations to provide safety and health training and education to employers and workers in the workplace. It expects each organization awarded a grant to develop a training and/or education program that addresses a safety and health topic, to recruit workers and employers for the training, and to conduct the training. Grantees are also expected to follow up with people who have been trained to find out what changes were made to reduce hazards in their workplaces as a result of the training.

The OSHA inspection process consists of an opening conference, a walkthrough, and a closing conference with the employer. Results can take up to 6 months, after which OSHA may issue citations. These may include fines and will include dates by which hazard must be abated. When an OSHA inspection is conducted in the workplace, workers have the right to have a worker representative accompany the inspector on the inspection. Workers can talk to the inspector privately. They may point out hazards, describe injuries, illnesses, or near misses that resulted from those hazards and describe any concerns they have about a safety or health issue. Workers also can find out about inspection results and abatement measures and get involved in any meetings or hearings related to the inspection. Workers may also object to the date set for the violation to be corrected and be notified if the employer files a contest.

OSHA provides free consultation services for employers needing help in establishing and maintaining a safe and healthy workplace. The consultation services help an employer survey their workplace hazards and evaluate existing safety and health management systems without attaching any penalties or citations. Employers with outstanding safety achievements can be recognized by the OSHA Voluntary Protection Program (VPP) at three levels: Star, Merit, and Demonstration. They may participate in OSHA's Strategic Partnership Program for employers who are more interested in cooperative relationships with employer groups and employees, such as trade unions and professional associations. Through its Susan Harwood Grant, OSHA also funds initiatives that aim to conduct safety and health training. OSHA also funds initiatives to conduct safety and health training through its Susan Harwood Grant.

9.8   Multiple-Choice Questions

9.1   Under OSHA, employers with _____ or more employees must maintain records of and report occupational injuries and occupational illnesses.

  A.   6
  B.   11
  C.   35
  D.   50

9.2   Which of the following best describes the primary purpose of the Occupational Safety and Health Administration?

  A.   Set and enforce the safety and health standards for almost all workers in the United States.
  B.   Ensure that employees of state agencies have safe and healthy working conditions.
  C.   Provide safe and healthy working conditions to all self-employed persons.
  D.   Ensure that family farms provide healthy and safe working environments.

9.3   The Occupational Safety and Health Act was intended to _____

    A.   Set national, state, and local safety and health standards.
    B.   Assure every person safe and healthful working conditions.
    C.   Provide safe and healthful working conditions to self-employed persons.
    D.   Prevent the occurrence of occupational illnesses among public employees.

9.4   All of the following are covered by the Occupational Safety and Health Act, EXCEPT _____.

    A.   Federal agents
    B.   Hospital nurses
    C.   Crane operators
    D.   Self-employed persons

9.5   According to OSHA, employers must report occupational injuries that result in any of the following, EXCEPT _____.

    A.   First aid treatment
    B.   Loss of consciousness
    C.   Restriction of motion
    D.   Transfer to another job

9.6   Which of the following would most likely NOT be considered a reportable injury according to OSHA?

    A.   Mike breaks his arm while playing in a softball game during a mandatory company picnic.
    B.   John sprains his ankle after becoming tangled in his car's seat belt in the company parking lot.
    C.   Leah breaks her wrist after slipping in a puddle on a stairwell inside the company building.
    D.   Tom injures his back during a traffic accident as he delivers lumber in a company truck.

9.7   According to the Occupational Safety and Health Act, employers are responsible for _____.

    A.   Transferring workers who are cited for OSHA violations
    B.   Examining workplace conditions for OSHA compliance
    C.   Scheduling annual consultations with OSHA representatives
    D.   Replacing old equipment on an annual basis to comply with OSHA

9.8   OSHA does not approve individual States to have their own safety and health program.

    •   True
    •   False

9.9   OSHA does not cover _____ businesses.

    A.   Private
    B.   Self-employed
    C.   Small
    D.   Large

9.10   Workers may file complaint with OSHA if they believe that one of the following situations exist in your workplace.

    A.   A violation safe conditions
    B.   A violation of the job hazard analysis

C. Imminent danger

D. No hazard

9.11 The OSHA inspection does not consist of:

A. Opening conference

B. Walkthrough

C. Closing conference

D. Post-closing conference

9.12 OSHA has standards that cover:

A. Maritime

B. Construction

C. Agriculture

D. All of the above

9.13 The Section (5)(a)(1) of the OSH Act is known as:

A. The General Duty Clause

B. Preamble

C. The best practice

D. The OSH Act

9.14 Workers have the right to refuse to do a job if they believe in _____ faith that they are exposed to an imminent danger.

A. Good

B. Great

C. Ghastly

D. None of the above

9.15 What may employers who are accepted into the Safety and Health Achievement Recognition Program (SHARP) receive from OSHA?

A. Start designation in the VPP program

B. Exemption from programmed inspections

C. Exemption from accident inspections

D. Susan Harwood Training Grant

9.9 Practice Problems

9.1 List some historical milestones of OSHA.

9.2 List OSHA's assigned missions.

9.3 List the four elements that must be present to prove a violation under the General Duty Clause.

9.4 List the employer categories that OSHA does not cover.

9.5 List the situations or results of injury or illness that must be recorded if it is work-related.

9.6 List the priorities of OSHA's inspection scheduling.

9.7 List the five types of violations that may be cited by OSHA.

9.8 List some OSHA Services and Programs.

9.10   Critical Thinking and Discussion Topic

A construction worker directed a worker to remove a plank of wood from a construction site. The plank was essential to carry the safe loading of construction activities. After removing the plank, an accident happened and another worker was severely injured. OSHA team inspected the site and fined the company. The company fired the worker as he listened to his colleague and did not ask the site supervisor about removing the plank. Was firing the worker lawful under OSHA standard?

CHAPTER 10

General Safety and Health Provisions

(Photo courtesy by BECOSAN from Pexels.)

10.1   General

OSHA obligates contractor or subcontractor for any part of the contract not require any laborer or mechanic employed to work in surroundings or under working conditions which are unsanitary, hazardous, or dangerous to his health or safety. It is the responsibility of the employer to initiate and maintain such programs. The use of any machinery, tool, material, or equipment which is not in compliance with any applicable requirement of this standard is prohibited. Such machine, tool, material, or equipment is to be identi-fied as unsafe by tagging or locking the controls to render them inoperable or be physi-cally removed from its place of operation. Part 1926, Subpart C, Standard Number 1926.20 includes such regulations for Safety and Health Regulations for Construction. It states that:

- The employer shall permit only those employees qualified by training or experience to operate equipment and machinery.

- Standards requiring the employer to provide personal protective equipment (PPE), including respirators and other types of PPE, because of hazards to employees impose a separate compliance duty with respect to each employee covered by the requirement. The employer must provide PPE to each employee required to use the PPE, and each failure to provide PPE to an employee may be considered a separate violation.

- Standards in this part requiring training on hazards and related matters, such as standards requiring that employees receive training or that the employer train employees, provide training to employees, or institute or implement a training program, impose a separate compliance duty with respect to each employee covered by the requirement. The employer must train each affected employee in the manner required by the standard, and each failure to train an employee may be considered a separate violation.

10.2   Safety Training and Education

Part 1926, Subpart C, Standard Number 1926.21 establishes and supervises programs for the education and training of employers and employees in the recognition, avoidance, and prevention of unsafe conditions in employments. It states that:

- The employer should avail himself of the safety and health training programs the Secretary provides.

- The employer shall instruct each employee in the recognition and avoidance of unsafe conditions and the regulations applicable to his work environment to control or eliminate any hazards or other exposure to illness or injury.

- Employees required to handle or use poisons, caustics, and other harmful substances shall be instructed regarding the safe handling and use, and be made aware of the potential hazards, personal hygiene, and personal protective measures required.

- In jobsite areas where harmful plants or animals are present, employees who may be exposed shall be instructed regarding the potential hazards, and how to avoid injury, and the first aid procedures to be used in the event of injury.

- Employees required to handle or use flammable liquids, gases, or toxic materials shall be instructed in the safe handling and use of these materials and made aware of the specific requirements contained in applicable areas of OSHA. The factor that determines whether a liquid is flammable or not is its flash point. Flash point is the lowest temperature at which a liquid can give off vapor to form an ignitable mixture in air near the surface of the liquid. The lower the flash point, the easier it is to ignite the material.

Example 10.1
The factor that determines whether a liquid is flammable or not is its:

    A. Container

    B. Liquid

    C. Flash point

    D. Flash

Answer: C

10.3 First Aid and Medical Attention

Part 1926, Subpart C, Standard Number 1926.23 states that first aid services and provisions for medical care shall be made available by the employer for every employee covered by these regulations. Regulations prescribing specific requirements for first aid, medical attention, and emergency facilities are contained in Subpart D of Part 1926.

10.4 Fire Protection and Prevention

Part 1926, Subpart C, Standard Number 1926.24 states that the employer shall be responsible for the development and maintenance of an effective fire protection and prevention program at the jobsite throughout all phases of the construction, repair, alteration, or demolition work. The employer shall ensure the availability of the fire protection and suppression equipment required by Subpart F of Part 1926.

10.5 Housekeeping

Part 1926, Subpart C, Standard Number 1926.25 states that during the course of construction, alteration, or repairs, form and scrap lumber with protruding nails, and all other debris, shall be kept cleared from work areas, passageways, and stairs, in and around buildings or other structures. Combustible scrap and debris shall be removed at regular intervals during the course of construction. Safe means shall be provided to facilitate such removal. Containers shall be provided for the collection and separation of waste, trash, oily and used rags, and other refuse. Containers used for garbage and other oily, flammable, or hazardous wastes, such as caustics, acids, harmful dusts, etc., shall be equipped with covers. Garbage and other waste shall be disposed of at frequent and regular intervals.

Example 10.2
Combustible scrap and debris shall be removed _____ during the course of construction.

    A. At regular intervals

    B. At daily basis

    C. At weekly basis

    D. Only when the bins are full

Answer: A

10.6   Personal Protective Equipment

Part 1926, Subpart C, Standard Number 1926.28 states that the employer is responsible for requiring the wearing of appropriate personal protective equipment in all operations where there is an exposure to hazardous conditions or where this part indicates the need for using such equipment to reduce the hazards to the employees.

10.7   Acceptable Certifications

Part 1926, Subpart C, Standard Number 1926.29 states that:

- Pressure vessels. Current and valid certification by an insurance company or regulatory authority shall be deemed as acceptable evidence of safe installation, inspection, and testing of pressure vessels provided by the employer.

- Boilers. Boilers provided by the employer shall be deemed to be in compliance with the requirements of this part when evidence of current and valid certification by an insurance company or regulatory authority attesting to the safe installation, inspection, and testing is presented.

- Other requirements. Regulations prescribing specific requirements for other types of pressure vessels and similar equipment are contained in Subparts F and O of this part.

Example 10.3
The safe installation, inspection, and testing of pressure vessels can be certified by: [Select all that apply]

   A. An insurance company

   B. A regulatory authority

   C. A competent person

   D. A licensed engineer

Answers: *A and B*

10.8   Means of Egress

Part 1926, Subpart C, Standard Number 1926.34 states that:

In every building or structure exits shall be so arranged and maintained as to provide free and unobstructed egress from all parts of the building or structure at all times when it is occupied. No lock or fastening to prevent free escape from the inside of any building shall be installed except in mental, penal, or corrective institutions where supervisory personnel is continually on duty and effective provisions are made to remove occupants in case of fire or other emergency.

Exit marking.   Exits shall be marked by a readily visible sign. Access to exits shall be marked by readily visible signs in all cases where the exit or way to reach it is not immediately visible to the occupants.

Maintenance and workmanship. Means of egress shall be continually maintained free of all obstructions or impediments to full instant use in the case of fire or other emergency.

10.9 Employee Emergency Action Plans

Part 1926, Subpart C, Standard Number 1926.35 states that:

This section applies to all emergency action plans required by a particular OSHA standard. The emergency action plan shall cover those designated actions employers and employees must take to ensure employee safety from fire and other emergencies.

Elements. The following elements, at a minimum, shall be included in the plan:

- Emergency escape procedures and emergency escape route assignments
- Procedures to be followed by employees who remain to operate critical plant operations before they evacuate
- Procedures to account for all employees after emergency evacuation has been completed
- Rescue and medical duties for those employees who are to perform them
- The preferred means of reporting fires and other emergencies
- Names or regular job titles of persons or departments who can be contacted for further information or explanation of duties under the plan

Alarm system. The employer shall establish an employee alarm system. If the employee alarm system is used for alerting fire brigade members, or for other purposes, a distinctive signal for each purpose shall be used.

Evacuation. The employer shall establish in the emergency action plan the types of evacuation to be used in emergency circumstances.

Training. Before implementing the emergency action plan, the employer shall designate and train a sufficient number of persons to assist in the safe and orderly emergency evacuation of employees. The employer shall review the plan with each employee covered by the plan at the following times:

- Initially when the plan is developed
- Whenever the employee's responsibilities or designated actions under the plan change
- Whenever the plan is changed

The employer shall review with each employee upon initial assignment those parts of the plan which the employee must know to protect the employee in the event of an emergency. The written plan shall be kept at the workplace and made available for employee review. For those employers with 10 or fewer employees the plan may be communicated orally to employees and the employer need not maintain a written plan.

10.10 Summary

OSHA obligates the employer to initiate and maintain any type of safety programs. The employer shall instruct each employee in the recognition and avoidance of unsafe conditions and the regulations applicable to his work environment. Appropriate personal protective equipment, first aid services, and provisions for medical care shall be made available by the employer for every employee. Ensure that managers understand their safety and health responsibilities, as described under the management commitment and employee involvement element of the guidelines.

10.11 Multiple-Choice Questions

10.1 Who is responsible to initiate and maintain safety programs necessary to comply with OSHA standard?

A. Employer
B. Employee
C. Both employer and employee
D. OSHA

10.2 What type of employer does not require written emergency action plan?

A. All employers require it.
B. Employers more than 10 employees.
C. Employers with 10 or fewer employees.
D. Written emergency action plan is not required at all.

10.3 Before implementing the emergency action plan, how many people are to be trained to assist in the safe and orderly emergency evacuation of employees?

A. At least 10
B. Fifty percent of the employees
C. Sufficient number of employees
D. There is no recommendation about the number

10.4 Contractor or subcontractor of a construction project is allowed to require laborer to work in:

A. Unsanitary areas
B. Hazardous areas
C. Dangerous areas
D. None of the above

10.5 Employers who implement effective health and safety programs should expect all these result except:

A. Improved safety
B. Improved employee morale
C. Decreased workers' compensation costs
D. Decreased, through slight, productivity

10.6 An effective occupational safety and health program will include all of the following elements except:

A. Management commitment and employee involvement
B. Systematic identification and random evaluation
C. Worksite hazard analysis and safety training
D. Hazard prevention and control programs

10.12   Practice Problems

10.1   List a few safety and education training regulations for general safety and health provisions.

10.2   What is the regulation for first aid and medical attention?

10.3   What is the regulation for fire protection and prevention?

10.4   What is the regulation for housekeeping?

10.5   What is the regulation for exit marking?

10.6   What is the regulation for PPE for general safety and health provisions?

10.7   List down the elements to be considered for emergency action plans.

10.13   Critical Thinking and Discussion Topic

Assume that you are a recent graduate and is hired by ABC Construction Company as a site project supervisor. On the first day of your site visit, you intended to inspect a bulldozer for its safety issues. However, the bulldozer operator responded to get out of his work. The operator also told you that he had never had an accident or injury cases in his 25-year career. His office file also shows that he has excellent safety record. As a supervisor, how would you handle this situation?

(Photo courtesy of Pixabay from Pexels.)

CHAPTER 11

Occupational Health and Environmental Controls

(Photo courtesy of 123RF.)

11.1   General

Occupational health is a field of health care made up of multiple disciplines dedicated to the well-being and safety of employees in the workplace. Occupational health services include employee wellness, pre-placement testing, ergonomics, occupational therapy, occupational medicine, and more. Occupational health deals with all aspects of health and safety in the workplace and has a strong focus on primary prevention of hazards. The main focus in occupational health is on three different objectives:

1. The maintenance and promotion of workers' health and working capacity

2. The improvement of working environment and work to become conducive to safety and health

3. Development of work organizations and working cultures in a direction which supports health and safety at work and in doing so also promotes a positive social climate and smooth operation and may enhance productivity of the undertakings

11.2   Medical Services and First Aid

Part 1926, Subpart D, Standard Number 1926.50 states that the employer must ensure the availability of easily accessible medical personnel for advice and consultation on matters of occupational health. In the absence of an infirmary, clinic, hospital, or physician, that is reasonably accessible in terms of time and distance to the worksite, which is available for the treatment of injured employees, a person who has a valid certificate in first-aid training shall be available at the worksite to render first aid.

Part 1926, Subpart D, Standard Number 1926.50 states that:

- Proper equipment for prompt transportation of the injured person to a physician or hospital, or a communication system for contacting necessary ambulance service, shall be provided.

- In areas where 911 emergency dispatch services are not available, the telephone numbers of the physicians, hospitals, or ambulances shall be conspicuously posted.

- In areas where 911 emergency dispatch services are available and an employer uses a communication system for contacting necessary emergency-medical service, the employer must:

  - Ensure that the communication system is effective in contacting the emergency-medical service; and

  - When using a communication system in an area that does not automatically supply the caller's latitude and longitude information to the 911 emergency dispatcher, the employer must post in a conspicuous location at the worksite.

- Where the eyes or body of any person may be exposed to injurious corrosive materials, suitable facilities for quick drenching or flushing of the eyes and body shall be provided within the work area for immediate emergency use.

11.3   Sanitation

Regarding the sanitation of the workplace, Part 1926, Subpart D, Standard Number 1926.51 has regulations for potable water, nonpotable water, toilets, food handling, etc. These are discussed below.

11.3.1   Potable Water

Potable water means water that meets the standards for drinking purposes of the State or local authority having jurisdiction, or water that meets the quality standards prescribed by the U.S. Environmental Protection Agency's National Primary Drinking Water Regulations.

- An adequate supply of potable water shall be provided in all places of employment.
- Portable containers used to dispense drinking water shall be capable of being tightly closed, and equipped with a tap. Water shall not be dipped from containers.
- Any container used to distribute drinking water shall be clearly marked as to the nature of its contents and not used for any other purpose.
- The common drinking cup is prohibited.
- Where single service cups (to be used but once) are supplied, both a sanitary container for the unused cups and a receptacle for disposing of the used cups shall be provided.

11.3.2   Nonpotable Water

- Outlets for nonpotable water, such as water for industrial or firefighting purposes only, shall be identified by signs to indicate clearly that the water is unsafe and is not to be used for drinking, washing, or cooking purposes.
- There shall be no cross-connection, open or potential, between a system furnishing potable water and a system furnishing nonpotable water.

11.3.3   Toilets at Construction Jobsites

Toilets (Fig. 11.1) shall be provided for employees according to Table 11.1: Under temporary field conditions, provisions shall be made to assure not less than one toilet facility is available at jobsites, not provided with a sanitary sewer. The requirements of sanitation facilities shall not apply to mobile crews having transportation readily available to nearby toilet facilities.

11.3.4   Food Handling

All employees' food service facilities and operations shall meet the applicable laws, ordinances, and regulations of the jurisdictions in which they are located. All employee food service facilities and operations shall be carried out in accordance with sound hygienic principles. In all places of employment where all or part of the food service is provided, the food dispensed shall be wholesome, free from spoilage, and shall be processed, prepared, handled, and stored in such a manner as to be protected against contamination.

11.3.5   Washing Facilities

The employer shall provide adequate washing facilities with sanitary condition for employees engaged in the application of paints, coating, herbicides, or insecticides, or in other operations where contaminants may be harmful to the employees. Such

FIGURE 11.1   Portable toilets in a construction site.
(Photo courtesy of 123RF.)

Number of Employees	Number of Toilet
20 or less	1 toilet seat
20 or more	1 toilet seat and 1 urinal per 40 workers
200 or more	1 toilet seat and 1 urinal per 50 workers

TABLE 11.1   Recommended Number of Toilet Facility

facilities shall be in near proximity to the worksite and shall be so equipped as to enable employees to remove such substances. Lavatories shall be made available in all places of employment. Each lavatory shall be provided with hot and cold running water, or tepid running water. Hand soap or similar cleansing agents shall be provided.

No employee shall be allowed to consume food or beverages in a toilet room or in any area exposed to a toxic material. Every enclosed workplace shall be so constructed, equipped, and maintained, so far as reasonably practicable, as to prevent the entrance or harborage of rodents, insects, and other vermin. A continuing and effective extermination program shall be instituted where their presence is detected.

11.4   Occupational Noise Exposure

Part 1926, Subpart D, Standard Number 1926.52 states that protection against the effects of noise exposure shall be provided when the sound levels exceed those shown in

Duration per Day, hours	Sound Level, dBA Slow Response
8	90
6	92
4	95
3	97
2	100
1.5	102
1	105
0.5	110
0.25 or less	115

TABLE 11.2 Permissible Noise Exposure

Table 11.2 of this section when measured on the A-scale of a standard sound level meter at slow response. When employees are subjected to sound levels exceeding those listed in Table 11.2, feasible administrative or engineering controls shall be utilized. If such controls fail to reduce sound levels within the levels of the table, personal protective equipment shall be provided and used to reduce sound levels within the levels of the table. Exposure to impulsive or impact noise should not exceed 140 dB peak sound pressure level.

11.5   Illumination

Part 1926, Subpart D, Standard Number 1926.56 states that construction areas, ramps, runways, corridors, offices, shops, and storage areas shall be lighted to not less than the minimum illumination intensities listed in Table 11.3 while any work is in progress.

Foot-Candles	Area of Operation
5	General construction area lighting
3	General construction areas, concrete placement, excavation and waste areas, access ways, active storage areas, loading platforms, refueling, and field maintenance areas
5	Indoors: warehouses, corridors, hallways, and exit ways
5	Tunnels, shafts, and general underground work areas (Exception: minimum of 10 foot-candles is required at tunnel and shaft heading during drilling, mucking, and scaling. Bureau of Mines–approved cap lights shall be acceptable for use in the tunnel heading)
10	General construction plant and shops (e.g., batch plants, screening plants, mechanical and electrical equipment rooms, carpenter shops, rigging lofts and active store rooms, mess halls, and indoor toilets and workrooms)
30	First aid stations, infirmaries, and offices

TABLE 11.3 Recommended Illumination Intensities

11.6 Ventilation

Part 1926, Subpart D, Standard Number 1926.57 states that exhaust fans, jets, ducts, hoods, separators, and all necessary appurtenances, including refuse receptacles, shall be so designed, constructed, maintained, and operated as to ensure the required protection by maintaining a volume and velocity of exhaust air sufficient to gather dusts, fumes, vapors, or gases from said equipment or process, and to convey them to suitable points of safe disposal, thereby preventing their dispersion in harmful quantities into the atmosphere where employees work.

The exhaust system shall be in operation continually during all operations which it is designed to serve. If the employee remains in the contaminated zone, the system shall continue to operate after the cessation of said operations, the length of time to depend upon the individual circumstances and effectiveness of the general ventilation system.

The air outlet from every dust separator, and the dusts, fumes, mists, vapors, or gases collected by an exhaust or ventilating system shall discharge to the outside atmosphere. Collecting systems which return air to work area may be used if concentrations which accumulate in the work area air do not result in harmful exposure to employees. Dust and refuse discharged from an exhaust system shall be disposed of in such a manner that it will not result in harmful exposure to employees.

11.7 Methylenedianiline

4,4'-Methylenedianiline (MDA) is an industrial chemical that is not known to occur naturally. It is also commonly known as diaminodiphenylmethane or MDA. It occurs as a colorless to pale yellow solid and has a faint odor. MDA is used mainly for making polyurethane foams, which have a variety of uses, such as insulating materials in mailing containers. It is also used for making coating materials, glues, fiber, dyes, and rubber.

When 4,4'-methylenedianiline enters the environment, the following actions happen to it:

- 4,4'-Methylenedianiline is found in tiny particles in air which will settle to land or water in rain or snow.
- Most of the 4,4'-methylenedianiline in water will attach itself to particles and sink to the bottom sediment.
- 4,4'-Methylenedianiline in water or sediment will be broken down by bacteria and other microorganisms.
- It does not build up in the food chain.
- 4,4'-Methylenedianiline becomes strongly attached to soil and will not easily move into groundwater.
- It may take as long as 10 days for bacteria and microorganisms in soil to break down 4,4'-methylenedianiline.

A construction worker might be exposed to 4,4'-methylenedianiline in any of the following ways:

- Working in an industry that makes or uses 4,4'-methylenedianiline.
- Touching consumer goods such as polyurethane foams that contain it.

- Living near a hazardous waste site where 4,4'-methylenedianiline is disposed of.
- Being treated by a kidney dialysis machine. Tiny amounts are released from the polyurethane parts of the machine when it is sterilized by radiation or heat.

Part 1926, Subpart D, Standard Number 1926.60 states that the employer shall assure that no employee is exposed to an airborne concentration of MDA in excess of ten parts per billion (10 ppb) as an 8-hour time-weighted average and a STEL (short-term exposure limit as determined by any 15-minute sample period) of one hundred parts per billion (100 ppb).

11.8 Lead

Pure lead (Pb) is a heavy metal at room temperature and pressure. A basic chemical element, it can combine with various other substances to form numerous lead compounds. Lead has been poisoning workers for thousands of years. Lead can severely damage the central nervous system, cardiovascular system, reproductive system, hematological system, and kidneys. When absorbed into the body in high enough doses, lead can be toxic. In addition, workers' lead exposure can harm their children's development. In the construction industry, inhalation is the most common route of lead absorption into the body. In construction, lead is used frequently for roofs, cornices, tank linings, and electrical conduits. In plumbing, soft solder, used chiefly for soldering tinplate and copper pipe joints, is an alloy of lead and tin. Soft solder has been banned for many uses in the United States. In addition, the Consumer Product Safety Commission bans the use of lead-based paint in residences. Because lead-based paint inhibits the rusting and corrosion of iron and steel, however, lead continues to be used on bridges, railways, ships, lighthouses, and other steel structures, although substitute coatings are available. Construction projects vary in their scope and potential for exposing workers to lead and other hazards. Projects such as removing paint from a few interior residential doors may involve limited exposure. Others projects, however, may involve removing or stripping substantial quantities of lead-based paints on large bridges and other structures. You can avoid using lead-containing materials by selecting other materials. Epoxy-covered zinc-containing primers can be used instead of lead-containing coating. Also you can use equipment that decreases the risk of lead emission.

Short-term (acute) overexposure—as short as days—can cause acute encephalopathy, a condition affecting the brain that develops quickly into seizures, coma, and death from cardiorespiratory arrest. Short-term occupational exposures of this type are highly unusual but not impossible. Extended, long-term (chronic) overexposure can result in severe damage to the central nervous system, particularly the brain. It can also damage the blood-forming, urinary, and reproductive systems. There is no sharp dividing line between rapidly developing acute effects of lead and chronic effects that take longer to develop.

Workers potentially at risk for lead exposure include those involved in:

- Iron work
- Demolition work
- Painting
- Lead-based paints abatement
- Plumbing

- Heating and air conditioning maintenance and repair
- Electrical work and carpentry, renovation, and remodeling work

Plumbers, welders, and painters are among those workers most exposed to lead. Significant lead exposures also can arise from removing paint from surfaces previously coated with lead-based paint such as bridges, residences being renovated, and structures being demolished or salvaged. With the increase in highway work, bridge repair, residential lead abatement, and residential remodeling, the potential for exposure to lead-based paint has become more common. Workers at the highest risk of lead exposure are those involved in:

- Abrasive blasting
- Welding, cutting, and burning on steel structures

Other operations with the potential to expose workers to lead include:

- Lead burning
- Using lead-containing mortar
- Power tool cleaning without dust collection systems
- Rivet busting
- Cleanup activities where dry expendable abrasives are used
- Movement and removal of abrasive blasting enclosures
- Manual dry scraping and sanding
- Manual demolition of structures
- Heat-gun applications
- Power tool cleaning with dust collection systems
- Melting and casting lead and Babbitt metal
- Soldering
- Reclaiming lead-acid batteries
- Grinding or sanding lead-containing materials
- Machining lead
- Cutting or heating lead-containing materials

Example 11.1
Lead can _____ damage your nervous, urinary, blood-forming, and reproductive systems.

   A. Moderately

   B. Slightly

   C. Severely

   D. Haphazardly

Answer: C

Early signs of lead poisoning can be overlooked as everyday medical complaints. These include:

- Loss of appetite
- Metallic taste
- Irritability
- Moodiness
- Joint and muscle aches
- Trouble sleeping
- Lack of concentration
- Fatigue
- Decreased sex drive
- Headaches

Example 11.2
Lead can cause severe damage to the body even before the symptoms appear. What are some early signs of lead poisoning? Select all that apply.

A. Metallic taste

B. Kidney failure

C. Joint and muscle ache

D. Trouble sleeping

E. Lack of concentration

F. Tremors

G. Convulsions or seizures

Answers: *A, C, D, and E*

Brief intense exposure or prolonged overexposure can result in severe damage to your blood-forming, nervous, urinary, and reproductive systems. Some noticeable medical problems include:

- Anemia
- Kidney failure
- Stomach pains
- High blood pressure
- Convulsions or seizures
- Constipation or diarrhea
- Tremors
- Nausea

- Wrist or foot drop
- Reduced fertility

Part 1926, Subpart D, Standard Number 1926.62 covers lead in a variety of forms, including metallic lead, all inorganic lead compounds, and organic lead soaps. It states that the employer shall assure that no employee is exposed to lead at concentrations greater than fifty micrograms per cubic meter of air (50 µg/m³) averaged over an 8-hour period. If an employee is exposed to lead for more than 8 hours in any work day the employees' allowable exposure, as a time weighted average (TWA) for that day, shall be reduced according to the following formula:

Allowable employee exposure (in µg/m³) = 400 divided by hours worked in the day.

OSHA's lead in construction standard applies to all construction work where an employee may be exposed to lead. All work related to construction, alteration, or repair, including painting and decorating, is included. Under this standard, construction includes, but is not limited to:

- Demolition or salvage of structures where lead or materials containing lead are present
- Removal or encapsulation of materials containing lead
- New construction, alteration, repair, or renovation of structures, substrates, or portions or materials containing lead
- Installation of products containing lead
- Lead contamination from emergency cleanup
- Transportation, disposal, storage, or containment of lead or materials containing lead where construction activities are performed
- Maintenance operations associated with these construction activities

Your employer is responsible for assessing each employee's exposure level. If the initial exposure is assessed to be at or above the action level (30 µg/m³), your employer must obtain samples that indicate the level of exposure for each work shift and for each task in each work area. The degree of daily exposure to lead for each monitored employee can be assessed through these samples. If you have a blood lead level of 40 µg/100 g, you must be tested until the level goes below 40 µg/100 g for two consecutive blood tests.

The results of all assessments that indicate the exposure level of employees to lead must include the following information:

- All observations, information, and calculations that show an employee's exposure to lead
- Measurements of any previous airborne lead
- Any complaints made by an employee of symptoms that indicate lead exposure
- Objective information about the materials that are used or the processes that have to be carried out

Example 11.3

If you are exposed to lead beyond the action level, your employer is required to perform medical monitoring. What are some of the medical monitoring actions that the employer must complete? (Select all that apply)

The employer:

A. Performs medical monitoring every 6 months.

B. Can discontinue lead monitoring when blood lead levels have fallen below 30 μg/m³ after three consecutive tests.

C. Notifies you in writing within 5 days of the test if your blood lead level exceeds 40 μg/100 g.

D. Can remove the equipment or change the processes that expose you to lead so monitoring can be eliminated.

E. Provides annual medical examinations to all employees whose blood lead levels have been at or above 40 μg/100 g during the previous year.

Answers: *A, C, D, and E*

Example 11.4

If you have a blood lead level of 40 μg/100 g, you must be tested until the level goes below 40 μg/100 g for two consecutive blood tests.

- True
- False

Answer:

True—If you have a blood lead level of 40 μg/100 g, you must be tested until the level goes below 40 μg/100 g for two consecutive blood tests.

11.9   Hazardous Waste Operations and Emergency Response (HAZWOPER)

Hazardous waste operations and emergency response (HAZWOPER) requires that workers be trained to perform their anticipated job duties without endangering themselves or others. Hazardous waste operation means any operation conducted within the scope of this standard. Hazardous waste site or Site means any facility or location within the scope of this standard at which hazardous waste operations take place. Health hazard means a chemical or a pathogen where acute or chronic health effects may occur in exposed employees. It also includes stress due to temperature extremes.

Emergency response or responding to emergencies means a response effort by employees from outside the immediate release area or by other designated responders (i.e., mutual-aid groups, local fire departments, etc.) to an occurrence which results, or is likely to result, in an uncontrolled release of a hazardous substance. Responses to incidental releases of hazardous substances where the substance can be absorbed, neutralized, or otherwise controlled at the time of release by employees in the immediate release area, or by maintenance personnel are not considered to be emergency responses within the scope of this standard. Responses to releases of hazardous substances where there is no potential safety or health hazard (i.e., fire, explosion, or chemical exposure) are not considered to be emergency responses.

An unexpected release of hazardous substances, or a substantial threat of a hazardous substance release, can pose a significant health and safety risk to workers. Unexpected releases can be caused by operation failures and unrelated outside events (e.g., natural disasters, terrorism). Workers can encounter hazardous substances through waste dumped in the environment—a serious safety and health issue that continues to endanger life and environmental quality. Employers must adequately prepare emergency response and cleanup workers to clearly understand their role(s) in managing unexpected releases of hazardous substances, so that they can act quickly and respond in a safe manner during an emergency.

The Superfund Amendments Reauthorization Act (SARA) of 1996 required OSHA to issue regulations protecting workers engaged in hazardous waste operations. OSHA's Hazardous Waste Operations and Emergency Response (HAZWOPER) standards (in construction 29 CFR 1926.65) established health and safety requirements for employers engaged in these operations, as well as responses to emergencies involving releases of hazardous substances. HAZWOPER requires that employers follow specific work policies, practices, and procedures to protect their workers potentially exposed to hazardous substances. The standards provide employers with the information and training criteria necessary to ensure workplace health and safety during hazardous waste, emergency response, and cleanup operations involving hazardous substances. HAZWOPER aims to prevent and minimize the possibility of worker injury and illness resulting from potential exposures to hazardous substances.

Exposures to hazardous substances pose a wide range of acute (i.e., immediate) and chronic (i.e., long-term) health effects. These may include chemical burns, sensitization, irritation, and other toxic effects that may lead to death. Hazardous substance releases can also result in fires, explosions, high-energy events, and/or toxic atmospheres depending on the physical properties and health hazards of the released substance(s). OSHA's Chemical Hazards and Toxic Substances Safety and Health Topic page provides more information on safety and health hazards from exposure to hazardous substances. Under the OSHA law, each employer is responsible for the safety and health of its workers and for providing a safe and healthful workplace. Employers must protect workers from anticipated hazards associated with participation in response and recovery operations for hazardous substances.

Example 11.5

In the construction industry, _____ is the most common route of lead absorption into the body.

    A. Inhalation

    B. Absorption

    C. Ingestion

    D. Injection

Answer: *A*

11.10 Summary

Workers in the construction industry are at an increased risk of lead exposure, because lead is used in everything from steel and iron structures to walls and lead pipes. Specific measures must be taken to protect workers from the deadly hazards posed by lead. Such measures include medical monitoring, medical surveillance where indicated, exposure assessments, regular monitoring of exposure levels, and additional monitoring where indicated. Employers are responsible for supervising workers to ensure compliance with all control measures, as well as necessary personal hygiene and house-keeping practices, to minimize employees' lead exposure. This may require supplying employees with protective clothing, a professional laundering service (or disposable clothes and shoe covers), and respiratory protection (including proper training) where needed. The two basic types of respirators that can provide protection against lead are air-purifying respirators and atmosphere-supplying respirators.

Lead can be very toxic, even deadly, if it is absorbed by the body in sufficient quantities, most commonly by either unintentional inhalation or ingestion. Because our bodies are slow to remove lead from our systems, someone who inhales small doses of lead over a long period of time can end up with lead poisoning. When lead enters the body it circulates in the bloodstream and accumulates in various organs, possibly causing irreversible harm to body tissues. If the amount of lead stored in the body continues to increase, the person can suffer numerous adverse health effects, including severe damage to kidneys, nervous, urinary, blood-forming, reproductive systems, decreased fertility, and danger to the unborn babies of pregnant workers, since lead particles can pass through the placenta. Workers must learn to recognize the early and later symptoms of lead poisoning, which range from headaches and fatigue to seizures and tremors.

11.11 Multiple-Choice Questions

11.1 To ensure occupational health, employer must provide:

    A. Hand soap
    B. Hot and cold water
    C. Personal clothing
    D. Lavatories

11.2 There are 160 employees in a construction site. How many toilet seats are to be provided at the minimum level at the jobsite to ensure occupational health?

    A. 2
    B. 3
    C. 4
    D. 5

11.3 No employee is allowed to consume food or beverages in a: [select all that apply]

    A. Toilet room
    B. Toxic area
    C. Construction site
    D. Temporary shelter

11.4 In a construction site, the work-hour per day is 6 hours. What is the maximum allowable sound level in dBA slow response at the jobsite?

A. 90
B. 92
C. 95
D. 97

11.5 In a construction site, if the number of work-hour increases, the maximum allowable sound level:

A. Increases
B. Decreases
C. Remains the same
D. No recommendation is provided

11.6 In construction, lead is used frequently for: [select all that apply]

A. Roofs
B. Cornices
C. Tank linings
D. Electrical conduits

11.7 In a construction site with lead exposure, the daily work-hour is 8. According to Part 1926, Subpart D, what is the allowable exposure limit for lead in $\mu g/m^3$ at this site?

A. 30
B. 50
C. 80
D. 400

11.8 Lead has an action level of _____ micrograms per cubic meter.

A. 10
B. 20
C. 30
D. 40

11.9 Early signs of lead poisoning include:

A. Loss of appetite
B. Metallic taste
C. Irritability
D. All of the Above

11.10 Workers can accidentally consume lead particles while eating or drinking contaminated food or beverages, or by eating drinking, or smoking with contaminated hands.

- True
- False

11.11 All contaminated clothes that have to be launched cleaned, or disposed of should be placed in _____ containers and sealed off.

A. Closed
B. Open
C. Slotted
D. Permeable

11.12 _____ compounds were often applied to steel and iron structures in the form of paint primer.

A. Asbestos
B. Lead
C. Water
D. Silica

11.13 There is no material available which can be used as a substitute to lead.

- True
- False

11.12   Exercises

11.1 According to Part 1926, Subpart D, list some requirements of potable water.

11.2 According to Part 1926, Subpart D, list some requirements of nonpotable water.

11.3 According to Part 1926, Subpart D, list some requirements of the number of toilet seats and urinals based on the number of employees.

11.4 According to Part 1926, Subpart D, list the allowable sound level based on daily work-hours.

11.5 What happens to 4,4'-methylenedianiline when it enters the environment?

11.6 How might be a construction worker exposed to 4,4'-methylenedianiline?

11.7 According to Part 1926, Subpart D, what is the exposure limit for 4,4'-methylenedianiline?

11.8 How might a construction worker be exposed to lead?

11.9 According to Part 1926, Subpart D, what is the allowable exposure limit for lead?

11.13   Critical Thinking and Discussion Topic

During working in a site, you heard a big noise and your ears started bothering you. You went to your company health office and he carefully inspected your ears. He advised that there was no issue in your ears and it would go away within hours. You did not like this consultation and went to your supervisor and explained everything. Your supervisor replied that you should go back to work as the health officer inspected and found no issue. However, you are still not happy with your ears. Soon after this situation, what should you do?

(Photo courtesy of Life of Pix from Pexels.)

Personal Protective and Lifesaving Equipment

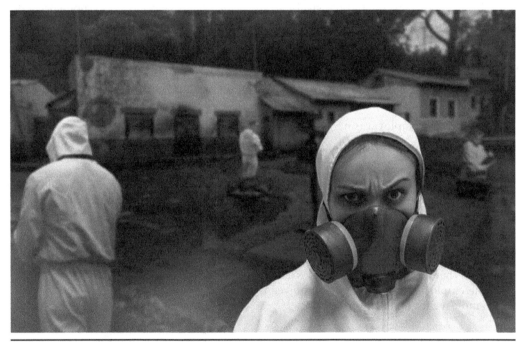

(Photo courtesy of 123RF.)

12.1   General

The basics of personal protective equipment (PPE) have been discussed in Chapter 2. The regulations by OSHA about the PPE and a few lifesaving equipment are discussed in the current chapter.

12.2   General Criteria for Personal Protective Equipment

Part 1926, Subpart E, Standard Number 1926.95 states that protective equipment, including personal protective equipment for eyes, face, head, and extremities, protective clothing, respiratory devices, and protective shields and barriers, shall be provided, used, and maintained in a sanitary and reliable condition wherever it is necessary by reason of hazards of processes or environment, chemical hazards, radiological hazards, or mechanical irritants encountered in a manner capable of causing injury or impairment in the function of any part of the body through absorption, inhalation, or physical contact. Where employees provide their own protective equipment, the employer shall be responsible to assure its adequacy, including proper maintenance, and sanitation of such equipment. All personal protective equipment shall be of safe design and construction for the work to be performed. Except as provided below, the protective equipment, including personal protective equipment (PPE), used to comply with this part, shall be provided by the employer at no cost to employees.

a. The employer is not required to pay for non-specialty safety-toe protective footwear (including steel-toe shoes or steel-toe boots) and non-specialty prescription safety eyewear, provided that the employer permits such items to be worn off the jobsite.

b. When the employer provides metatarsal guards and allows the employee, at his or her request, to use shoes or boots with built-in metatarsal protection, the employer is not required to reimburse the employee for the shoes or boots.

c. The employer is not required to pay for:

• Everyday clothing, such as long-sleeve shirts, long pants, street shoes, and normal work boots; or

• Ordinary clothing, skin creams, or other items, used solely for protection from weather, such as winter coats, jackets, gloves, parkas, rubber boots, hats, raincoats, ordinary sunglasses, and sunscreen.

d. The employer must pay for replacement PPE, except when the employee has lost or intentionally damaged the PPE.

Where an employee provides adequate protective equipment he or she owns, the employer may allow the employee to use it and is not required to reimburse the employee for that equipment.

Example 12.1

_____ personal protective equipment (PPE) should be of safe design and construction, and should be maintained in a clean and reliable fashion.

A. Used

B. New

C. Some

D. All

Answer: D

12.3   Head Protection

Employees working in areas where there is a possible danger of head injury from impact, or from falling or flying objects, or from electrical shock and burns, shall be protected by protective helmets. Part 1926, Subpart E, Standard Number 1926.100 states that the employer must provide each employee with head protection that meets the specifications contained in American National Standards Institute (ANSI) Z89.1. The employer must ensure that head protection is provided for each employee exposed to high-voltage electric shock and burns. OSHA will deem any head protection device that the employer demonstrates is at least as effective as a head protection device.

12.4   Hearing Protection

Part 1926, Subpart E, Standard Number 1926.101 states that wherever it is not feasible to reduce the noise levels or duration of exposures to those specified in 1926.52, ear protective devices shall be provided and used. Ear protective devices inserted in the ear shall be fitted or determined individually by competent persons. Plain cotton is not an acceptable protective device.

12.5   Eye and Face Protection

The employer is expected to ensure that each employee uses appropriate eye or face protection when exposed to eye or face hazards from flying particles, molten metal, liquid chemicals, acids or caustic liquids, chemical gases or vapors, or potentially injurious light radiation. When there is a hazard from flying objects, the employer must ensure that each employee uses eye protection that provides side protection. It is also the duty of employer to ensure that each employee who wears prescription lenses while engaged in operations that involve eye hazards wears eye protection that incorporates the prescription in its design, or wears eye protection that can be worn over the prescription lenses without disturbing the proper position of the prescription lenses or the protective lenses. Part 1926, Subpart E, Standard Number 1926.102 states that protectors shall meet the following minimum requirements:

- They shall provide adequate protection against the particular hazards for which they are designed.
- They shall be reasonably comfortable when worn under the designated conditions.
- They shall fit snugly and shall not unduly interfere with the movements of the wearer.
- They shall be durable.
- They shall be capable of being disinfected.
- They shall be easily cleanable.

All protective goggles shall bear a label identifying the following data:

- The laser wavelengths for which use is intended
- The optical density of those wavelengths
- The visible light transmission

12.6 Respiratory Protection

Part 1910, Subpart I, Standard Number 1910.134 states that a respirator shall be provided to each employee when such equipment is necessary to protect the health of such employee. The employer shall provide the respirators which are applicable and suitable for the purpose intended. The employer shall be responsible for the establishment and maintenance of a respiratory protection program. The program shall cover each employee required by this section to use a respirator.

12.7 Safety Belts, Lifelines, and Lanyards

Some requirements of safety belts, lifelines, and lanyards as specified by Part 1926, Subpart E, Standard Number 1926.104 are as follows:

- Lifelines, safety belts, and lanyards shall be used only for employee safeguarding.
- Lifelines shall be secured above the point of operation to an anchorage or structural member capable of supporting a minimum dead weight of 5400 pounds.
- Lifelines used on rock-scaling operations, or in areas where the lifeline may be subjected to cutting or abrasion, shall be a minimum of 7/8-inch wire core manila rope. For all other lifeline applications, a minimum of 3/4-inch manila or equivalent, with a minimum breaking strength of 5000 pounds, shall be used.
- Safety belt lanyard shall be a minimum of 1/2-inch nylon, or equivalent, with a maximum length to provide for a fall of no greater than 6 feet. The rope shall have a nominal breaking strength of 5400 pounds.
- All safety belt and lanyard hardware shall be drop forged or pressed steel, cadmium plated. Surface shall be smooth and free of sharp edges.
- All safety belt and lanyard hardware, except rivets, shall be capable of withstanding a tensile loading of 4000 pounds without cracking, breaking, or taking a permanent deformation.

12.8 Safety Nets

Some requirements of safety nets as specified by Part 1926, Subpart E, Standard Number 1926.105 are listed below:

- Safety nets shall be provided when workplaces are more than 25 feet above the ground or water surface, or other surfaces where the use of ladders, scaffolds, catch platforms, temporary floors, safety lines, or safety belts is impractical.

- Where safety net protection is required by this part, operations shall not be undertaken until the net is in place and has been tested.

- Nets shall extend 8 feet beyond the edge of the work surface where employees are exposed and shall be installed as close under the work surface as practical but in no case more than 25 feet below such work surface. Nets shall be hung with sufficient clearance to prevent user's contact with the surfaces or structures below. Such clearances shall be determined by impact load testing.

- The mesh size of nets shall not exceed 6 inches by 6 inches. All new nets shall meet accepted performance standards of 17,500 foot-pounds minimum impact resistance as determined and certified by the manufacturers, and shall bear a label of proof test. Edge ropes shall provide a minimum breaking strength of 5000 pounds.

Example 12.2

The mesh size of nets shall not exceed _____ inches by _____ inches.

    A. 6, 6

    B. 3, 3

    C. 9, 9

    D. 12, 12

Answer: *A*

12.9   Working Over or Near Water

Employees working over or near water, where the danger of drowning exists must have U.S. Coast Guard–approved life jacket or buoyant work vests. Prior to and after each use, the buoyant work vests or life preservers are to be inspected for defects which would alter their strength or buoyancy. Part 1926, Subpart E, Standard Number 1926.106 states that ring buoys with at least 90 feet of line shall be provided and readily available for emergency rescue operations. Distance between ring buoys shall not exceed 200 feet. At least one lifesaving skiff shall be immediately available at locations where employees are working over or adjacent to water.

Example 12.3

It is necessary for employers to provide personal equipment to employees if any of these conditions exist, except:

    A. When hazards exist or are likely to be present in work environment

    B. When employees might come into contract with hazard during work

    C. When employers cannot eliminate workplace hazards

    D. When employers cannot provide safety training

Answer: *D*

12.10   Summary

PPE is the last line of defense to protect workers from hazards on the jobsite. Employers are expected to enforce engineering controls and other safety measures to prevent accidents and injuries. In the case that these steps fail or are not possible, the PPE is there to prevent injury when hazards occur. Employers must provide PPE free of cost to workers exposed to possible hazards. All PPE should meet the American National Standards Institute (ANSI) specifications.

12.11   Multiple-Choice Questions

12.1 All protective goggles shall bear a label identifying some data. Which of the following data is not a requirement?

A. The laser wavelengths for which use is intended
B. The expiry date of the goggle
C. The optical density of those wavelengths
D. The visible light transmission

12.2 Personal protective equipment must be:

A. Used and maintained in a sanitary and reliable condition
B. Can be provided by employees by themselves
C. Of safe design and construction for the work to be performed
D. All of the above

12.3 The employer is expected to ensure that each employee uses appropriate eye or face protection when exposed to eye or face hazards from: [select all that apply]

A. Flying particles
B. Molten metal
C. Liquid chemicals, acids or caustic liquids, chemical gases or vapors
D. Potentially injurious light radiation

12.4 Which of the following statements is not true for the safety nets?

A. Safety nets shall be provided when workplaces are more than 25 feet above the ground
B. Nets shall extend 8 feet beyond the edge of the work surface where employees are exposed
C. The mesh size of nets shall not exceed 12 inches by 12 inches
D. All new nets shall meet accepted performance standards of 17,500 foot-pounds minimum impact resistance
E. Edge ropes shall provide a minimum breaking strength of 5000 pounds

12.5 Safety nets shall be provided when workplaces are more than _____ feet above the ground.

A. 5
B. 10
C. 20
D. 25

12.6 Edge ropes in safety nets shall provide a minimum breaking strength of _____ pounds.

   A. 5000
   B. 4000
   C. 3000
   D. 2000

12.12 Practice Problems

12.1 List the items employer is not required to pay for.

12.2 List the minimum requirements of the eye and face protectors based on Part 1926, Subpart E, Standard Number 1926.10.

12.3 List some requirements of safety belts, lifelines, and lanyards as specified by Part 1926, Subpart E, Standard Number 1926.104.

12.4 What are the requirements for working over or near water?

12.13 Critical Thinking and Discussion Topic

Carpenter: The company supplied hand gloves and goggles are disgusting. They often get in the way of our works.

Supervisor: These are the PPE recommended by the experts. Get used to it and one day, you will find it worthy.

Carpenter: I will feel better and will perform better without these PPEs. If any accident happens, it is my responsibility.

Supervisor: But if any injury happens, I will be asked.

Carpenter: I am not telling it anybody; it is my responsibility.

What should be your opinion as a supervisor at this stage of conversation?

(Photo courtesy of 123RF.)

CHAPTER **13**

Fire Protection and Prevention

(Photo courtesy of Pixabay from Pexels.)

13.1   General

The occurrence of something burning is called a fire (Fig. 13.1). Fires occur in the presence of the following elements:

- Oxygen—air, compressed oxygen
- Burning medium—wood, combustible materials, paper, gasoline, etc.
- Heat source—flames, sparking elements, heaters

Figure 13.1   Fire in a site.
(Photo courtesy of Pixabay from Pexels.)

Following are the different types of fires:

- Class A: Ordinary combustible (rags, paper)
- Class B: Combustible/flammable liquids (petroleum, diesel)
- Class C: Electrical fires (equipment, breakers)
- Class D: Flammable/combustible metal fires (magnesium, potassium)

Class A fires involve ordinary combustible materials such as wood, paper, rags, rubbish, and other solids. Class B fires occur due to flammable and combustible liquids such as gasoline, fuel oil, paint thinner, hydraulic fluids, flammable cleaning solvents, and other hydrocarbon fuels. Class C fires involve energized electrical equipment such as power outlets, circuit breakers, defective wiring, and overloaded circuits. Class D fires occur in combustible metals such as magnesium, aluminum powder, and alkali metals.

Employers should:

- Plan how to prevent fires.
- Train employees on fire prevention.
- Promote strong fire prevention practices, and methods.
- Perform regular inspection of workplaces.

Example 13.1

Table 13.1 shows a list of sources of fire and marks the type of fire that may be produced from the fire sources.

	Class A	Class B	Class C
Lumber	X		
Cleaning solvents		X	
Cloth rags	X		
Papers	X		
Hydraulic fluids		X	
Power outlets			X
Gasoline		X	
Bad wiring			X
Circuit breaks			X

TABLE 13.1   Marking the Type of Fire Produced from Fire Sources

13.2   Fire Protection Equipment

13.2.1   Fire Extinguishers

A fire extinguisher (Fig. 13.2a) is a device used for putting out fires. There are four different types of fire extinguishers classified according to the type of fire they extinguish. The four different types of fire extinguishers are Class A, B, C, and D. The classes relate to the type of fire they extinguish. Older fire extinguishers used geometric shapes to identify their type, but more current extinguishers use a labeling system that incorporates both words and pictures to distinguish the type of fire they are best suited for. Class A and Class B fire extinguishers have a numerical rating that is designed to determine the extinguishing potential for each size and type of extinguisher. The types of extinguishers and their uses are as follows:

- Class A—Pressurized water cans, clean agent/halogen, and wet chemical for specific applications
- Class B—Carbon dioxide, dry chemical, wet chemical, clean agent/halogen
- Class C—Dry chemical, carbon dioxide, clean agent/halogen
- Class D—Dry powder
- Multiclass extinguishers—Carbon dioxide/dry chemical

Class A Extinguishers.   Class A extinguishers are water-based or wet chemical solutions that are used on paper, cloth, wood, trash, and other common combustible fires. These extinguishers utilize a cooling and soaking stream that is effective on Class A fires. The numerical rating for this class of fire extinguisher refers to the amount of water the fire extinguisher holds and to the amount of fire it will extinguish.

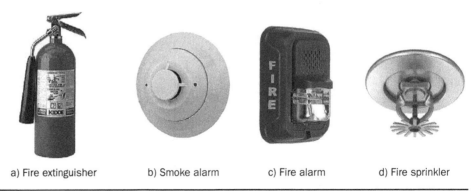

a) Fire extinguisher   b) Smoke alarm   c) Fire alarm   d) Fire sprinkler

FIGURE 13.2   Fire protection equipment.

Class B Extinguishers.   Class B extinguishers are pressurized with nonflammable carbon dioxide gas, dry chemical, wet chemical, or clean agent/halogen. Carbon dioxide reduces, or smothers, the oxygen content to a point where combustion cannot continue. Carbon dioxide is a clean, noncontaminating, odorless gas, and can safely be applied to clothing, equipment, and valuable documents without causing extreme damage. Class B extinguishers are used on fires involving flammable liquids including grease, gasoline, oil, paint thinner, hydraulic fluids, flammable cleaning solvents, and other hydrocarbon fuels. Carbon dioxide is extremely cold when disbursed from the extinguisher. The numerical rating for this class of fire extinguisher denotes the area in square feet of a flammable liquid fire that a person can expect to extinguish.

Class C Extinguishers.   Class C fire extinguishers are used on fires involving energized electrical equipment. Such fires must be extinguished using a nonconductive extinguishing agent such as carbon dioxide or a dry chemical or a clean agent/halogen. Carbon dioxide is most effective in extinguishing electrical fires, as it does not leave a residue that can harm sensitive electronics. This class of fire extinguishers does not have a numerical rating. Class C extinguishers have only a letter rating because there is no readily measurable quantity for Class C fires. The presence of the letter "C" indicates that the extinguishing agent is nonconductive.

Class D Extinguishers.   Class D extinguishers are designed for use on flammable metals and are often specific to the metal in question. Metals such as magnesium, potassium, titanium, and sodium burn at high temperatures and give off sufficient oxygen to support combustion. These metals react violently with water or other chemicals and must be handled with great care. The most common extinguishers for Class D fires use a dry powder designed specifically for this purpose. A common method of extinguishing small flammable metals fires is to cover the fire in dry sand. No picture designator is used on Class D extinguishers and this type of extinguisher generally has no rating.

Multiclass Fire Extinguishers.   Many fire extinguishers can be used on more than one class of fire and are called multipurpose extinguishers. Multiclass fire extinguishers are labeled with more than one class designator, such as A-B, B-C, or A-B-C. Multiclass fire

Fire Extinguishers	Description
a) Class A b) Class B c) Class C d) Class D e) Multiclass	A. Labeled with more than one class, and contains dry chemicals and a nonflammable gas as a propellant B. Water-based solution extinguishes burning paper, cloth, and wood trash with a cooling and soaking stream C. Designed for use on flammable metals and are often specific to the metal in question D. Nonconductive extinguishing agent used on fires involving energized electrical equipment E. Carbon dioxide gas used on fires involving gasoline, paint thinner, hydraulic fluids, and flammable cleaning solvents

TABLE 13.2   Matching the Class of Fire Extinguishers to the Correct Description

extinguishers typically contain dry chemicals and an extinguishing agent that uses a compressed, nonflammable gas as a propellant.

Extinguishers must be placed in an easily accessible location and should be in good operating condition. Extinguishers should be placed adjacent to a normal path of travel. At a minimum, fire extinguishers must be placed at all points of egress on construction projects and in close proximity of combustible/flammable materials stored on the site. The proper class must be marked on the extinguisher, so that it can be used according to the class of fire.

Example 13.2

Extinguishers must be in easily accessible locations, and they must be marked with the proper class so that they can be used on the right kind of fire. Match the class of fire extinguishers to the correct description shown in Table 13.2.

Answers: *eA, aB, dC, cD, bE*

13.2.2   Smoke Alarms

In case of a building fire, the first step is to warn the occupants and to evacuate the building as soon as possible. Early fire warnings can be given by means of active smoke and fire alarms (Fig. 13.2b and 13.2c, respectively) installed in strategic locations throughout a building. The two primary types of smoke alarms in use are ionization and photoelectric alarms. Ionization smoke detectors activate more quickly in fast, flaming fires that consume combustible materials rapidly and spread quickly. The photoelectric type of smoke detector is quicker to respond in slow, smoldering fires. These types of detectors provide early detection of smoke. When installed correctly, they provide accurate and dependable smoke detection. A combination of both types of detectors provides the greatest protection against fast moving fires and smoldering fires.

13.2.3   Fire Sprinklers

Fire sprinklers (Fig. 13.2d) are designed to provide 24-hour protection by detecting and controlling fires before they become a threat to lives or property. Fire sprinklers are designed to react quickly and independently of one another so that only those detectors in the affected area activate. Most fires are controlled by one or two sprinklers disbursing a minimal amount of water, which reduces the fire and water damage significantly.

13.3 Injuries and First Aid

The majority of fire-related deaths (50–80%) are caused by smoke inhalation. Actual flames and burns are second to smoke inhalation as the cause of deaths in fires. The National Traumatic Occupational Fatalities surveillance system recorded 1587 fire and flame-related occupational fatalities among the civilian workforce in the United States between 1980 and 1995. Of these fatalities, 433 resulted from 127 incidents that involved two or more victims.

Although smoke inhalation is the primary cause of deaths in fires, it is second to burns in the cause of injuries. Smoke from a fire may contain poison gases or may be hot enough to burn a victim's throat and lungs, resulting in serious breathing problems and even death. Symptoms of heavy smoke inhalation include breathing trouble, coughing, drowsiness, an upset stomach, vomiting, unconsciousness, and death.

It is important to evacuate from a smoky room as quickly as possible. If available, use a piece of wet cloth to cover your mouth and nostrils as you crawl as close to ground level as possible to safety. Once you're in fresh air, rest while taking deep breaths, and do not enter the smoky area until the fire is completely extinguished, all smoke has been removed, and fire officials have cleared the area.

For all burns beyond mild first-degree burns, seek medical attention immediately. Improper treatment can exacerbate damage. Minor first-degree burns can be treated by flushing the area with cold running water. Apply a clean, water-cooled cloth over the area to relieve pain. Do not apply ointment. Seek medical attention if the pain persists or if the burn appears worse.

Even if there is no visible evidence on the surface of the skin, electrical burns can cause deep tissue damage. Commence expired air resuscitation (EAR) or cardiopulmonary resuscitation (CPR) if pulse and breathing are absent. Immediately seek medical attention.

Example 13.3

True or False?

A. Most fire-related deaths are caused by flames and burns.

B. Symptoms of heavy smoke inhalation include drowsiness, an upset stomach, and vomiting.

C. Smoke is the primary cause of injuries in fire.

D. When evacuating a smoke-filled room, hold a wet cloth over your mouth and nostrils.

E. The proper first aid treatment for the second-degree burn is to rinse it and then apply ointment.

F. Electrical burns can cause deep tissue damage even if there is no visible evidence on the surface of the skin.

Answers:
False: A, C, E
True: B, D, F

13.4 Fire Protection Specifications

13.4.1 General Requirements

OSHA standards require a fire prevention plan to prevent fires and protect all employees from fire hazards. A fire prevention plan details:

- Names of employees responsible for controlling fire hazards from fuel sources
- Names of employees responsible for fire prevention equipment or equipment to control fires
- Safeguarding controls and maintenance of safeguards to prevent accidental fires caused by heat causing machinery, equipment, and materials
- Handling and storage of combustible or flammable materials
- Naming specific fire sources and locations and available equipment for fighting fires

Part 1926, Subpart F, Standard Number 1926.150 states that:

- The employer shall be responsible for the development of a fire protection program to be followed throughout all phases of the construction and demolition work.
- Access to all available firefighting equipment shall be maintained at all times.
- All firefighting equipment, provided by the employer, shall be conspicuously located.
- All firefighting equipment shall be periodically inspected and maintained in operating condition. Defective equipment shall be immediately replaced.
- As warranted by the project, the employer shall provide a trained and equipped firefighting organization (Fire Brigade) to assure adequate protection to life.

13.4.2 Water Supply

A temporary or permanent water supply, of sufficient volume, duration, and pressure, required to properly operate the firefighting equipment shall be made available as soon as combustible materials accumulate. Where underground water mains are to be provided, they shall be installed, completed, and made available for use as soon as practicable.

13.4.3 Portable Firefighting Equipment

For fire extinguishers and small hose lines:

- A fire extinguisher, rated not less than 2A, shall be provided for each 3000 square feet of the protected building area, or major fraction thereof. Travel distance from any point of the protected area to the nearest fire extinguisher shall not exceed 100 feet.

- One 55-gallon open drum of water with two fire pails may be substituted for a fire extinguisher having a 2A rating.

- A 1/2-inch diameter garden-type hose line, not to exceed 100 feet in length and equipped with a nozzle, may be substituted for a 2A-rated fire extinguisher, providing it is capable of discharging a minimum of 5 gallons per minute with a minimum hose stream range of 30 feet horizontally.

- One or more fire extinguishers, rated not less than 2A, shall be provided on each floor. In multistory buildings, at least one fire extinguisher shall be located adjacent to stairway.

- Extinguishers and water drums, subject to freezing, shall be protected from freezing.

- A fire extinguisher, rated not less than 10B, shall be provided within 50 feet of wherever more than 5 gallons of flammable or combustible liquids or 5 pounds of flammable gas are being used on the jobsite.

- Carbon tetrachloride and other toxic vaporizing liquid fire extinguishers are prohibited.

- Portable fire extinguishers shall be inspected periodically and maintained in accordance with maintenance and use of portable fire extinguishers.

- Fire extinguishers which have been listed or approved by a nationally recognized testing laboratory shall be used to meet the requirements of this subpart.

For fire hose and connections (Fig. 13.3):

- One hundred feet, or less, of 1 1/2-inch hose, with a nozzle capable of discharging water at 25 gallons or more per minute, may be substituted for a

FIGURE 13.3 Firefighting hose.
(Photo courtesy of Pixabay from Pexels.)

fire extinguisher rated not more than 2A in the designated area, provided that the hose line can reach all points in the area.

- If fire hose connections are not compatible with local firefighting equipment, the contractor shall provide adapters, or equivalent, to permit connections.

- During demolition involving combustible materials, charged hose lines, supplied by hydrants, water tank trucks with pumps, or equivalent, shall be made available.

13.4.4 Fixed Firefighting Equipment

For sprinkler protection:

- If the facility being constructed includes the installation of automatic sprinkler protection, the installation shall closely follow the construction and be placed in service as soon as applicable laws permit following completion of each story.

- During demolition or alterations, existing automatic sprinkler installations shall be retained in service as long as reasonable.

- The operation of sprinkler control valves shall be permitted only by properly authorized persons. Modification of sprinkler systems to permit alterations or additional demolition should be expedited so that the automatic protection may be returned to service as quickly as possible.

- Sprinkler control valves shall be checked daily at close of work to ascertain that the protection is in service.

For standpipes:

- In all structures in which standpipes are required, or where standpipes exist in structures being altered, they shall be brought up as soon as applicable laws permit, and shall be maintained as construction progresses in such a manner that they are always ready for fire protection use.

- The standpipes shall be provided with Siamese fire department connections on the outside of the structure, at the street level, which shall be conspicuously marked. There shall be at least one standard hose outlet at each floor.

For fire alarm devices:

An alarm system, for example, telephone system, siren, etc., shall be established by the employer whereby employees on the site and the local fire department can be alerted for an emergency. The alarm code and reporting instructions shall be conspicuously posted at phones and at employee entrances.

For fire cutoffs:

- Fire walls and exit stairways, required for the completed buildings, shall be given construction priority.

- Fire doors, with automatic closing devices, shall be hung on openings as soon as practicable.

- Fire cutoffs shall be retained in buildings undergoing alterations or demolition until operations necessitate their removal.

13.5   Fire Prevention

Part 1926, Subpart F, Standard Number 1926.151 gives the detailed specifications for the fire prevention plan. The major specifications are listed below.

13.5.1   Ignition Hazards

- Electrical wiring and equipment for light, heat, or power purposes shall be installed in compliance with the requirements of Subpart K of this part.
- Internal combustion engine powered equipment shall be so located that the exhausts are well away from combustible materials. When the exhausts are piped to outside the building under construction, a clearance of at least 6 inches shall be maintained between such piping and combustible material.
- Smoking shall be prohibited at or in the vicinity of operations which constitute a fire hazard, and shall be conspicuously posted: "No Smoking or Open Flame" (Fig. 13.4).
- Portable battery-powered lighting equipment, used in connection with the storage, handling, or use of flammable gases or liquids, shall be of the type approved for the hazardous locations.
- The nozzle of air, inert gas, and steam lines or hoses, when used in the cleaning or ventilation of tanks and vessels that contain hazardous concentrations of flammable gases or vapors, shall be bonded to the tank or vessel shell. Bonding devices shall not be attached or detached in hazardous concentrations of flammable gases or vapors.

13.5.2   Temporary Buildings

- No temporary building shall be erected where it will adversely affect any means of exit.

FIGURE 13.4   No smoking or open flame sign.

- Temporary buildings, when located within another building or structure, shall be of either noncombustible construction or of combustible construction having a fire resistance of not less than 1 hour.

- Temporary buildings, located other than inside another building and not used for the storage, handling, or use of flammable or combustible liquids, flammable gases, explosives, or blasting agents, or similar hazardous occupancies, shall be located at a distance of not less than 10 feet from another building or structure. Groups of temporary buildings, not exceeding 2000 square feet in aggregate, shall, for the purposes of this part, be considered a single temporary building.

13.5.3 Open Yard Storage

- Combustible materials shall be piled with due regard to the stability of piles and in no case higher than 20 feet.

- Driveways between and around combustible storage piles shall be at least 15 feet wide and maintained free from accumulation of rubbish, equipment, or other articles or materials. Driveways shall be so spaced that a maximum grid system unit of 50 feet by 150 feet is produced.

- The entire storage site shall be kept free from accumulation of unnecessary combustible materials. Weeds and grass shall be kept down and a regular procedure provided for the periodic cleanup of the entire area. In an open yard storage area, you should not have to go further than 100 feet to reach the nearest fire extinguishing unit.

- When there is a danger of an underground fire that land shall not be used for combustible or flammable storage.

- Method of piling shall be solid wherever possible and in orderly and regular piles. No combustible material shall be stored outdoors within 10 feet of a building or structure.

13.5.4 Indoor Storage

- Noncompatible materials, which may create a fire hazard, shall be segregated by a barrier having a fire resistance of at least 1 hour.

- Material shall be piled to minimize the spread of fire internally and to permit convenient access for firefighting. Stable piling shall be maintained at all times. Aisle space shall be maintained to safely accommodate the widest vehicle that may be used within the building for firefighting purposes.

- When you're stacking material in an indoor storage area, leave at least 18 inches between the top of the pile and the nearest sprinkler.

- For indoor storage, materials cannot be stored within 36 inches of a fire door opening.

Example 13.4

The construction and arrangement of temporary structures and open yard and indoor storage can have an important effect on fire safety and prevention. You are well aware of this as your company's safety officer. Type in the correct number so that these sentences meet the fire safety guidelines.

A. A temporary combustible structure covering 2000 square feet should be at least _____ feet away from any other building.

B. Do not pile combustible materials in open yard storage higher than _____ feet.

C. In an open yard storage area, the driveways between combustible storage spaces should be at least _____ feet wide.

D. In an open yard storage area, you should not have to go further than _____ feet to reach the nearest fire extinguishing unit.

E. That stack of combustible material outdoors should be at least _____ feet away from all buildings or structures

F. When you're stacking material in an indoor storage area, leave at least _____ inches between the top of the pile and the nearest sprinkler.

Answers:
A. 10
B. 20
C. 15
D. 100
E. 10
F. 18

Example 13.5

True or False?

A. A temporary combustible structure can store flammables or explosives as long as the facility is at least 10 feet away from any other buildings.

B. Combustible materials should not be stacked or piled higher than 10 feet in an open yard storage area.

C. For indoor storage, materials cannot be stored within 36 inches of a fire door opening.

D. Employers must hire the local fire departments to assist in creating the company's fire prevention plan.

E. Fire extinguishers should be placed within 100 feet of the protected area.

Answers:
True: C, E
False: A, B, D

13.6   Flammable Liquids

The major specifications from the Part 1926, Subpart F, Standard Number 1926.152 are summarized below.

13.6.1   General Requirements

Only approved containers and portable tanks shall be used for storage and handling of flammable liquids (Fig. 13.5). Approved safety cans or Department of Transportation–approved containers shall be used for the handling and use of flammable liquids in quantities of 5 gallons or less, except that this shall not apply to those flammable liquid materials which are highly viscid (extremely hard to pour), which may be used and handled in original shipping containers. For quantities of 1 gallon or less, the original container may be used, for storage, use, and handling of flammable liquids. Flammable liquids shall not be stored in areas used for exits, stairways, or normally used for the safe passage of people.

13.6.2   Indoor Storage of Flammable Liquids

No more than 25 gallons of flammable liquids shall be stored in a room outside of an approved storage cabinet. For storage of liquefied petroleum gas, quantities of flammable liquid in excess of 25 gallons shall be stored in an acceptable or approved cabinet.

13.6.3   Storage Outside Buildings

- Storage of containers (not more than 60 gallons each) shall not exceed 1100 gallons in any one pile or area. Piles or groups of containers shall be separated by a 5-foot clearance. Piles or groups of containers shall not be nearer than 20 feet to a building.
- Within 200 feet of each pile of containers, there shall be a 12-foot-wide access way to permit approach of fire control apparatus.
- Storage areas shall be kept free of weeds, debris, and other combustible material not necessary to the storage.

Figure 13.5   Flammable liquid.

13.7 Summary

While the majority of fire-related deaths are caused by smoke inhalation, actual flames and burns follow this. Although fire sprinklers are designed to provide 24-hour protection by detecting and controlling fires before they become a threat to lives or property, a combination of both types of fire detector (ionization and photoelectric) provide the greatest protection against fast moving and smoldering fires. Some of the newer fire extinguishers can be used on more than one type of fire. Multiclass fire extinguishers are labeled with more than one class designator, whereas Class C extinguishers have only a letter rating because there is no readily measurable quantifier for this type of fire. In the event of a fire, a safe and speedy response depends on how well employees and employers are prepared for an emergency. Thus, proper response requires planning and cooperation among workers, and includes the planning of escape routes, guidelines to prevent fires from spreading, and safe evacuation procedures.

13.8 Multiple-Choice Questions

13.1. Some common workplace fire hazards are:

A. Contains types of chemicals
B. Cardboard and paper items could still fuel a fire
C. Faulty electrical wiring
D. All of the above

13.2. Fire classifications are based on:

A. The location of the fire
B. How long the fire has been burning
C. The combustible materials involved in the fire
D. The amount of training the fire fighter has

13.3. What is a common extinguishing agent for:

A. Flammable solvent
B. Wood, cloth, and paper
C. Electrical fires
D. Sand

13.4. The most common extinguishing agent for flammable liquid fires is

A. Water fog
B. Mist
C. Dry chemical
A. Sand

13.5. The two primary types of smoke alarms in use are _____ and photoelectric alarms.

A. Radiation
B. Ionization
C. Pulverization
D. None of the above

13.6. A class A fire is a combustible/flammable liquid fire.

• True
• False

13.7. Many fire extinguishers can be used on more than one class of fire and are called multipur-pose extinguishers.

- True
- False

13.8. There are only two types of fire extinguishers for fighting all types of fires.

- True
- False

13.9. Fire _____ are designed to react quickly and independently of one another so that only those detectors in the affected area activate.

A. Alarms
B. Watch
C. Mains
D. Sprinklers

13.10. Class _____ extinguishers are pressurized with nonflammable carbon dioxide gas, dry chemical, wet chemical, or clean agent/halogen.

A. A
B. B
C. C
D. D

13.11. Which of the following elements are required for sustaining a fire?

A. Oxygen
B. Fuel (burning medium)
C. Heat source
D. All of the above

13.12. What is the primary cause of death due to fires?

A. Smoke inhalation
B. Burns
B. Arc flash
D. Falls

13.13. In the event of a fire, a safe and _____ responds on how well employees and employers are prepared for emergencies.

A. Speedy
B. Slow
C. Measured
D. Paced

13.14. Class _____ fire extinguishers are used on fires involving energized electrical equipment.

A. A
B. B
C. C
D. D

13.9 Practice Problems

13.1. List the types of fires.

13.2. List the types of fire extinguishers.

13.3. List the primary types of smoke alarms.

13.4. What are the elements of fire prevention plan?

13.5. What are the requirements for indoor storage of flammable liquids?

13.10 Critical Thinking and Discussion Topic

While working on a construction site, your finger gets burnt from a fire. The nearby health services are little away and if you go there you will not be able to come back to work. Also, today's schedule is pretty tight. If today's work cannot be completed, it may hamper the work scheduled for tomorrow. This is why, your supervisor advised you to put some antiseptic cream and finish the work scheduled for today. As a compensation, you are offered to take a day off the next day. What will you do in this situation?

CHAPTER 14

Signs, Signals, and Barricades

(Photo courtesy of 123RF.)

14.1 Accident Prevention Signs and Tags

Signs and symbols must be visible at all times when work is being performed and must be removed or promptly covered when the hazards no longer exist. For instance, directional signs, other than automotive traffic signs, are white with a black panel and a white directional symbol. Any additional wording on the sign must be in black letters on the white background. Part 1926, Subpart G, Standard Number 1926.200 states that signs and symbols (Fig. 14.1) required shall be visible at all times when work is being performed, and shall be removed or covered promptly when the hazards no longer exist.

a) Danger sign   b) Caution sign   c) Exit sign   d) Safety instruction sign   e) Directional sign

FIGURE **14.1**   Accident prevention signs.

Danger signs:

- Danger signs shall be used only where an immediate hazard exists, and shall follow the specifications.
- Danger signs shall have red as the predominating color for the upper panel; black outline on the borders; and a white lower panel for additional sign wording.

Caution signs:

- Caution signs shall be used only to warn against potential hazards or to caution against unsafe practices.
- Caution signs shall have yellow as the predominating color; black upper panel and borders: yellow lettering of "caution" on the black panel; and the lower yellow panel for additional sign wording. Black lettering shall be used for additional wording.
- The standard color of the background shall be yellow; and the panel, black with yellow letters. Any letters used against the yellow background shall be black.

Exit signs: Exit signs, when required, shall be lettered in legible red letters, not less than 6 inches high, on a white field and the principal stroke of the letters shall be at least three-fourths inch in width.

Safety instruction signs: Safety instruction signs, when used, shall be white with green upper panel with white letters to convey the principal message. Any additional wording on the sign shall be black letters on the white background.

Directional signs: Directional signs, other than automotive traffic signs, shall be white with a black panel and a white directional symbol. Any additional wording on the sign shall be black letters on the white background.

Traffic control signs and devices:

- At points of hazard, construction areas shall be posted with legible traffic control signs and protected by traffic control devices.
- The design and use of all traffic control devices, including signs, signals, markings, barricades, and other devices, for protection of construction workers shall conform to the Manual on Uniform Traffic Control Devices (MUTCD).

Accident prevention tags:

- Accident prevention tags shall be used as a temporary means of warning employees of an existing hazard, such as defective tools, equipment, etc. They shall not be used in place of, or as a substitute for, accident prevention signs.
- For accident prevention tags, employers shall follow specifications.

Example 14.1
Accident prevention signs and symbols must be:

A. Visible at all times whether work is performed or not

B. Visible at all times when work is being performed

C. Removed when the hazards no longer exist

D. Promptly covered when the hazards no longer exist

Answers: B, C, and D

14.2 Signaling

Part 1926, Subpart G, Standard Number 1926.201 states that signaling by flaggers and the use of flaggers, including warning garments worn by flaggers, shall conform to the MUTCD. Regulations for crane and hoist signaling will be found in applicable American National Standards Institute standards.

14.3 Summary

Lesson signs and symbols must be visible at all times when work is being performed and must be removed or promptly covered when the hazards no longer exist. For instance, directional signs other than automotive traffic signs are white with a black panel and a white directional symbol. Any additional wording on the sign must be in black letters on the white background.

14.4 Multiple-Choice Questions

14.1. Regulations for crane and hoist signaling is specified by:
- A.   American National Standards Institute
- B.   Manual on Uniform Traffic Control Devices
- C.   Crane and Hoist Society
- D.   OSHA

14.2. The colors used in the upper panel and the lower panel of a danger sign are respectively:
- A.   Red, White
- B.   Red, Yellow
- C.   Black, Red
- D.   Red, Black

14.3. The background color of the upper panel of a safety instruction sign is most nearly:

A. Yellow
B. Orange
C. White
D. Green

14.4. Any additional wording other than the principal message on the Safety instruction sign shall be:

A. Red letters on the white background
B. White letters on the black background
C. Black letters on the white background
D. Black letters on the orange background

14.5. The predominating color of caution sign is:

A. Orange
B. Yellow
C. Red
D. Green

14.6. Any letters used against the yellow background of caution sign shall be:

A. Black
B. White
C. Red
D. Yellow

14.7. Directional signs, other than automotive traffic signs, shall be _____ with a black panel and white directional symbol. Any additional wording on the sign must be in black letters on the white background.

A. Yellow
B. White
C. Black
D. Red

14.8. Every caution sign must have _____ as the dominant color, a black upper panel and borders with yellow lettering of the world "caution" on the black panel, and a lower yellow panel for additional sign wording. Black lettering must be used for additional wording.

A. Yellow
B. Red
C. Blue
D. Black

14.9. Each danger sign must have _____ as the dominant color for the upper panel, black outline on the borders, and a white lower panel for additional sign wording.

A. Yellow
B. Purple
C. Red
D. Blue

14.5 Practice Problems

14.1. List some regulations for danger and caution signs.

14.2. List some regulations for exit and directional signs.

14.3. List some regulations for safety instructions, traffic control, and accident prevention signs.

14.4. List some regulations for signaling.

14.6 Critical Thinking and Discussion Topic

While working on a construction site, you found that you needed four precaution sign boards to meet the safety regulations. However, you had only three at that time. It might take several hours to go back to warehouse and bring one more sign board. Your supervisor advised you to put the three sign boards instead of four and started to work. Your supervisor argued that people will not understand the spacing between two consecutive sign boards exceeds the regulations. What will you do in this situation?

(Photo courtesy of frans-van-heerden from Pexels.)

CHAPTER 15

Materials Storage, Handling, Use, and Disposal

(Photo by Tiger Lily from Pexels.)

15.1 General

Every industry needs a continuous inflow and outflow of resources and materials. Therefore, there is an ever-present need to handle and store materials that are so vital to industry. However, improper handling and storage of materials can be hazardous if precautions are not taken. Two of the major hazards involved in handling and storing materials include:

- Bulkiness
- Weight of materials

Often, handling heavy and bulky objects results in back and spinal injuries. Workers that lift these objects may suffer from acute and chronic back pains. Body movement is a common factor related to back injuries. Bending, twisting, and turning are some of the common body movements that lead to back and spinal injuries. Another common factor that can potentially cause severe injuries is falling objects. Materials that have been improperly stacked can present a great danger.

When workers move materials they must be aware of the following:

- Improper lifting may cause strains and sprains.
- Falling materials may cause bruises, fractures, or even death.

It is very important that efforts be made by both the employer and worker to ensure that dangers from improper material handling are minimized, if not eliminated, from the workplace. Inspections of the workplace must be carried out on a periodical basis to ensure that they are free from any related hazards.

Example 15.1
Bending, twisting, and turning are some of the common body movements that can lead to back and spinal injuries.

- True
- False

Answer: *True*

15.2   Methods of Hazards Prevention in Handling and Storing Materials

If the worker has to manually handle an object, he or she must ask for assistance when a load is:

- Bulky to the extent that it cannot be grasped or lifted properly
- Bulky to the extent that it cannot be seen around or over
- One that cannot be handled safely

Blocks.   When placing blocks under raised loads, workers must make sure that the raised loads are kept in a raised position until their hands have been removed from beneath them. The blocks must be large and sturdy enough to be able to support the load. Block materials with cracks, splintered pieces, and rot must not be used.

Handles, Holders, and Protective Equipment.   All loads ideally should be moved via mechanical means when possible. When loads are to be moved manually, using handles and holders may minimize chances of injuries to the fingers and hands. In the case of loads with sharp and rough edges, workers must wear gloves. It also may be advisable for a worker to wear steel-toed shoes when carrying heavy or bulky loads, so as to minimize the risk of foot injuries in case of accidentally dropping the load.

Load Weight and Mechanical Moving Equipment.   Workers must never overload mechanical moving equipment. All types of material handling equipment have maximum weight specifications which must be adhered to. As such, the type of equipment used to move a load from one point to another must be dictated by the specifications of the load itself.

Stored Materials.   Workers must ensure that stored materials do not create hazards. For example, workers must ensure that storage spaces are not left to accumulate flammable materials, cause explosions or tripping hazards, or easily harbor rats and other pests. In addition, storage containers must have adequate capacity to handle the height and weight of stored items, as well as being accessible and in good condition.

Bound Material.   All materials stored in tiers shall be stacked, racked, blocked, interlocked, or otherwise secured to prevent sliding, falling, or collapse. Furthermore, maximum safe load limits that have been specified by building inspectors must not be exceeded or otherwise violated, and signs displaying load limits must be posted in all storage areas except for slab on grade.

Height Limitations.   Always adhere to height limitations when stacking materials. Lumber that is manually handled must not be stacked at a height of more than 16 feet; 20 feet if a forklift is being used. Painting stripes on poles and walls is a good way to indicate the maximum height allowed.

Stacking Lumber.   If used lumber is being stacked, workers must ensure that all nails have been removed before stacking lumber. Furthermore, workers must ensure that the lumber stacks are on level and solidly supported bracing. Lumber must be stacked such that it is stable and self-supporting.

Bricks.   Loose bricks must not be stacked to heights of more than 7 feet. When a stack of loose bricks exceeds 4 feet, they must be tapered back 2 inches for every foot of height over and above the 4-foot level.

Masonry Blocks.   When masonry blocks are stacked to a height of 6 feet or higher, the stacks must be tapered back one-half block for each tier over the 6-foot level.

Bags and Bundles.   It is advisable that when bags and bundles are stacked, interlocking rows are used. Bagged materials must be stacked by stepping back the layers and cross-keying the bags at least every 10 bags high. When workers remove bags from the stack, they must start with the topmost layer working their way down. Noncompatible materials shall be segregated in storage. Baled paper and rags must be kept at a minimum of 18 inches from walls, sprinklers, and partitions. Finally, it is advised that boxed materials be banded, or at least held in place using cross-ties or shrink plastic fiber.

Drums, Barrels, and Kegs.   Drums, barrels, and kegs must be stored symmetrically. If they are stored on their sides, the bottom tiers must be blocked accordingly to prevent them from rolling. When barrels are stacked on end, planks must be placed between each tier to make a firm, flat stacking surface. If the stack reaches two or more tiers, the lowest tier must be secured on either side to prevent the barrels from shifting

15.3   General OSHA Requirements for Storage

Part 1926, Subpart H, Standard Number 1926.250 states the following:
General:

- All materials stored in tiers shall be stacked, racked, blocked, interlocked, or otherwise secured to prevent sliding, falling, or collapse.
- The weight of stored materials on floors within buildings and structures shall not exceed maximum safe load limits. Employers shall conspicuously post maximum safe load limits of floors within buildings and structures, in pounds per square foot, in all storage areas, except when the storage area is on a floor or slab on grade. Posting is not required for storage areas in all single-family residential structures and wood-framed multifamily residential structures.
- Aisles and passageways shall be kept clear to provide for the free and safe movement of material handling equipment or employees. Such areas shall be kept in good repair.
- When a difference in road or working levels exist, means such as ramps, blocking, or grading shall be used to ensure the safe movement of vehicles between the two levels.

Material storage:

- Material stored inside buildings under construction shall not be placed within 6 feet of any hoistway or inside floor openings, nor within 10 feet of an exterior wall which does not extend above the top of the material stored.
- Each employee required to work on stored material in silos, hoppers, tanks, and similar storage areas shall be equipped with personal fall arrest equipment.
- Noncompatible materials shall be segregated in storage.
- Bagged materials shall be stacked by stepping back the layers and cross-keying the bags at least every 10 bags high.
- Materials shall not be stored on scaffolds or runways in excess of supplies needed for immediate operations.
- Brick stacks shall not be more than 7 feet in height. When a loose brick stack reaches a height of 4 feet, it shall be tapered back 2 inches in every foot of height above the 4-foot level.
- When masonry blocks are stacked higher than 6 feet, the stack shall be tapered back one-half block per tier above the 6-foot level.
- Lumber:
  - Used lumber shall have all nails withdrawn before stacking.
  - Lumber shall be stacked on level and solidly supported sills.
  - Lumber shall be so stacked as to be stable and self-supporting.
  - Lumber piles shall not exceed 20 feet in height provided that lumber to be handled manually shall not be stacked more than 16 feet high.

- Structural steel, poles, pipe, bar stock, and other cylindrical materials, unless racked, shall be stacked and blocked so as to prevent spreading or tilting.

Housekeeping:

- Storage areas shall be kept free from accumulation of materials that constitute hazards from tripping, fire, explosion, or pest harborage. Vegetation control will be exercised when necessary.

Dockboards (bridge plates):

- Portable and powered dockboards shall be strong enough to carry the load imposed on them.
- Portable dockboards shall be secured in position, either by being anchored or equipped with devices which will prevent their slipping.
- Handholds, or other effective means, shall be provided on portable dockboards to permit safe handling.
- Positive protection shall be provided to prevent railroad cars from being moved while dockboards or bridge plates are in position.

Example 15.2

All materials stored in tiers shall be stacked, racked, blocked, interlocked, or otherwise secured to prevent:

    A. Sliding

    B. Falling

    C. Collapsing

    D. All of the above

Answer: D

15.4   Rigging Equipment for Material Handling

All loads ideally should be moved via mechanical means when possible. Rigging methods and equipment are used to attach heavy loads to lifting devices such as cranes, derricks, or chain falls. Rigging equipment may include: rope, chains, or synthetic webbing slings, which connect the loads to lifting machinery by means of hooks or shackles. Part 1926, Subpart H, Standard Number 1926.251 states the following regulations.

15.4.1   General Regulations

- Rigging equipment for material handling shall be inspected prior to use on each shift and as necessary during its use to ensure that it is safe. Defective rigging equipment shall be removed from service.
- Employers must ensure that rigging equipment:

- Has permanently affixed and legible identification markings as prescribed by the manufacturer that indicate the recommended safe working load
- Not be loaded in excess of its recommended safe working load as prescribed on the identification markings by the manufacturer
- Not be used without affixed, legible identification markings

- Rigging equipment, when not in use, shall be removed from the immediate work area so as not to present a hazard to employees.
- Special custom design grabs, hooks, clamps, or other lifting accessories, for such units as modular panels, prefabricated structures, and similar materials, shall be marked to indicate the safe working loads and shall be proof-tested prior to use to 125% of their rated load.
- Scope: This section applies to slings used in conjunction with other material handling equipment for the movement of material by hoisting, in employments covered by this part. The types of slings covered are those made from alloy steel chain, wire rope, metal mesh, natural or synthetic fiber rope (conventional three-strand construction), and synthetic web (nylon, polyester, and polypropylene).
- Inspections: Each day before being used, the sling and all fastenings and attachments shall be inspected for damage or defects by a competent person designated by the employer. Additional inspections shall be performed during sling use, where service conditions warrant. Damaged or defective slings shall be immediately removed from service.

Example 15.3

All loads ideally should be moved via _____ means when possible.

  A. Physical
  B. Mechanical
  C. Personal
  D. None of the above

Answer: B

15.4.2 Alloy Steel Chains

- Welded alloy steel chain slings shall have permanently affixed durable identification stating size, grade, rated capacity, and sling manufacturer.
- Hooks, rings, oblong links, pear-shaped links, welded or mechanical coupling links, or other attachments, when used with alloy steel chains, shall have a rated capacity at least equal to that of the chain.
- Job or shop hooks and links, or makeshift fasteners, formed from bolts, rods, etc., or other such attachments, shall not be used.

- Employers must not use alloy steel-chain slings with loads in excess of the rated capacities (i.e., working load limits) indicated on the sling by permanently affixed and legible identification markings prescribed by the manufacturer.

- Inspections:
  - In addition to the inspection required by other paragraphs of this section, a thorough periodic inspection of alloy steel chain slings in use shall be made on a regular basis, to be determined on the basis of (a) frequency of sling use; (b) severity of service conditions; (c) nature of lifts being made; and (d) experience gained on the service life of slings used in similar circumstances. Such inspections shall in no event be at intervals greater than once every 12 months.
  - The employer shall make and maintain a record of the most recent month in which each alloy steel chain sling was thoroughly inspected, and shall make such record available for examination.

15.4.3 Wire Rope

- Employers must not use improved plow-steel wire rope and wire-rope slings with loads in excess of the rated capacities (i.e., working load limits) indicated on the sling by permanently affixed and legible identification markings prescribed by the manufacturer.

- Protruding ends of strands in splices on slings and bridles shall be covered or blunted.

- Wire rope shall not be secured by knots, except on haul back lines on scrapers.

- The following limitations shall apply to the use of wire rope:
  - An eye splice made in any wire rope shall have not less than three full tucks. However, this requirement shall not operate to preclude the use of another form of splice or connection which can be shown to be as efficient and which is not otherwise prohibited.
  - Except for eye splices in the ends of wires and for endless rope slings, each wire rope used in hoisting or lowering, or in pulling loads, shall consist of one continuous piece without knot or splice.
  - Eyes in wire rope bridles, slings, or bull wires shall not be formed by wire rope clips or knots.
  - Wire rope shall not be used if, in any length of eight diameters, the total number of visible broken wires exceeds 10% of the total number of wires, or if the rope shows other signs of excessive wear, corrosion, or defect.

- Slings shall not be shortened with knots or bolts or other makeshift devices.

- Sling legs shall not be kinked.

- Slings used in a basket hitch shall have the loads balanced to prevent slippage.

- Slings shall be padded or protected from the sharp edges of their loads.

- Hands or fingers shall not be placed between the sling and its load while the sling is being tightened around the load.
- Shock loading is prohibited.
- A sling shall not be pulled from under a load when the load is resting on the sling.
- Minimum sling lengths.
  - Cable laid and 6 × 19 and 6 × 37 slings shall have a minimum clear length of wire rope 10 times the component rope diameter between splices, sleeves, or end fittings.
  - Braided slings shall have a minimum clear length of wire rope 40 times the component rope diameter between the loops or end fittings.
  - Cable laid grommets, strand laid grommets, and endless slings shall have a minimum circumferential length of 96 times their body diameter.
- Safe operating temperatures. Fiber core wire rope slings of all grades shall be permanently removed from service if they are exposed to temperatures in excess of 200°F (93.33°C). When nonfiber core wire rope slings of any grade are used at temperatures above 400°F (204.44°C) or below minus 60°F (15.55°C), recommendations of the sling manufacturer regarding use at that temperature shall be followed.
- End attachments:
  - Welding of end attachments, except covers to thimbles, shall be performed prior to the assembly of the sling.
  - All welded end attachments shall not be used unless proof tested by the manufacturer or equivalent entity at twice their rated capacity prior to initial use. The employer shall retain a certificate of the proof test, and make it available for examination.
- Wire rope slings shall have permanently affixed, legible identification markings stating size, rated capacity for the type(s) of hitch(es) used and the angle upon which it is based, and the number of legs if more than one.

15.4.4 Natural Rope and Synthetic Fiber

- Employers must not use natural- and synthetic-fiber rope slings with loads in excess of the rated capacities (i.e., working load limits) indicated on the sling by permanently affixed and legible identification markings prescribed by the manufacturer.
- All splices in rope slings provided by the employer shall be made in accordance with fiber rope manufacturers' recommendations.
  - In manila rope, eye splices shall contain at least three full tucks, and short splices shall contain at least six full tucks (three on each side of the centerline of the splice).

- In layed synthetic fiber rope, eye splices shall contain at least four full tucks, and short splices shall contain at least eight full tucks (four on each side of the centerline of the splice).

- Strand end tails shall not be trimmed short (flush with the surface of the rope) immediately adjacent to the full tucks. This precaution applies to both eye and short splices and all types of fiber rope. For fiber ropes under 1-inch diameter, the tails shall project at least six rope diameters beyond the last full tuck. For fiber ropes 1-inch diameter and larger, the tails shall project at least 6 inches beyond the last full tuck. In applications where the projecting tails may be objectionable, the tails shall be tapered and spliced into the body of the rope using at least two additional tucks (which will require a tail length of approximately six rope diameters beyond the last full tuck).

- For all eye splices, the eye shall be sufficiently large to provide an included angle of not greater than 60° at the splice when the eye is placed over the load or support.

- Knots shall not be used in lieu of splices.

- Safe operating temperatures: Natural and synthetic fiber rope slings, except for wet frozen slings, may be used in a temperature range from minus 20°F (−28.88°C) to plus 180°F (82.2°C) without decreasing the working load limit. For operations outside this temperature range and for wet frozen slings, the sling manufacturer's recommendations shall be followed.

- Splicing: Spliced fiber rope slings shall not be used unless they have been spliced in accordance with the following minimum requirements and in accordance with any additional recommendations of the manufacturer:

  - In manila rope, eye splices shall consist of at least three full tucks, and short splices shall consist of at least six full tucks, three on each side of the splice centerline.

  - In synthetic fiber rope, eye splices shall consist of at least four full tucks, and short splices shall consist of at least eight full tucks, four on each side of the centerline.

  - Strand end tails shall not be trimmed flush with the surface of the rope immediately adjacent to the full tucks. This applies to all types of fiber rope and both eye and short splices. For fiber rope under 1 inch (2.54 cm) in diameter, the tail shall project at least six rope diameters beyond the last full tuck. For fiber rope 1 inch (2.54 cm) in diameter and larger, the tail shall project at least 6 inches (15.24 cm) beyond the last full tuck. Where a projecting tail interferes with the use of the sling, the tail shall be tapered and spliced into the body of the rope using at least two additional tucks (which will require a tail length of approximately six rope diameters beyond the last full tuck).

  - Fiber rope slings shall have a minimum clear length of rope between eye splices equal to 10 times the rope diameter.

  - Knots shall not be used in lieu of splices.

- ○ Clamps not designed specifically for fiber ropes shall not be used for splicing.
- ○ For all eye splices, the eye shall be of such size to provide an included angle of not greater than 60° at the splice when the eye is placed over the load or support.
- End attachments: Fiber rope slings shall not be used if end attachments in contact with the rope have sharp edges or projections.
- Removal from service: Natural and synthetic fiber rope slings shall be immediately removed from service if any of the following conditions are present:
  - ○ Abnormal wear
  - ○ Powdered fiber between strands
  - ○ Broken or cut fibers
  - ○ Variations in the size or roundness of strands
  - ○ Discoloration or rotting
  - ○ Distortion of hardware in the sling
- Employers must use natural- and synthetic-fiber rope slings that have permanently affixed and legible identification markings that state the rated capacity for the type(s) of hitch(es) used and the angle upon which it is based, type of fiber material, and the number of legs if more than one.

15.4.5   Synthetic Webbing (Nylon, Polyester, and Polypropylene)

- The employer shall have each synthetic web sling marked or coded to show:
  - ○ Name or trademark of manufacturer
  - ○ Rated capacities for the type of hitch
  - ○ Type of material
- Rated capacity shall not be exceeded.
- Webbing: Synthetic webbing shall be of uniform thickness and width and selvage edges shall not be split from the webbing's width.
- Fittings: Fittings shall be:
  - ○ Of a minimum breaking strength equal to that of the sling
  - ○ Free of all sharp edges that could in any way damage the webbing
- Attachment of end fittings to webbing and formation of eyes: Stitching shall be the only method used to attach end fittings to webbing and to form eyes. The thread shall be in an even pattern and contain a sufficient number of stitches to develop the full breaking strength of the sling.
- Environmental conditions: When synthetic web slings are used, the following precautions shall be taken:

- Nylon web slings shall not be used where fumes, vapors, sprays, mists, or liquids of acids or phenolics are present.

- Polyester and polypropylene web slings shall not be used where fumes, vapors, sprays, mists, or liquids of caustics are present.

- Web slings with aluminum fittings shall not be used where fumes, vapors, sprays, mists, or liquids of caustics are present.

- Safe operating temperatures: Synthetic web slings of polyester and nylon shall not be used at temperatures in excess of 180°F (82.2°C). Polypropylene web slings shall not be used at temperatures in excess of 200°F (93.33°C).

- Removal from service: Synthetic web slings shall be immediately removed from service if any of the following conditions are present:

  - Acid or caustic burns

  - Melting or charring of any part of the sling surface

  - Snags, punctures, tears, or cuts

  - Broken or worn stitches

  - Distortion of fittings

15.4.6   Shackles and Hooks

- Employers must not use shackles with loads in excess of the rated capacities (i.e., working load limits) indicated on the shackle by permanently affixed and legible identification markings prescribed by the manufacturer.

- The manufacturer's recommendations shall be followed in determining the safe working loads of the various sizes and types of specific and identifiable hooks. All hooks for which no applicable manufacturer's recommendations are available shall be tested to twice the intended safe working load before they are initially put into use. The employer shall maintain a record of the dates and results of such tests.

15.5   Handling Other Equipment

There are a number of methods an employer can use to reduce the frequency and severity of conveyor-related injuries:

- Emergency buttons or pull cords designed to stop the conveyor must be installed preferably near worker stations.

- Accessible conveyor belts must have emergency stop cables that run along the entire length of the belt.

- Emergency stop systems must be designed in such a manner that they have to be reset before the conveyor can start again. This ensures that conveyor can run only after an employee has been removed from danger.

Example 15.4
It is important for workers to place raised loads in a way so as to not create a new hazard.

- True
- False

Answer: True—When placing blocks under raised loads, workers must make sure that the raised loads are kept in a raised position until their hands have been removed from beneath them. The blocks must be large and sturdy enough to be able to support the load. Block materials with cracks, splintered places, and rot must not be used.

15.6   Disposal of Waste Materials

Part 1926, Subpart H, Standard Number 1926.252 states that:

- Whenever materials are dropped more than 20 feet to any point lying outside the exterior walls of the building, an enclosed chute of wood, or equivalent material, shall be used. For the purpose of this paragraph, an enclosed chute is a slide, closed in on all sides, through which material is moved from a high place to a lower one.

- When debris is dropped through holes in the floor without the use of chutes, the area onto which the material is dropped shall be completely enclosed with barricades not less than 42 inches high and not less than 6 feet back from the projected edge of the opening above. Signs warning of the hazard of falling materials shall be posted at each level. Removal shall not be permitted in this lower area until debris handling ceases above.

- All scrap lumber, waste material, and rubbish shall be removed from the immediate work area as the work progresses.

- Disposal of waste material or debris by burning shall comply with local fire regulations.

- All solvent waste, oily rags, and flammable liquids shall be kept in fire resistant covered containers until removed from worksite.

Example 15.5
Workers must _____ overload mechanical moving equipment.

   A. Always
   B. Sometimes
   C. Never
   D. All of the above

Answer: C

15.7 Summary

Some materials cannot be stacked due to shape, size, or fragility constraints. In most cases, these can be safely stored on shelves or in bins. All materials stored in tiers must be stacked, racked, blocked, interlocked, or otherwise secured to prevent sliding, falling, or collapse. Storage containers must also have adequate capacity to handle the height and weight of stored items, as well as be accessible and in good condition. Fitting loads with handles and holders may minimize the chances of injuries to the fingers and hands.

Bending, twisting, and turning are some of the common body movements that can lead to back and spinal injuries. Frequent handling of heavy and bulky objects often results in such injuries. Workers that lift these types of objects may suffer from acute and chronic back pains.

It is imperative that your company's management play an active role in the effective implementation of a safety and health program designed for handling materials and their storage. When management is closely involved with such programs, it can persuade supervisors and employees alike of its importance, and motivate them to take the program seriously.

Employees must also be aware of the reactive qualities of different materials and keep them properly segregated. All passage ways must be maintained clear of obstructions. By implementing ergonomically designed systems and training employees, employers can greatly reduce the number of personnel injuries.

15.8 Multiple-Choice Questions

15.1. Material stored inside buildings under construction shall not be placed within _____ feet of any hoistway or inside floor openings.

    A. 2
    B. 4
    C. 6
    D. 8

15.2. Material stored inside buildings under construction shall not be placed _____ feet of an exterior wall which does not extend above the top of the material stored.

    A. 4
    B. 7
    C. 10
    D. 12

15.3. Bagged materials shall be stacked by stepping back the layers and cross-keying the bags at least every _____ bags high.

    A. 4
    B. 7
    C. 10
    D. 12

15.4. Brick stacks shall not be more than _____ feet in height.

    A. 4
    B. 7
    C. 10
    D. 12

15.5. When a loose brick stack reaches a height of 4 feet, it shall be tapered back _____ in every foot of height above the 4-foot level.

    A. 1 inches
    B. 2 inches
    C. 3 inches
    D. 4 inches

15.6. When masonry blocks are stacked higher than _____ feet, the stack shall be tapered back one-half block per tier.

    A. 2
    B. 6
    C. 10
    D. 14

15.7. Lumber piles shall not exceed _____ feet in height.

    A. 12
    B. 16
    C. 18
    D. 20

15.8. Lumber to be handled manually shall not be stacked more than _____ feet high.

    A. 16
    B. 14
    C. 12
    D. 10

15.9. Fiber rope slings shall have a minimum clear length of rope between eye splices equal to _____ times the rope diameter.

    A. 4
    B. 6
    C. 8
    D. 10

15.10. Whenever materials are dropped more than _____ feet to any point lying outside the exterior walls of the building, an enclosed chute of wood, or equivalent material, shall be used.

    A. 14
    B. 16
    C. 18
    D. 20

15.9   Practice Problems

15.1. List some general requirements for material storage.

15.2. List some general requirements for material handling.

15.3. List some general requirements for alloy steel chains.

15.4. List some general requirements for wire rope.

15.5. List some general requirements for natural rope and synthetic fiber.

15.6. List some general requirements for synthetic webbing (nylon, polyester, and polypropylene).

15.7. List some general requirements for shackles and hooks.

15.8. List some general requirements for disposal of waste materials.

15.10   Critical Thinking and Discussion Topic

In a construction site, the amount of waste products is not huge. The site manager is worried that the waste will find its way from the construction site to the adjoining properties. Your logic is the amount of waste products is very small and it should not bother the site or the environment. What will you do in this situation?

(Photo courtesy of andrea-piacquadio from Pexels.)

CHAPTER 16

Tools–Hand and Power

(Photo courtesy of Pexels.)

16.1 General

Hand and power tools are a part of our everyday lives. These tools help us in performing tasks that otherwise would be difficult or impossible. However, even simple tools can be hazardous and have the potential for causing severe injuries when used or maintained improperly. Special attention to hand and power tool safety is necessary to reduce or eliminate these hazards. Workers using hand and power tools may be exposed to these hazards:

- Falling or flying objects which can be abrasive, or may splash
- Harmful dusts, fumes, mists, vapors, and gases

- Frayed or damaged electrical cords, hazardous connections, and improper grounding

Basic tool safety rules include the following:

- Perform maintenance regularly.
- Use the right tool for the job.
- Inspect tools before use.
- Operate according to manufacturers' instructions.
- Use the proper personal protective equipment (PPE).
- Use required and provided guards.

Example 16.1

Workers using hand and power tools may be exposed to:

A. Falling of Flying objects

B. Harmful dusts

C. Damaged electrical cords

D. All of the above

Answer: D

16.2   General Requirements for Tools

Injuries can be prevented by keeping tools in good condition, using the right tool for the job, using the tool properly, and wearing required personal protective equipment. The employer is responsible for the safe condition of tools and equipment used by employees.

- Keep your work area well lit, clean, and dry.
- Stand where you have firm footing and good balance while you use any tools.
- Arrange the work and use portable tools so that the tool will move away from your hands and body if it slips.
- Make sure that the material you are working on is held securely—use clamps or a vise if you need to.
- Use the right tool for the job. Don't force a small tool to do heavy-duty work.
- Regularly inspect tools, cords, and accessories. Repair or replace problem equipment immediately.
- Never use a dull blade or cutting edge.
- Keep electric cables and cords clean and free from kinks. Never carry a tool by its cords.

- Use all guards and safety devices (i.e., three-prong plugs, double-insulated tools, and safety switches) that are designed to be used with the equipment.
- Dress right. Never wear clothing or jewelry that could become entangled in machinery or power tools.
- Use protective equipment when necessary. This might include safety glasses, hearing protection, and respiratory protection.
- Make adjustments and accessory changes when machinery is turned off and unplugged.
- Maintain your tools. Keep them sharp, oiled, and stored in a safe, dry place.
- Install or repair equipment only if you're qualified. A faulty job may cause fires or seriously injure you or other workers.
- Direct the tool away from other employees working in close proximity.

Good tool habits soon become second nature. Follow the tool safety guidelines at your workplace and the equipment you operate will serve you efficiently and safely.

Example 16.2
While using saw blades, knives, or other sharp tools, employees should direct the tool _____ other employees working in close proximity.

  A. Toward

  B. Diagonal toward

  C. Away from

  D. Tilted toward

Answer: C

Example 16.3
The _____ is responsible for the safe condition of tools and equipment used by employees.

  A. Employer

  B. OSHA compliance officer

  C. City Manager

  D. Construction Case Manager

  Answer: A

Some general regulations based on the Part 1926, Subpart I, Standard Number 1926.300 are stated in the following text.

16.2.1   Guarding

The point of operations must be guarded or otherwise protected. The point of operations is where the work is actually performed on the materials.

- When power-operated tools are designed to accommodate guards, they shall be equipped with such guards when in use.

- Belts, gears, shafts, pulleys, sprockets, spindles, drums, fly wheels, chains, or other reciprocating, rotating, or moving parts of equipment shall be guarded if such parts are exposed to contact by employees or otherwise create a hazard.

- Types of guarding. One or more methods of machine guarding shall be provided to protect the operator and other employees in the machine area from hazards such as those created by point of operation, ingoing nip points, rotating parts, flying chips and sparks. Examples of guarding methods are barrier guards, two-hand tripping devices, electronic safety devices, etc.

- For point of operation guarding:
  - Point of operation is the area on a machine where work is actually performed upon the material being processed.

  - The point of operation of machines whose operation exposes an employee to injury shall be guarded. The guarding device shall be in conformity with any appropriate standards therefore, or, in the absence of applicable specific standards, shall be so designed and constructed as to prevent the operator from having any part of his body in the danger zone during the operating cycle.

  - Special hand tools for placing and removing material shall be such as to permit easy handling of material without the operator placing a hand in the danger zone. Such tools shall not be in lieu of other guarding required by this section, but can only be used to supplement protection provided.

Example 16.4

The point of _____ is where the work is actually performed on the materials and must be guarded or otherwise protected.

    A. Operations

    B. Routine

    C. Placement

    D. Maintenance

Answer: *A*

16.2.2   Personal Protective Equipment

Employees using hand and power tools and exposed to the hazard of falling, flying, abrasive, and splashing objects, or exposed to harmful dusts, fumes, mists, vapors, or gases shall be provided with the particular personal protective equipment necessary to protect them from the hazard.

16.2.3   Switches

- All handheld powered platen sanders, grinders with wheels 2-inch diameter or less, routers, planers, laminate trimmers, nibblers, shears, scroll saws, and jigsaws with blade shanks one-fourth of an inch wide or less may be equipped with only a positive "on-off" control.

- All handheld powered drills, tappers, fastener drivers, horizontal, vertical, and angle grinders with wheels greater than 2 inches in diameter, disc sanders, belt sanders, reciprocating saws, saber saws, and other similar operating powered tools shall be equipped with a momentary contact "on-off" control and may have a lock-on control provided that turnoff can be accomplished by a single motion of the same finger or fingers that turn it on.

- All other handheld powered tools, such as circular saws, chain saws, and percussion tools without positive accessory holding means, shall be equipped with a constant pressure switch that will shut off the power when the pressure is released.

Example 16.5

_____ tools must be fitted with appropriate guards and safety switches. They are extremely hazardous when used improperly.

A. Hand

B. Foot

C. Power

D. None of the above

Answer: C

16.3   Hand Tools

Some hand tools (Fig. 16.1) used in constructions are listed below.

Knives.   Using knives as pries, screwdrivers, can openers, awls, or punches can easily damage the blades. A sharp blade needs less pressure to cut and has less chance of getting hung up and slipping. Always move the blade away from you as you cut.

Screwdrivers.   Using screwdrivers as pries, can openers, punches, chisels, wedge, etc., can cause chipped, rounded bent, dull tips; bent shafts; and split or broken handles. If the screwdriver tip doesn't fit the screw, you'll apply more force and the screwdrivers can easily slip. Redress the tips of flat head screwdrivers to keep them sharp and square edged. Screwdrivers with shorter shafts give you better control. Screwdrivers with thicker handles apply more torque, with less effort on your part.

Hammers and Mallets.   Nail hammers are designed to drive nails. Ball-peen hammers are designed for striking cold chisels and metal punches. Mallets have a striking head of plastic, wood, or rawhide and are designed for striking wood chisels, punches, or dies. Sledgehammers are for striking concrete or stone. You can damage a hammer by trying to use it for the wrong purpose. Don't use a hammer with a mushroomed striking surface or a loose handle. You can damage other tools by trying to force them by hitting them with a hammer.

(a)   (b)
(Photo courtesy of Rodolfo Quirós.)   (Photo courtesy of Skitterphoto.)

Figure 16.1   Construction workers working with hand tools.

Pliers.   Don't substitute pliers for a wrench. The face of the pliers is not designed to grip a fastener, and the pliers can easily slip off of the nut or bolt. Pliers are designed for gripping so you can more easily bend or pull material. They'll provide a strong grip if you protect them from getting bent out of shape and keep the gripping surface from being damaged.

Cutters.   Use cutters or snipes to remove banding wire or strapping. Trying to use a pry bar to snap open banding can cause injuries. Keep the cutting edges sharp and protect them from getting nicked or gouged.

Wrenches.   Use adjustable open-ended wrenches for light-duty jobs when the proper sized wrench isn't available. Position yourself so you will be pulling the wrench toward you, with the open end facing you—this lessens the chance of the wrench slipping off of the fastener when you apply force. Select an open-ended wrench to fit the fastener for medium-duty jobs. With the snug fit, these wrenches can apply more force than an adjustable open-ended wrench. Again, pull the wrench with the open end facing you to avoid slippage. Box and socket wrenches should be used when a heavy pull is required. Because they completely encircle the fastener, they apply even pressure with a minimal chance of slipping. Some box wrenches are designed for heavy-duty use, and they do have a striking surface. But, in general, don't try to increase the torque by hitting the wrench with a hammer or by adding a cheater bar to the wrench's handle—this can break or damage the wrench. If the fastener is too tight, use some penetrating oil to lubricate it.

Wood Saws.   For cutting wood, use a cross-cut saw to cut across the grain, and use a ripping saw to cut with the grain. Select a saw with coarse teeth for sawing green wood, thick lumber, or for making coarse cuts. Fine-toothed saws can be used to make fine cuts in dry wood. After use, wipe the saw with a lightly oiled rag to keep the teeth clean. Protect the saw from getting bent or damaged in storage.

Metalworking Hand Tools.   Hack saws should have the blade installed with the teeth facing forward, and apply pressure on the forward stroke. Use a light pressure to avoid twisting and breaking the blade. Metal files need to be kept clean and protected from damage. Hitting the file against a hard object to clean it can damage the file—use a file card for cleaning.

The employer is responsible for the safe condition of tools and equipment used by employees. Employers shall not issue or permit the use of unsafe hand tools. Employees should be trained in the proper use and handling of tools and equipment. Hand tool hazards are often caused by misuse and improper maintenance. Part 1926, Subpart I, Standard Number 1926.301 states that:

- Employers shall not issue or permit the use of unsafe hand tools.
- Wrenches, including adjustable, pipe, end, and socket wrenches shall not be used when jaws are sprung to the point that slippage occurs.
- Impact tools, such as drift pins, wedges, and chisels, shall be kept free of mushroomed heads.
- The wooden handles of tools shall be kept free of splinters or cracks and shall be kept tight in the tool.

16.4   Power-Operated Hand Tools

Power tools must be fitted with appropriate guards and safety switches. They are extremely hazardous when used improperly. Different types of power tools are determined by their power source:

- Electric
- Pneumatic
- Liquid fuel
- Hydraulic
- Powder-actuated

Some commonly used power-operated tools are listed below.

Saws.   The circular saw is a heavy-duty tool with interchangeable blades for all types of woodcutting. The saber saw is somewhat smaller and used for smaller woodcutting jobs and curved cuts. A chainsaw may be either gasoline or electricity powered. Before cutting, inspect the material to be cut for nails or foreign objects. Make sure blade guards are in place and working properly. Saws are noisy and the sound may drown out warning shouts or instructions. Wear goggles or goggles and a face shield to protect yourself from flying debris or sawdust. Inspect the blade regularly. First, turn the saw off and unplug it. Don't use dull or loose blades. Don't overload the motor by pushing too hard or cutting material that is too heavy. Be sure you have form footing and balance when using any saw. Slips or falls can be deadly when you're holding a power tool.

Drills.   Variable speed drills are versatile tools used for boring holes, turning screws, buffing, and grinding. Select the correct drill bit for the job to be done. Use only a sharp bit. Make sure the material being drilled is secured or clamped firmly. Hold the drill firmly and at the correct angle. Don't force it to work or lean on it with all your strength. Always remove the bit from the drill when you're finished. Use a droll bit sharpener to maintain the cutting edge on drill bits.

Grinding Wheels.   Bench grinders are useful for sharpening, shaping, and smoothing metal, wood, plastic, or stone. Keep machine guards in place and wear hearing and eye protection. Before use, make sure that wheels are firmly held on spindles and work rests are tight. Stand to one side while starting the motor, until the operating speed is reached—this prevents injury if a defective wheel breaks apart. Use light pressure when you start grinding—too much on a cold wheel may cause failure.

Sanders.   Two types of sanders are orbital and belt. Arrange the cord so that it won't be damaged by the abrasive belt. Keep both hands on the tool for good control. Hold onto the sander when you plug it in. Clean dust and chips from the motor and vent holes regularly and lubricate when necessary.

Impact Wrenches.   They operate on electricity or compressed air and deliver extra power and torque for fastening and loosening bolts and drilling. Don't force a wrench to take on a job bigger than it's designed to handle. Don't use standard hand sockets or driver parts with an impact tool. Don't reverse direction of rotation while the trigger is depressed.

Soldering Irons or "Guns."   They can be dangerous because of the heat they generate. Handle with care—they can easily cause serious burns. Always assume that a soldering iron is hot. Rest a heated iron on a rack or metal surface. Never swing an iron to remove solder. Hold small soldering jobs with pliers, never in your hand. Wait until the tool is cool before you put it away.

Propane and Gas Torches.   These commonly used tools pose fire and heat hazards. Never use a flame to test for propane or gas leaks. Never store the fuel tanks in an unventilated area, and never use a tank with a leaking valve. Use torches in well-ventilated areas. Avoid breathing the vapor and fumes they generate.

Glue Guns.   A glue gun can be a real time saver. However, because it generates temperatures as high as 450°F, avoid contact with the hot nozzle and glue.

Shop Vacuums.   Clean filters regularly and never use your vacuum to pick up flammable liquids or smoldering materials.

Table Saw.   This saw has a large circular blade used to make a variety of cuts in wood or other material. Never reach over the saw to push stock. Stand slightly to one side, never in line with the saw. A "kickback" occurs when material being cut is thrown back toward the operator. This is one of the greatest hazards in running a table saw. To avoid it:

- Never use a dull blade.
- Don't cut "freehand" or attempt to rip badly warped wood.
- Use the splitter guard.
- Don't drop wood on an unguarded saw.

Radial-Arm Saw.   Often called the number one multipurpose saw in the shop, this saw blade is mounted on a moveable head, and slides in tracks or along a shaft. Most have built-in safety devices such as key switches to start them, blade guards, anti-kickback pawls, and blade brakes. The saw and motor should always be returned to the rear of the table against the column after a cut is made. The lower blade guard on a radial arm saw is designed to prevent the operator from coming into contact with the rotating blade. This guard must automatically return to the covering position when withdrawn from the work.

Drill Press.   The stationary drill press is a larger, more powerful version of a portable drill. Clamp or securely fasten the material being drilled whenever possible. Make sure any attachments are fastened tightly.

Power Sanders.   Always select the correct grade of abrasive for the job. Move the work around to avoid heating and burning a portion of the disc, belt, or wood. Remember to use the dust collector if the sander has one.

Shapers.   A shaper is used mainly for grooving and fluting woods. It can be dangerous because of its high speed and because the cutters are difficult to guard completely. When using a shaper, avoid loose clothing, wear eye protection, and make sure that cutters are sharp and securely fastened.

Welding Machines.   The high-intensity arc of even a small welding machine can cause severe burns. Flame-resistant clothing and hand and eye protection are needed to protect against hot sparks and molten metal. Keep the area around the welding operation clean—hot sparks can start fires.

Example 16.6

The lower blade guard on a radial arm saw is NOT designed to prevent the operator from coming into contact with the rotating blade.

- True
- False

Answer: *False*

Some major regulations based on the Part 1926, Subpart I, Standard Number 1926.302 are listed in the following text.

16.4.1   Electric Power-Operated Tools

- Electric power-operated tools shall either be of the approved double-insulated type or grounded.
- The use of electric cords for hoisting or lowering tools shall not be permitted.

16.4.2   Pneumatic Power Tools

- Pneumatic power tools shall be secured to the hose or whip by some positive means to prevent the tool from becoming accidentally disconnected.
- Safety clips or retainers shall be securely installed and maintained on pneumatic impact (percussion) tools to prevent attachments from being accidentally expelled.

- All pneumatically driven nailers, staplers, and other similar equipment provided with automatic fastener feed, which operate at more than 100 psi pressure at the tool shall have a safety device on the muzzle to prevent the tool from ejecting fasteners, unless the muzzle is in contact with the work surface.
- Compressed air shall not be used for cleaning purposes except where reduced to less than 30 psi and then only with effective chip guarding and personal protective equipment which meets the requirements of Subpart E of this part. The 30 psi requirement does not apply for concrete form, mill scale, and similar cleaning purposes.
- The manufacturer's safe operating pressure for hoses, pipes, valves, filters, and other fittings shall not be exceeded.
- The use of hoses for hoisting or lowering tools shall not be permitted.
- All hoses exceeding 1/2-inch inside diameter shall have a safety device at the source of supply or branch line to reduce pressure in case of hose failure.
- Airless spray guns of the type which atomize paints and fluids at high pressures (1000 pounds or more per square inch) shall be equipped with automatic or visible manual safety devices which will prevent pulling of the trigger to prevent release of the paint or fluid until the safety device is manually released.
- In lieu of the above, a diffuser nut which will prevent high-pressure, high-velocity release, while the nozzle tip is removed, plus a nozzle tip guard which will prevent the tip from coming into contact with the operator, or other equivalent protection, shall be provided.
- The blast cleaning nozzles shall be equipped with an operating valve which must be held open manually. A support shall be provided on which the nozzle may be mounted when it is not in use.

Example 16.7

It is important to drive fasteners into very hard or brittle material that might chip or splatter, or make the fasteners ricochet.

- True
- False

False—Avoid driving into materials easily penetrable unless materials are backed by substance that will prevent the pin of fastener from passing through. Also don't drive fasteners into very hard or brittle material that might chip or splatter, or make the fasteners ricochet.

Example 16.8

Ensure that a _____ tool is fastened securely to the air hose to prevent a disconnection.

A. Hand

B. Electric

C. Running

D. Pneumatic

Answer: D

16.4.3   Fuel-Powered Tools

- All fuel-powered tools shall be stopped while being refueled, serviced, or maintained, and fuel shall be transported, handled, and stored in accordance with Subpart F of OSHA Part 1926.

- When fuel-powered tools are used in enclosed spaces, the requirements for concentrations of toxic gases and use of personal protective equipment should be applied.

16.4.4   Hydraulic-Powered Tools

- The fluid used in hydraulic-powered tools shall be fire-resistant fluids approved under Schedule 30 of the U.S. Bureau of Mines and shall retain its operating characteristics at the most extreme temperatures to which it will be exposed.

- The manufacturer's safe operating pressures for hoses, valves, pipes, filters, and other fittings shall not be exceeded.

16.4.5   Powder-Actuated Tools

- Only employees who have been trained in the operation of the particular tool in use shall be allowed to operate a powder-actuated tool.

- The tool shall be tested each day before loading to see that safety devices are in proper working condition. The method of testing shall be in accordance with the manufacturer's recommended procedure.

- Any tool found not in proper working order, or that develops a defect during use, shall be immediately removed from service and not used until properly repaired.

- Personal protective equipment shall be in accordance with Subpart E of OSHA Part 1926.

- Tools shall not be loaded until just prior to the intended firing time. Neither loaded nor empty tools are to be pointed at any employees. Hands shall be kept clear of the open barrel end.

- Loaded tools shall not be left unattended.

- Fasteners shall not be driven into very hard or brittle materials including, but not limited to, cast iron, glazed tile, surface-hardened steel, glass block, live rock, face brick, or hollow tile.

- Driving into materials easily penetrable shall be avoided unless such materials are backed by a substance that will prevent the pin or fastener from passing completely through and creating a flying missile hazard on the other side.

- No fastener shall be driven into a spalled area caused by an unsatisfactory fastening.

- Tools shall not be used in an explosive or flammable atmosphere.

- All tools shall be used with the correct shield, guard, or attachment recommended by the manufacturer.
- Powder-actuated tools used by employees shall meet all other applicable requirements of American National Standards Institute, A10.3-1970, Safety Requirements for Explosive-Actuated Fastening Tools.

16.5   Abrasive Wheels and Tools

Abrasive wheels (Fig. 16.2) are powered wheels used in workshop grinder machines (fixed or portable), and are made up of small abrasive particles (grit) that have been stuck together by a bonding material to form wheel structures of different thickness. These are used for surface modification or preparations (external or internal) by periphery grinding, abrasive cutoff and face grinding standard designs; they (disc) can also be used for cutting materials. Some are used for polishing, deburring, sanding, and finishing. Part 1926, Subpart I, Standard Number 1926.303 states that:

Power.   All grinding machines shall be supplied with sufficient power to maintain the spindle speed at safe levels under all conditions of normal operation.

Guarding.   Grinding machines shall be equipped with safety guards in conformance with the requirements of American National Standards Institute. The safety guard shall cover the spindle end, nut, and flange projections. The safety guard shall be mounted so as to maintain proper alignment with the wheel, and the strength of the fastenings shall exceed the strength of the guard, except:

Figure 16.2   Abrasive wheel.
(Photo courtesy of Trainer Bubble.)

- Safety guards on all operations where the work provides a suitable measure of protection to the operator may be so constructed that the spindle end, nut, and outer flange are exposed; and where the nature of the work is such as to entirely cover the side of the wheel, the side covers of the guard may be omitted.
- The spindle end, nut, and outer flange may be exposed on machines designed as portable saws.

Use of abrasive wheels

- Floor stand- and bench-mounted abrasive wheels, used for external grinding, shall be provided with safety guards (protection hoods). The maximum angular exposure of the grinding wheel periphery and sides shall be not more than 90 degrees, except that when work requires contact with the wheel below the horizontal plane of the spindle, the angular exposure shall not exceed 125 degrees. In either case, the exposure shall begin not more than 65 degrees above the horizontal plane of the spindle. Safety guards shall be strong enough to withstand the effect of a bursting wheel.
- Floor- and bench-mounted grinders shall be provided with work rests which are rigidly supported and readily adjustable. Such work rests shall be kept at a distance not to exceed one-eighth inch from the surface of the wheel.
- Cup-type wheels used for external grinding shall be protected by either a revolving cup guard or a band-type guard.
- When safety guards are required, they shall be so mounted as to maintain proper alignment with the wheel, and the guard and its fastenings shall be of sufficient strength to retain fragments of the wheel in case of accidental breakage. The maximum angular exposure of the grinding wheel periphery and sides shall not exceed 180 degrees.
- When safety flanges are required, they shall be used only with wheels designed to fit the flanges. Only safety flanges, of any type and design and properly assembled so as to ensure that the pieces of the wheel will be retained in case of accidental breakage, shall be used.
- All abrasive wheels shall be closely inspected and ring-tested before mounting to ensure that they are free from cracks or defects.
- Grinding wheels shall fit freely on the spindle and shall not be forced on. The spindle nut shall be tightened only enough to hold the wheel in place.
- All employees using abrasive wheels shall be protected by eye protection equipment, except when adequate eye protection is afforded by eye shields which are permanently attached to the bench or floor stand.

Other requirements.   All abrasive wheels and tools used by employees shall meet other applicable requirements of American National Standards Institute.

Work rests.   On offhand grinding machines, work rests shall be used to support the work. They shall be of rigid construction and designed to be adjustable to compensate for wheel wear. Work rests shall be kept adjusted closely to the wheel with a maximum opening of 1/8 inch (0.3175 cm) to prevent the work from being jammed between the wheel and the rest, which may cause wheel breakage. The work rest shall be securely clamped after each adjustment. The adjustment shall not be made with the wheel in motion.

16.6   Woodworking Tools

Part 1926, Subpart I, Standard Number 1926.304 states that:

Disconnect switches.   All fixed, power-driven woodworking tools shall be provided with a disconnect switch that can be either locked or tagged in the off position.

Speeds.   The operating speed shall be etched or otherwise permanently marked on all circular saws over 20 inches in diameter or operating at over 10,000 peripheral feet per minute. Any saw so marked shall not be operated at a speed other than that marked on the blade. When a marked saw is retensioned for a different speed, the marking shall be corrected to show the new speed.

Self-feed.   Automatic feeding devices shall be installed on machines whenever the nature of the work will permit. Feeder attachments shall have the feed rolls or other moving parts covered or guarded so as to protect the operator from hazardous points.

Guarding.   All portable, power-driven circular saws shall be equipped with guards above and below the baseplate or shoe. The upper guard shall cover the saw to the depth of the teeth, except for the minimum arc required to permit the base to be tilted for bevel cuts. The lower guard shall cover the saw to the depth of the teeth, except for the minimum arc required to allow proper retraction and contact with the work. When the tool is withdrawn from the work, the lower guard shall automatically and instantly return to the covering position.

Other requirements.   All woodworking tools and machinery shall meet other applicable requirements of American National Standards Institute, 01.1-1961, Safety Code for Woodworking Machinery.

Radial saws.   In radial saw (Fig. 16.3a), the upper hood shall completely enclose the upper portion of the blade down to a point that will include the end of the saw arbor. The upper hood shall be constructed in such a manner and of such material that it will protect the operator from flying splinters, broken saw teeth, etc., and will defect saw-dust away from the operator. The sides of the lower exposed portion of the blade shall be guarded to the full diameter of the blade by a device that will automatically adjust itself to the thickness of the stock and remain in contact with stock being cut to give maximum protection possible for the operation being performed.

Hand-fed crosscut table saws.   Each circular crosscut table saw (Fig. 16.3b) shall be guarded by a hood which shall meet all the requirements of hoods for circular ripsaws.

Hand-fed ripsaws.   Each circular hand-fed ripsaw (Fig. 16.3c) shall be guarded by a hood which shall completely enclose the portion of the saw above the table and that portion of the saw above the material being cut. The hood and mounting shall be arranged so that the hood will automatically adjust itself to the thickness of and remain in contact

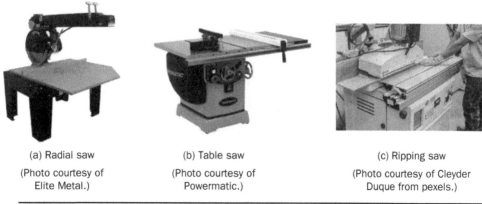

(a) Radial saw

(Photo courtesy of
Elite Metal.)

(b) Table saw

(Photo courtesy of
Powermatic.)

(c) Ripping saw

(Photo courtesy of Cleyder
Duque from pexels.)

Figure 16.3 Different types of saws.

with the material being cut but it shall not offer any considerable resistance to insertion of material to saw or to passage of the material being sawed. The hood shall be made of adequate strength to resist blows and strains incidental to reasonable operation, adjusting, and handling, and shall be so designed as to protect the operator from flying splinters and broken saw teeth. It shall be made of material that is soft enough so that it will be unlikely to cause tooth breakage. The hood shall be so mounted as to ensure that its operation will be positive, reliable, and in true alignment with the saw; and the mounting shall be adequate in strength to resist any reasonable side thrust or other force tending to throw it out of line.

16.7   Jacks-Lever and Ratchet, Screw, and Hydraulic

Part 1926, Subpart I, Standard Number 1926.305 states that:

General requirements.   The manufacturer's rated capacity shall be legibly marked on all jacks and shall not be exceeded. All jacks shall have a positive stop to prevent over travel.

Blocking.   When it is necessary to provide a firm foundation, the base of the jack shall be blocked or cribbed. Where there is a possibility of slippage of the metal cap of the jack, a wood block shall be placed between the cap and the load.

Operation and maintenance.   After the load has been raised, it shall be cribbed, blocked, or otherwise secured at once. Hydraulic jacks exposed to freezing temperatures shall be supplied with an adequate antifreeze liquid. All jacks shall be properly lubricated at regular intervals. Each jack shall be thoroughly inspected at times which depend upon the service conditions. Inspections shall be not less frequent than the following:

- For constant or intermittent use at one locality, once every 6 months
- For jacks sent out of shop for special work, when sent out and when returned
- For a jack subjected to abnormal load or shock, immediately before and immediately thereafter

Repair or replacement parts shall be examined for possible defects. Jacks which are out of order shall be tagged accordingly, and shall not be used until repairs are made.

16.8   Air Receivers

Part 1926, Subpart I, Standard Number 1926.306 states the following.

16.8.1   General Requirements

Application.   This section applies to compressed air receivers, and other equipment used in providing and utilizing compressed air for performing operations such as cleaning, drilling, hoisting, and chipping. On the other hand, however, this section does not deal with the special problems created by using compressed air to convey materials or the problems created when men work in compressed air such as in tunnels and caissons. This section is not intended to apply to compressed air machinery and equipment used on transportation vehicles such as steam railroad cars, electric railway cars, and automotive equipment.

New and existing equipment.   All new air receivers installed after the effective date of these regulations shall be constructed in accordance with the 1968 edition of the A.S.M.E. Boiler and Pressure Vessel Code Section VIII. All safety valves used shall be constructed, installed, and maintained in accordance with the A.S.M.E. Boiler and Pressure Vessel Code, Section VIII Edition 1968.

16.8.2   Installation and Equipment Requirements

Installation.   Air receivers shall be so installed that all drains, handholes, and manholes therein are easily accessible. Under no circumstances shall an air receiver be buried underground or located in an inaccessible place.

Drains and traps.   A drain pipe and valve shall be installed at the lowest point of every air receiver to provide for the removal of accumulated oil and water. Adequate automatic traps may be installed in addition to drain valves. The drain valve on the air receiver shall be opened and the receiver completely drained frequently and at such intervals as to prevent the accumulation of excessive amounts of liquid in the receiver.

Gages and valves.   Every air receiver shall be equipped with an indicating pressure gage (so located as to be readily visible) and with one or more spring-loaded safety valves. The total relieving capacity of such safety valves shall be such as to prevent pressure in the receiver from exceeding the maximum allowable working pressure of the receiver by more than 10%.

No valve of any type shall be placed between the air receiver and its safety valve or valves.

Safety appliances, such as safety valves, indicating devices and controlling devices, shall be constructed, located, and installed so that they cannot be readily rendered inoperative by any means, including the elements.

All safety valves shall be tested frequently and at regular intervals to determine whether they are in good operating condition.

16.9   Mechanical Power-Transmission Apparatus

Part 1926, Subpart I, Standard Number 1926.307 specifies the specification used for different power-transmission apparatus such as flywheels, cranks and connecting rods, tail rods or extension piston rods shafting, pulleys, belt, rope, and chain drives, cone-pulley belts, keys, setscrews, and other projections, belt shifters, clutches, shippers, poles, perches, and fasteners, etc.

16.10   Overall Working Safety

Proper care and safety when using tools and machinery is vital. Some overall safety rules are listed below.

1. Respect your equipment, know the dangers it presents, and take safety precautions necessary to work without injury.
2. Take out only the tools that you will need for the job. Piles of extra tools can get in the way or get lost.
3. Always wear appropriate personal protective equipment.
4. Maintain tools and equipment with regular servicing and good housekeeping practices. Putting tools away after use keeps them from getting damaged or disappearing.
5. If you don't know how to use a particular tool, don't be afraid to admit it. Find someone who does and learn from an experienced worker.
6. Carry your tools safely. Use a tool box or a tool chest to move tools around. If you need to carry tools, especially on a ladder, wear a tool belt.

Example 16.9
Basic tool safety includes:

A. Use the wrong tool for any job.

B. Maintenance of tools isn't suggested.

C. Inspect tools before use.

D. Disregard operating guidance from the manufacturers' instructions.

Answer: C

Example 16.10
_____ should be trained in the proper use and handling of tools and equipment.

A. Trainers

B. Employees

C. OSHA Compliance Officers

D. Manufacturers

Answer: B

16.11   Summary

We regularly use hand and power tools as part of our everyday lives. They help us in performing tasks that otherwise might be difficult or impossible to accomplish. However, even simple tools can be hazardous and have the potential for causing severe injuries when used or maintained improperly. So, paying special attention to power tool safety is necessary to help reduce or eliminate hazards.

This chapter discussed the various types of tools and their general safety precautions. Electric power tools must be double insulated and have a three-wire cord plugged into a grounded receptacle. When using power tools, you should use gloves and safety shoes, keep work areas well lit, and ensure that cords don't present a tripping. Don't use power tools in wet locations unless approved for those conditions.

Abrasive wheels and tools may throw off flying fragments and should have guards and be inspected before use. Never remove guards or use a power tool without proper guarding in place. The point of operation must be guarded or otherwise protected. The lower blade guard on a radial arm saw is designed to prevent the operator from coming into contact with the rotating blade. This guard must automatically return to the covering position when withdrawn from the work.

Pneumatic tools are powered by compressed air and they include nail guns, staplers, chippers, drills, and sanders. Hazards include getting hit by a tool attachment or by a fastener that worker is using with the tool. They should have a safety device on the muzzle to prevent the tool from ejecting fasteners, unless the muzzle is in contact with the work surface. Avoid driving into materials easily penetrable unless materials are backed by a substance that will prevent the pin or fastener from passing through. Do not use powder-actuated tools in explosive or flammable environments.

16.12   Multiple-Choice Questions

16.1   Workers using hand and power tools may be exposed to hazard(s): [Select all that apply]

    A.   Falling or flying objects
    B.   Harmful dusts, fumes, mists, vapors, and gases
    C.   Cutting
    D.   Noise

16.2   Electric power-operated tools shall either be of the approved:

    A.   Single-insulated type
    B.   Double-insulated type
    C.   Wrapped up with tape
    D.   Grounded

16.3   When should fuel-powered tools be stopped? [Select all that apply]

    A.   When refueled
    B.   When serviced
    C.   When maintained
    D.   When in service

16.4   All of the following are general tool safety rules except:

    A.   Keeping tools in good condition
    B.   Arranging the work so a tool will move forward from you if it slips

C.   Wearing required personal protective equipment

D.   Making sure the material being worked on is held securely

16.5   If tools are used in areas where flammable liquids are stored or used:

A.   It's safe to use any type of hand tool.

B.   Use electrical power tools made from nonsparking alloy.

C.   Use electrical power tools that are approved for the hazard.

D.   No extra precautions are needed.

16.6   If using a hand tool could create chips, pieces, or sparks:

A.   Stand so the tool moves away from you.

B.   Wear safety glasses.

C.   Wear a cap to keep clean.

D.   None of the above.

16.7   A flat-head screwdriver:

A.   Should also be used as a pry or wedge

B.   Can be damaged if it isn't used properly

C.   Is still safe to use if it doesn't fit the screw

D.   Is safer if the head has round corners

16.8   Mallets are designed for striking:

A.   Nails

B.   Metal punches

C.   Wood chisels

D.   Concrete

16.9   Pliers can easily slip off of a nut or bolt:

A.   If you aren't using enough force

B.   If they aren't kept promptly oiled

C.   If you don't inspect them before each use

D.   When you try to use them as a wrench

16.10   For a heavy pull on a nut:

A.   Use an open-ended wrench with the opening facing you.

B.   Use an open-ended wrench with the opening facing away from you.

C.   Add a "cheater bar" to an open-ended wrench to get more torque.

D.   Use a box wrench or a socket wrench.

16.11   To cut wood along the grain:

A.   Use a ripping saw.

B.   Use a cross-cut saw.

C.   Use a fine-tooth saw if the wood is green.

D.   Use any saw, even if it's slightly damaged.

16.12   Use light pressure when you start using a grinding wheel:

A.   So that you get the feel of the tool

B.   If the work rest is loose

C.   Because too much pressure on a cold wheel could cause it to fail

D.   None of the above

16.13 When you use a variable speed drill, do all of the following except:

    A.   Use sharp bits.

    B.   Be sure the material being drilled is secured in place.

    C.   Hold the drill at the correct angle.

    D.   Leave the bit in the drill when you're done using it.

16.13 Practice Problems

16.1   List some basic tool safety rules.

16.2   List some examples of guarding methods.

16.3   Name some hand tools used in constructions.

16.4   Name some power-operated tools used in constructions.

16.5   List some safety cautions recommended by OSHA for electric power-operated tools.

16.6   List some safety cautions recommended by OSHA for electric pneumatic power tools.

16.7   List some safety cautions recommended by OSHA for electric hydraulic power tools.

16.14 Critical Thinking and Discussion Topic

While working on a construction site, you heard that the scrappers being used have not been inspected for safety issues in the last 3 years. However, the scrappers are working well without any trouble or noise. The operator said that these are the best machines for being so smooth and defect free. As a site supervisor, what will you do in this situation?

CHAPTER 17

Welding and Cutting

(Photo courtesy of Pixabay from Pexels.)

17.1 General

Welding, cutting, and sometimes, brazing are some of the most common industrial processes. *Welding* involves joining two or more pieces of metal together to form one piece. Molten metal is generated by an intense heat source, such as oxygen and gas, or by an electrical arc. Some extremely hazardous materials used in welding operations include:

A. Fluorine compounds

B. Metals containing lead

C. Cadmium-containing filler metals

Unlike welding processes that join two pieces of metal, *cutting* processes involve separating or severing a piece of metal through intense heat generated to melt the metal. Cutting processes include oxygen and fuel-gas and electrical arc gouging.

Brazing is a metal-joining process in which two or more metal items are joined together by melting and flowing a filler metal into the joint. The filler metal has a lower melting point than the adjoining metal. Brazing differs from welding in that it does not involve melting the work pieces and from soldering in higher temperatures for a similar process. It also requires much more closely fitted parts than when soldering. The filler metal flows into the gap between close-fitting parts by capillary action. The filler metal is brought slightly above its melting (liquidus) temperature while being protected by an appropriate atmosphere, usually by a flux. It flows over the base metal (in a process known as wetting) and is then cooled to unite the work pieces together. A major advantage of brazing is the ability to combine the same or different metals with considerable strength.

Whenever welding, cutting, or brazing occurs, everyone involved in the operation must take precautions to prevent fires, explosions, or personal injuries. There are three basic types of welding operations:

- **Oxygen–fuel-gas welding** joins metal parts by generating extremely high heat during combustion.

- **Resistance welding** joins metals by generating heat through resistance created to the flow of electric current.

- **Arc welding** joins or cuts metal parts by heat generated from an electric arc that extends between the welding electrode and the electrode placed on the equipment being welded.

Gas welding, or oxy/fuel welding, as is commonly referred to, is slower and easier to control than arc welding. This method combines metals by heating—the heat source is a flame produced by the combustion of a fuel gas such as acetylene, methyl acetylene (MAPP gas), or hydrogen. Temperatures can be as high as 6000°F. Sometimes, this process involves the use of pressure and filler material. Gases commonly used are oxygen and either acetylene, hydrogen, propane, or propylene. These gases are commonly supplied in compressed gas cylinders, which can pose additional handling and transport hazards. A primary danger for oxygen–fuel-gas welding operations stems from welding with compressed gas cylinders (CGCs) containing oxygen and acetylene. If the CGCs are damaged, the gas can escape with great force and the vessel itself could explode. Rocketing occurs when a CGC breaks or is damaged. The cylinder can then act like a rocket with enough force to break through the walls of concrete. Look for these danger signals when handling CGCs:

- Leaking (you may be able to hear or smell escaping gas)
- Corrosion
- Cracks or burn marks
- Contaminated valves
- Worn or corroded hoses
- Broken gauges or regulators

Resistance welding is the joining of metals by applying pressure and passing current through the metal area to be joined for a long period of time. The key benefit of resistance welding is that no other materials are required to create the bond, making this process extremely cost-effective.

In the arc-welding process, the intense heat needed to melt the metal is produced by an electric arc. The arc is formed between the actual piece of work and a manually or mechanically guided electrode (stick or wire) along the joint. The electrode can either be a rod for simply carrying the current between the tip and the work, or it can be a specially prepared rod or wire that not only conducts the current but also fills and supplies the metal filler to the joint. Power sources for arc welding could be either alternating (AC) or direct (DC) current. The working cable connects to the work piece, and the electrode cable creates an arc across the gap when the energized circuit and the electrode tip touch the work piece and are removed (still in close contact). At the tip, the arc produces a temperature of about 6500°F. This heat melts both the base metal and the electrode, producing a pool of molten metal. Metals can react chemically with elements in the air (oxygen and nitrogen) at high temperatures. Oxides and nitrides that form destroy the strength of the weld. A gas, vapor, or slag shield is used to cover the arc and the molten pool in order to prevent or minimize contact or molten metal with air.

Example 17.1
Some extremely hazardous materials used in welding operations include:

    A. Fluorine compounds

    B. Metals containing lead

    C. Cadmium-containing filler metals

    D. All of the above

Answer: D

17.2   Hazards Related to Welding and Cutting

If proper preventive measures are not taken, these processes may be extremely hazardous, as shown in Fig. 17.1. Burns to the skin, flash burns to the eyes, and fire are some of the most immediate and serious hazards associated with welding, cutting, and brazing. The common hazards are associated with welding:

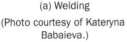

| (a) Welding | (b) Cutting a metal object |
| (Photo courtesy of Kateryna Babaieva.) | (Photo courtesy of Anamul Rezwan.) |

Figure 17.1   Welding and cutting operations.

- Eyes and skin can be damaged from exposure to ultraviolet and infrared rays produced by electric arcs and gas flames.
- Closed containers holding flammables or combustibles can explode under high heat (proper cleaning and purging procedures must be followed before hot work is started).
- Toxic gases, fumes, and dust may be released during welding and cutting operations.
- Welding or cutting near combustible or flammable materials creates a fire hazard.
- Metal splatter and electric shock cause injuries.

The use of compressed gases to create extremely hot flames may expose workers to very dangerous conditions. There are many welding techniques that have different types of hazards associated with them, including the risk of fire, explosion, gas, and fumes hazards. Similar hazards also prevail for cutting.

In regards to these hazards, supervisor's duties include, but are not limited to, the following:

- Safe handling of the cutting or welding equipment, and the safe use of the cutting or welding process.
- Determining the combustible materials and hazardous areas present or likely to be present in the work location.

The following measures can be done for protecting combustibles from ignition:

- Have the work moved to a location free from dangerous combustibles.
- If the work cannot be moved, have the combustibles moved to a safe distance from the work or have the combustibles properly shielded against ignition.
- See that cutting and welding are so scheduled that plant operations that might expose combustibles to ignition are not started during cutting or welding.
- Secure authorization for the cutting or welding operations from the designated management representative.
- Determine that the cutter or welder secures approval that conditions are safe before going ahead.
- Determine that fire protection and extinguishing equipment are properly located at the site.

17.3   Safe Practices for Welding and Cutting

The OSHA standard mandates all employers adopt and follow good work practices in all welding, cutting, and brazing techniques to prevent injuries, fires, and explosions. Some major regulations from the Part 1926, Subpart J, Standard Number 1926.350 are listed in the following text.

17.3.1   Transporting, Moving, and Storing Compressed Gas Cylinders

- When cylinders are hoisted, they shall be secured on a cradle, slingboard, or pallet. They shall not be hoisted or transported by means of magnets or choker slings.
- Cylinders shall be moved by tilting and rolling them on their bottom edges. They shall not be intentionally dropped, struck, or permitted to strike each other violently.
- When cylinders are transported by powered vehicles, they shall be secured in a vertical position.
- Valve protection caps shall not be used for lifting cylinders from one vertical position to another. Bars shall not be used under valves or valve protection caps to pry cylinders loose when frozen. Warm, not boiling, water shall be used to thaw cylinders loose.
- Unless cylinders are firmly secured on a special carrier intended for this purpose, regulators shall be removed and valve protection caps put in place before cylinders are moved.
- A suitable cylinder truck, chain, or other steadying device shall be used to keep cylinders from being knocked over while in use.
- When work is finished, when cylinders are empty, or when cylinders are moved at any time, the cylinder valve shall be closed.
- Compressed gas cylinders shall be secured in an upright position at all times except, if necessary, for short periods of time while cylinders are actually being hoisted or carried.

- Oxygen cylinders in storage shall be separated from fuel-gas cylinders or combustible materials (especially oil or grease), a minimum distance of 20 feet (6.1 meter) or by a noncombustible barrier at least 5 feet (1.5 meter) high having a fire-resistance rating of at least one-half hour.

- Inside of buildings, cylinders shall be stored in a well-protected, well-ventilated, dry location, at least 20 feet (6.1 meter) from highly combustible materials such as oil or excelsior. Cylinders should be stored in definitely assigned places away from elevators, stairs, or gangways. Assigned storage places shall be located where cylinders will not be knocked over or damaged by passing or falling objects, or subject to tampering by unauthorized persons. Cylinders shall not be kept in unventilated enclosures such as lockers and cupboards.

Example 17.2
Cylinders may be stored inside buildings if the storage meets all these conditions except:

A. Well-ventilated

B. Secure

C. Dry

D. Well-lit

Answer: D

17.3.2 Placing Cylinders

- Cylinders shall be kept far enough away from the actual welding or cutting operation so that sparks, hot slag, or flame will not reach them. When this is impractical, fire-resistant shields shall be provided.

- Cylinders shall be placed where they cannot become part of an electrical circuit. Electrodes shall not be struck against a cylinder to strike an arc.

- Fuel-gas cylinders shall be placed with valve end up whenever they are in use. They shall not be placed in a location where they would be subject to open flame, hot metal, or other sources of artificial heat.

- Cylinders containing oxygen or acetylene or other fuel gas shall not be taken into confined spaces.

17.3.3 Treatment of Cylinders

- Cylinders, whether full or empty, shall not be used as rollers or supports.

- No person other than the gas supplier shall attempt to mix gases in a cylinder. No one except the owner of the cylinder or person authorized by him shall refill a cylinder. No one shall use a cylinder's contents for purposes other than those intended by the supplier.

- No damaged or defective cylinder shall be used.

17.3.4 Use of Fuel Gas

The employer shall thoroughly instruct employees in the safe use of fuel gas, as follows:

- Before a regulator to a cylinder valve is connected, the valve shall be opened slightly and closed immediately. This action is generally termed cracking and is intended to clear the valve of dust or dirt that might otherwise enter the regulator. The person cracking the valve shall stand to one side of the outlet, not in front of it. The valve of a fuel-gas cylinder shall not be cracked where the gas would reach welding work, sparks, flame, or other possible sources of ignition.

- The cylinder valve shall always be opened slowly to prevent damage to the regulator. For quick closing, valves on fuel-gas cylinders shall not be opened more than 1.5 turns. When a special wrench is required, it shall be left in position on the stem of the valve while the cylinder is in use so that the fuel-gas flow can be shut off quickly in case of an emergency. In the case of manifolded or coupled cylinders, at least one such wrench shall always be available for immediate use. Nothing shall be placed on top of a fuel-gas cylinder, when in use, which may damage the safety device or interfere with the quick closing of the valve.

- When manifolds and headers are not in use, they should be capped properly.

- Fuel gas shall not be used from cylinders through torches or other devices which are equipped with shutoff valves without reducing the pressure through a suitable regulator attached to the cylinder valve or manifold.

- Before a regulator is removed from a cylinder valve, the cylinder valve shall always be closed and the gas released from the regulator.

- If, when the valve on a fuel-gas cylinder is opened, there is found to be a leak around the valve stem, the valve shall be closed and the gland nut tightened. If this action does not stop the leak, the use of the cylinder shall be discontinued, and it shall be properly tagged and removed from the work area. In the event that fuel gas should leak from the cylinder valve, rather than from the valve stem, and the gas cannot be shut off, the cylinder shall be properly tagged and removed from the work area. If a regulator attached to a cylinder valve will effectively stop a leak through the valve seat, the cylinder need not be removed from the work area.

- If a leak should develop at a fuse plug or other safety device, the cylinder shall be removed from the work area.

Example 17.3
When manifolds and headers are not in use, they should be _____.

A. Removed

B. Stored

C. Capped

D. Shielded

Answer: C

17.4 Arc Welding and Cutting

Part 1926, Subpart J, Standard Number 1926.351 states the following.

17.4.1 Manual Electrode Holders

- Only manual electrode holders which are specifically designed for arc welding and cutting, and are of a capacity capable of safely handling the maximum rated current required by the electrodes, shall be used.

- Any current-carrying parts passing through the portion of the holder which the arc welder or cutter grips in his hand, and the outer surfaces of the jaws of the holder, shall be fully insulated against the maximum voltage encountered to ground.

17.4.2 Welding Cables and Connectors

- All arc-welding and arc-cutting cables shall be of the completely insulated, flexible type, capable of handling the maximum current requirements of the work in progress, taking into account the duty cycle under which the arc welder or cutter is working.

- Only cable free from repair or splices for a minimum distance of 10 feet from the cable end to which the electrode holder is connected shall be used, except that cables with standard insulated connectors or with splices whose insulating quality is equal to that of the cable are permitted.

- When it becomes necessary to connect or splice lengths of cable one to another, substantial insulated connectors of a capacity at least equivalent to that of the cable shall be used. If connections are effected by means of cable lugs, they shall be securely fastened together to give good electrical contact, and the exposed metal parts of the lugs shall be completely insulated.

- Cables in need of repair shall not be used. When a cable becomes worn to the extent of exposing bare conductors, the portion thus exposed shall be protected by means of rubber and friction tape or other equivalent insulation.

Example 17.4

To be acceptable for use, cables must be free of repairs and splices for at least _____ feet from cable end to electrode holders.

   A. 10

   B. 15

   C. 20

   D. 25

Answer: A

17.4.3   Operating Instructions

Employers shall instruct employees in the safe means of arc welding and cutting as follows:

- When electrode holders are to be left unattended, the electrodes shall be removed and the holders shall be so placed or protected that they cannot make electrical contact with employees or conducting objects.
- Hot electrode holders shall not be dipped in water; to do so may expose the arc welder or cutter to electric shock.
- When the arc welder or cutter has occasion to leave his work or to stop work for any appreciable length of time, or when the arc-welding or arc-cutting machine is to be moved, the power supply switch to the equipment shall be opened.
- Any faulty or defective equipment shall be reported to the supervisor.

Whenever practicable, all arc-welding and arc-cutting operations shall be shielded by noncombustible or flameproof screens which will protect employees and other persons working in the vicinity from the direct rays of the arc.

17.5   Fire Prevention

In order to protect workers from fire-related hazards, OSHA has specified basic preventive measures along with certain special preventive measures for exceptionally dangerous tasks, like welding and cutting containers and welding in confined spaces. The OSHA standard's basic preventive measures for fire in welding, cutting, and brazing focus on eliminating fire hazards. These include:

- Before an object is welded or cut, it must be moved to a safe place.
- If the object to be welded or cut cannot be moved, then the area must be cleared of all moveable fire hazards.
- If certain fire hazards cannot be removed from the area, then workers are required to use guards to protect the immovable fire hazards from heat, sparks, and slag.
- If, however, the requirements mentioned above cannot be met, then employees must not perform any welding and cutting tasks.

Apart from these basic preventive measures for fires in welding and cutting, OSHA requires all workers to take certain special precautions. The OSHA standard requires workers to have proper fire extinguishing equipment available. All fire extinguishing equipment must be in proper working condition and must be kept close by for instant use. Preventive equipment may consist of buckets of sand, pails of water, fire extinguishers, or hoses. However, these may vary according to the nature of the work and the quantity of combustible material present in the vicinity.

Where there is a considerable amount of combustible material present closer than 35 feet from the work location, OSHA requires employers to station fire watchers whenever welding or cutting is performed. OSHA also requires employers to station fire watchers if considerable amounts of combustible materials are present that can be easily ignited by sparks, even if they are more than 35 feet away. The OSHA standard requires the fire watch to be maintained for at least half an hour after the welding or cutting task has been completed, and for a longer period of time if necessary.

Part 1926, Subpart J, Standard Number 1926.352 states that:

- When practical, objects to be welded, cut, or heated shall be moved to a designated safe location or, if the objects to be welded, cut, or heated cannot be readily moved, all movable fire hazards in the vicinity shall be taken to a safe place, or otherwise protected.

- If the object to be welded, cut, or heated cannot be moved and if all the fire hazards cannot be removed, positive means shall be taken to confine the heat, sparks, and slag, and to protect the immovable fire hazards from them.

- No welding, cutting, or heating shall be done where the application of flammable paints, or the presence of other flammable compounds, or heavy dust concentrations creates a hazard.

- Suitable fire extinguishing equipment shall be immediately available in the work area and shall be maintained in a state of readiness for instant use.

- When the welding, cutting, or heating operation is such that normal fire prevention precautions are not sufficient, additional personnel shall be assigned to guard against fire while the actual welding, cutting, or heating operation is being performed, and for a sufficient period of time after completion of the work to ensure that no possibility of fire exists. Such personnel shall be instructed as to the specific anticipated fire hazards and how the firefighting equipment provided is to be used.

- When welding, cutting, or heating is performed on walls, floors, and ceilings, since direct penetration of sparks or heat transfer may introduce a fire hazard to an adjacent area, the same precautions shall be taken on the opposite side as are taken on the side on which the welding is being performed.

- For the elimination of possible fire in enclosed spaces as a result of gas escaping through leaking or improperly closed torch valves, the gas supply to the torch shall be positively shut off at some point outside the enclosed space whenever the torch is not to be used or whenever the torch is left unattended for a substantial period of time, such as during the lunch period. Overnight and at the change of shifts, the torch and hose shall be removed from the confined space. Open-end fuel-gas and oxygen hoses shall be immediately removed from enclosed spaces when they are disconnected from the torch or other gas-consuming device.

- Hose connections must be clamped so they can withstand twice their pressure, which cannot be below 300 psi.

- Except when the contents are being removed or transferred, drums, pails, and other containers which contain or have contained flammable liquids shall be kept closed. Empty containers shall be removed to a safe area apart from hot work operations or open flames.

- Drums containers, or hollow structures, which contained toxic or flammable substances shall, before welding, cutting, or heating is undertaken on them, either be filled with water or thoroughly cleaned of such substances and ventilated and tested.

- Before heat is applied to a drum, container, or hollow structure, a vent or opening shall be provided for the release of any built-up pressure during the application of heat.

Example 17.5
Hose connections must be clamped so they can withstand twice their pressure, which cannot be below _____ psi.

   A. 600

   B. 500

   C. 400

   D. 300

Answer: D

17.6   Ventilation and Protection in Welding, Cutting, and Heating

Part 1926, Subpart J, Standard Number 1926.353 states the following.

17.6.1   Mechanical Ventilation

For purposes of this section, mechanical ventilation shall meet the following requirements:

- Mechanical ventilation shall consist of either general mechanical ventilation systems or local exhaust systems.
- General mechanical ventilation shall be of sufficient capacity and so arranged as to produce the number of air changes necessary to maintain welding fumes and smoke within safe limits, as defined in Subpart D of Part 1926
- Local exhaust ventilation shall consist of freely movable hoods intended to be placed by the welder or burner as close as practicable to the work. This system shall be of sufficient capacity and so arranged as to remove fumes and smoke at the source and keep the concentration of them in the breathing zone within safe limits as defined in Subpart D of Part 1926.
- Contaminated air exhausted from a working space shall be discharged into the open air or otherwise clear of the source of intake air.
- All air replacing the air that is withdrawn shall be clean and respirable.
- Oxygen shall not be used for ventilation purposes, comfort cooling, blowing dust from clothing, or for cleaning the work area.

17.6.2   Welding, Cutting, and Heating in Confined Spaces

- When sufficient ventilation cannot be obtained without blocking the means of access, employees in the confined space shall be protected by airline respirators in accordance with the requirements of Subpart E of Part 1926, and an employee on the outside of such a confined space shall be assigned to maintain communication with those working within it and to aid them in an emergency.

- Where a welder must enter a confined space through a manhole or other small opening, means shall be provided for quickly removing him in case of emergency. When safety belts and lifelines are used for this purpose they shall be so attached to the welder's body that his body cannot be jammed in a small exit opening. An attendant with a preplanned rescue procedure shall be stationed outside to observe the welder at all times and be capable of putting rescue operations into effect.

17.6.3   Welding, Cutting, or Heating of Metals of Toxic Significance

Welding, cutting, or heating in any enclosed spaces involving the following metals requires special care as recommended by OSHA:

- Zinc-bearing base or filler metals or metals coated with zinc-bearing materials
- Lead base metals
- Cadmium-bearing filler materials
- Chromium-bearing metals or metals coated with chromium-bearing materials

Employees must be protected by airline respirators for welding, cutting, or heating in any enclosed spaces of the following metals:

- Metals containing lead, other than as an impurity, or metals coated with lead-bearing materials
- Cadmium-bearing or cadmium-coated base metals
- Metals coated with mercury-bearing metals
- Beryllium-containing base or filler metals. Because of its high toxicity, work involving beryllium shall be done with both local exhaust ventilation and airline respirators

Employees performing such operations in the open air shall be protected by filter-type respirators, except that employees performing such operations on beryllium-containing base or filler metals shall be protected by airline respirators.

17.6.4   Inert-Gas Metal-Arc Welding

Since the inert-gas metal-arc-welding process involves the production of ultraviolet radiation of intensities of 5 to 30 times that produced during shielded metal-arc welding, the decomposition of chlorinated solvents by ultraviolet rays, and the liberation of toxic fumes and gases, employees shall not be permitted to engage in, or be exposed to the process until the following special precautions have been taken:

- The use of chlorinated solvents shall be kept at least 200 feet, unless shielded, from the exposed arc, and surfaces prepared with chlorinated solvents shall be thoroughly dry before welding is permitted on such surfaces.

- Employees in the area not protected from the arc by screening shall be protected by filter lenses meeting the requirements. When two or more welders are exposed to each other's arc, filter lens goggles of a suitable type, meeting the requirements of Subpart E of Part 1926, shall be worn under welding helmets. Hand shields to protect the welder against flashes and radiant energy shall be used when either the helmet is lifted or the shield is removed.

- Welders and other employees who are exposed to radiation shall be suitably protected so that the skin is covered completely to prevent burns and other damage by ultraviolet rays. Welding helmets and hand shields shall be free of leaks and openings, and free of highly reflective surfaces.

- When inert-gas metal-arc welding is being performed on stainless steel, the requirements shall be met to protect against dangerous concentrations of nitrogen dioxide.

Example 17.6
When working with stainless steel, workers must protect themselves from:

    A. Hydrogen cyanide

    B. Carbon monoxide

    C. Chlorine gas

    D. Nitrogen dioxide

Answer: D

17.6.5 General Welding, Cutting, and Heating

General welding, cutting, and heating may normally be done without mechanical ventilation or respiratory protective equipment, but where, because of unusual physical or atmospheric conditions, an unsafe accumulation of contaminants exists, suitable mechanical ventilation or respiratory protective equipment shall be provided. Employees performing any type of welding, cutting, or heating shall be protected by suitable eye protective equipment.

17.7 Welding, Cutting, and Heating in Way of Preservative Coatings

Before welding, cutting, or heating is commenced on any surface covered by a preservative coating whose flammability is not known, a test is to be made by a competent person to determine its flammability. Preservative coatings shall be considered to be highly flammable when scrapings burn with extreme rapidity. Part 1926, Subpart J, Standard Number 1926.354 states that precautions shall be taken to prevent ignition of highly flammable hardened preservative coatings. When coatings are determined to be highly flammable, they are to be stripped from the area to be heated to prevent ignition. The preservative coatings are to be removed a sufficient distance from the area to be heated to ensure that the temperature of the unstripped metal will not be appreciably raised. Artificial cooling of the metal surrounding the heating area may be used to limit the size of the area required to be cleaned.

17.8 Summary

All fire extinguishing equipment must be in proper working condition and must be kept close by for instant use. However, these devices may vary according to the nature of the work and the quantity of combustible material present in the vicinity. OSHA requires employers to station fire watchers if considerable amounts of combustible materials are present that can be easily ignited by sparks and requires the fire watch to be maintained for at least half an hour after the welding or cutting task has been completed or longer if necessary.

OSHA additionally requires all personnel to wear helmets and hand shields during all arc-welding or arc-cutting operations in order to protect themselves from direct radiant energy from the arc. It mandates all helpers and attendants to have proper eye protection. Also, if a welder is to enter a confined space, employers must station an attendant outside the space to monitor the welder at all times, and this attendant must be capable of putting rescue operations into effect.

All personnel who weld or supervise those individuals must know about marking of cylinders, storage of cylinders, and their operating procedures. OSHA mandates that all compressed gas distributors, for example, clearly mark all compressed gas cylinders. Markings must include either the chemical or trade name of the gas stored in the cylinder.

It is additionally important to store cylinders away from stairs, elevators, exit routes, or other areas where they can be toppled or damaged by passing people or falling objects, or where they could otherwise be tampered with. Due to the potential hazards associated with cylinders containing compressed gases, OSHA requires all personnel to store and operate these cylinders properly.

The OSHA standard requires employers to ensure that the service pipe system is always protected against the build up of excessive pressure and leaks. Protection may be provided using pressure relief devices, protective equipment, regulators, and hoses. OSHA mandates the use of pressure relief valves for preventing excess build up in fuel-gas, piping systems.

17.9 Multiple-Choice Questions

17.1 Welding fumes:

   A.   Are hot enough to melt steel
   B.   Are composed of microscopic particles
   C.   Can cause fires
   D.   Are harmless

17.2 _____ is a basic type of welding operation.

   A.   Arc flash
   B.   Resistance
   C.   Oxygen-argon gas
   D.   All of the above

17.3 Common welding hazards can include:

   A.   Microwave and ultraviolet radiation can damage the eye.
   B.   Vapors and mists are poisonous.
   C.   Welding on close containers can cause them to explode.
   D.   None of the above.

17.4 Oxygen cylinders:

   A. Are supposed to leak a little
   B. Can be dangerous if the gauges or regulators are broken
   C. Normally show burn marks after the first time they're used
   D. All of the above

17.5 An example of a compressed gas that can easily catch fire and burn is:

   A. Propane
   B. Nitrogen
   C. Argon
   D. Carbon dioxide

17.6 If you weld in an elevated area:

   A. You don't need fall protection such as a guardrail.
   B. Be sure to protect employees who might be below you.
   C. You don't need to put up a warning sign for hot surfaces when you're done.
   D. It's safer to do resistance welding if the surface is wet.

17.7 When handling an oxygen compressed gas cylinder:

   A. Wear oily gloves.
   B. Be absolutely certain it arrived in the horizontal position.
   C. Be sure to grease the valve before you open it.
   D. Identify its dangers by looking at the safety data sheet before you use it.

17.8 When you open the valve on a compressed gas cylinder:

   A. Make sure there's gas building up pressure in the regulator.
   B. Open the valve as quickly as you can.
   C. Point the outlet away from people and sources of ignition.
   D. Use any type of wrench to open a valve with hand wheels.

17.9 When a cylinder is being transported, the value must be:

   A. Installed
   B. Removed
   C. Open
   D. Capped

17.10 All cylinders must be stored away from:

   A. Radiators
   B. Fire
   C. Devices that produce heat
   D. All of the above

17.11 Devices that mix air or oxygen with flammable gasses cannot be used, unless approved by:

   A. An authorized person
   B. Employee
   C. Manufacturer
   D. Qualified welder or cutter

17.12 What is the recommended opening for valves?

   A. One-half turn
   B. Three turns
   C. One-and-a-half turns
   D. Full open

17.10   Practice Problems

17.1 Define welding, cutting, and brazing.

17.2 Discuss the types of welding.

17.3 List some dangers of handling compressed gas cylinders (CGCs).

17.4 List some hazards related to welding and cutting.

17.5 List the safe practices for transporting, moving, and storing compressed gas cylinders.

17.6 List the safe practices for treatment of gas cylinders.

17.7 List the safe operating instructions of arc welding.

17.8 List some OSHA standard's basic preventive measures for fire in welding, cutting, and brazing.

17.11   Critical Thinking and Discussion Topic

While working in a steel building construction, it was found that welding is going to be difficult for a steel connection for safety issue. One worker was advising to use bolts as he thinks bolts are stronger than welding and reliability of using bolts is better. Another worker said we needed to perform welding as it is in the design specifications. Which one do you prefer as a site manager?

CHAPTER 18

Electrical

(Photo courtesy of 123RF.)

18.1   General

This chapter addresses electrical safety requirements that are necessary for the practical safeguarding of employees involved in construction work and is divided into four major divisions and applicable definitions as follows:

- **Installation safety requirements.** Installation safety requirements are contained in 1926.402 through 1926.408. Included in this category are electric equipment and installations used to provide electric power and light on jobsites.

- **Safety-related work practices.** Safety-related work practices are contained in 1926.416 and 1926.417. In addition to covering the hazards arising from the use of electricity at jobsites, these regulations also cover the hazards arising from the accidental contact, direct or indirect, by employees with all energized lines, above or below ground, passing through or near the jobsite.

- **Safety-related maintenance and environmental considerations.** Safety-related maintenance and environmental considerations are contained in 1926.431 and 1926.432.

- **Safety requirements for special equipment.** Safety requirements for special equipment are contained in 1926.441.

To handle electricity safely, including working with electrical equipment, you need to understand how electricity acts, the hazards its presents, and how those hazards can be controlled.

Electricity is the flow of electrons through a conductor. An electron is a tiny particle of matter that orbits around the nucleus of an atom. Electrons of some atoms are easily moved out of their orbits. This ability of electrons to move or flow is the basis of electrical current. When you activate a switch to turn on an electric machine or tool, you allow current to flow from the generating source through conductors (wires) to the area of demand (motor).

A complete circuit is necessary for the controlled flow of electrons along a conductor. A complete circuit is made up of a source of electricity, a conductor, and a consuming device (load). The equation, Volts = Current × Resistance, is known as Ohm's Law. The factors discussed below relate to one another as described by this equation. This relationship makes it possible to change the qualities of an electrical current but keep an equivalent amount of power.

A force or pressure must be present before water will flow through a pipeline. Similarly, electrons flow through a conductor because electromotive force (EMF) is exerted. The unit of measure for EMF is the volt.

For electrons to move in a particular direction, a potential difference must exist between two points of the EMF source. For example, a battery has positive and negative poles.

The continuous movement for electrons past a given point is known as current. It is measured in amperes. The movement of electrons along a conductor meets with some opposition. This opposition is known as resistance. Resistance to the flow of electricity is measured in ohms. The amount of resistance provided by different materials varies widely. For example, most metals offer little resistance to the passage of electric current. However, porcelain, wood, pottery, and some other substances have a very high resistance to the flow of electricity. In fact, these substances can be used as insulators against the passage of electric current.

18.2   Hazards of Electricity

The primary hazards of electricity and its use are:

- Shock
- Burns
- Arc-blast
- Explosions
- Fires

18.2.1   Shock

Electric currents travel in closed circuits through some kind of conduction material. You get a shock when some part of your body becomes part of an electric circuit. An electric current enters the body at one point and exits the body at another location. High-voltage shocks can cause serious injury (especially burns) or death. You will get a shock if a part of your body completes an electrical circuit by:

- Touching a live wire and an electrical ground
- Touching a live wire and another wire at a different voltage

The severity of the shock a person receives depends on several factors:

- How much electric current flows through the body?
- What path the electric current takes through the body?
- How much time elapses while the body is part of the electric circuit?

A current as small as 60/1000 of an ampere can kill you if it passes through your chest. The typical household current operates at 15 amperes. The effects of an electric shock on the body can range from a tingle in the part touching the circuit to immediate cardiac arrest. A severe shock can cause more damage to the body than is readily visible. Relatively small burn marks may be all that are visible on the outside. However, a severely shocked person can suffer internal bleeding and severe destruction to tissues, muscles, and nerves. Finally, a person receiving an electric shock may suffer broken bones or other injuries that occur from falling after receiving a shock.

In its pure state, water is poor conductor of electricity. However if even small amounts of impurities are present in the water (salt in perspiration, for example), it becomes a ready electrical conductor. Therefore, if water is present anywhere in the work environment or on your skin, be extra careful around any source of electricity.

18.2.2   Burns

Burns can result when a person touches electrical wiring or equipment that is improperly used or maintained. Typically, such burn injuries occur on the hands.

18.2.3   Arc-Blast

Arc-blasts occur when high-amperage currents jump from one conductor to another through air, generally during opening or closing circuits, or when static electricity is discharged. Fire may occur if the arcing takes place in an atmosphere that contains an explosive mixture.

18.2.4   Explosions

Explosions occur when electricity provides a source of ignition for an explosive mixture in the atmosphere. Ignition can be due to overheated conductors or equipment, or normal arcing (sparking) at switch contacts. OSHA standards, the National Electrical Code, and related safety standards have precise requirements for electrical systems and equipment used in hazardous atmospheres.

18.2.5 Fires

Electricity is one of the most common causes of fire both in the home and workplace. Defective or misused electrical equipment is a major cause, with high-resistance connections being one of the primary sources of ignition. High-resistance connections occur where wires are improperly spliced or connected to other components such as receptacle outlets and switches.

High-resistance connections cause heat to build up. In some cases, enough heat can build up to start a fire.

18.3 Causes of Electrical Accidents

It is important for you to understand how to avoid electrical hazards when you work with electrical power tools, maintain electrical equipment, or install equipment for electrical operation. Accidents and injuries in working with electricity are caused by one or a combination of the following factors:

- Unsafe equipment and/or installation
- Unsafe workplace caused by environmental factors
- Unsafe work practices

18.4 Preventing Electrical Accidents

Protective methods to control electrical hazards include:

- Insulation
- Electrical protective devices
- Guarding
- Grounding
- PPE

18.4.1 Insulation

Insulation of glass, mica, rubber, or plastic is put on electrical conductors to protect you from electrical hazards. Before you begin work on any piece of electrical equipment, take a look at the insulation (on electrical cords, for example) to be sure that there are no exposed electrical wires.

18.4.2 Electrical Protective Devices

Electrical protective devices, including fuses, circuit breakers (CB), and ground-fault circuit breakers, and ground-fault circuit-interrupters (GFCIs), are critically important to electrical safety. These devices interrupt current flow when it exceeds the capacity of the conductor and should be installed where necessary. Current can exceed the capacity of the conductor when a motor is overloaded, for example, when you ask a 10 horsepower motor to do the work of a 12 horsepower motor. Or when a fault occurs, as when insulation fails in a circuit.

When a circuit is overloaded, the insulation becomes brittle over time. Eventually, it may crack and the circuit fails, or faults. Fault occurs in two ways. Most of the time a fault will occur between a conductor and an enclosure. This is called a ground fault. Infrequently, a fault will occur between two conductors. This is called a short circuit.

A device which prevents current from exceeding the conductor's capacity creates a weak link in the circuit. In the case of a fuse, the fuse is destroyed before another part of the system is destroyed. In the case of a circuit breaker, a set of contacts opens the circuit. Unlike a fuse, a circuit breaker can be reused by reclosing the contacts. Fuses and circuit breakers are designed to protect equipment and facilities, and in so doing, they also provide considerable protection against shock. However, the only electrical protective device whose sole purpose is to protect people is the GFCI. The GFCI is not an overcurrent device. It senses an imbalance in current flow over the normal path and opens the circuit. GFCIs are usually installed on circuits that are operated near water.

Example 18.1
_____ are overcurrent protective devices with a circuit opening fusible part that is heated and severed by the passage of overcurrent through that part.

    A. Circuits

    B. GFCI

    C. Assured Grounding Programs

    D. Fuses

Answer: D

Example 18.2
If working in damp locations, inspect electric cords and equipment to ensure that they are in good condition and free of defects and use a GFCI.

    • True

    • False

Answer: True

18.4.3   Guarding

Any live parts of electrical equipment operating at 50 volts or more must be guarded to avoid accidental contact. This protection can be accomplished in several different ways. The machinery or equipment can be located:

- In a room, enclosure, or vault accessible only to qualified personnel
- Behind substantial covers, panels, or partitions which prevent easy access
- On a balcony, platform, or gallery area which is elevated and not accessible to unqualified/unauthorized persons
- At least 8 feet above the floor of the work area

Any entrance to an area containing exposed live parts of electrical equipment must be marked with conspicuous warning signs. These signs should forbid entrance except by qualified persons.

18.4.4 Grounding

Grounding is necessary to protect you from electrical shock, safeguard against fire, and protect electrical equipment from damage. There are two kinds of grounding:

- Electrical circuit or system grounding, accomplished when one conductor of the circuit is intentionally connected to earth, protects the circuit, should lighting strike or other high-voltage contact occur. Grounding a system also stabilizes the voltage in the system so that the expected voltage levels are not exceeded under normal conditions.

- Electrical equipment grounding occurs when the equipment grounding conductor provides a path for dangerous fault current to return to the system ground at the supply source of the circuit, should the insulation fail.

When a tool or other piece of electrical equipment is grounded, a low-resistance path is intentionally created to the earth. This path has enough current-carrying capacity to prevent any buildup of voltages in the equipment which could pose a hazard to an employee using the equipment. Therefore, never remove the ground prong from a plug because the equipment no longer protects you from short circuit. If you're touching an ungrounded tool, you will become the path of least resistance of the ground.

Grounding does not guarantee that an employee will never receive a shock or be injured or killed by electricity in the workplace. However, this simple procedure will substantially reduce the likelihood of such accident. Be sure any equipment you work on is properly grounded.

18.5 Safe Work Practices for Handling Electricity

If your job requires you to work on or near exposed energized parts, you should be sure that any tools you use are in good repair, that you use good judgment when working near electrical lines, and that you use appropriate protective equipment.

18.5.1 Personal Protective Equipment

If you work on or near exposed energized parts, you must be trained as a qualified person. Your employer must provide you with protective equipment. You must use electrical protective equipment (see 29 CFR 1910.137) appropriate to the body parts that need protection and for the work to be done. Electrical protective equipment includes insulating blankets, matting, gloves, sleeves, overshoes, face protection, and hard hats among other equipment specially made to protect you from electricity.

18.5.2 Lockout/Tagout

Before any repair work or inspection of a piece of electrical equipment is begun by a qualified person, the current must be turned off at the switch box, and the switch padlocked in the OFF position.

The other step in this procedure is the tagging of the switch or controls of the machine or other equipment which is currently locked out of service. The tag indicates which circuits or pieces of equipment are out of service.

18.5.3 Work at Working Safely

Safety should be foremost in your mind when working with electrical equipment. You face hazards from the tools themselves and the electricity that powers them. It's up to you to wear protective equipment whenever it's specified, use all safety procedures, and work with tools correctly. The following general rules apply to every electrical equipment you use:

- Be sure your electrical equipment is maintained properly. Regularly inspect tools, cords, grounds, and accessories. Make repairs only if you are authorized to do so. Otherwise, arrange to have equipment repaired or replaced immediately.

- Be sure you use safety features like three-prong plugs, double-insulated tools, and safety switches. Be sure electrical cover and panels are in place and that you always follow proper procedure.

- Install or repair equipment only if you're qualified and authorized to do so. A faulty job may cause a fire or seriously injure other workers.

- Keep electric cables and cords clean and free from kinks. Never carry equipment by its cords.

- Use extension cords only when flexibility is necessary:

  - Never use them as substitutes for fixed wiring.

  - Never run them through holes on walls, ceilings, floors, doorways, or windows.

  - Never use them where they are concealed behind walls, ceilings, or floors.

- Don't touch water, damp surface, ungrounded metal, or any bare wires if you are not protected. Qualified employees must wear approved rubber gloves when working with live wires or ungrounded surfaces, and wear electrically insulated shoes or boots.

- Don't wear metal objects (rings, watches, etc.) when working with electricity. They might cause arcing.

- It you are working near overhead power lines of 50 kilovolts (kV) or less, you or any equipment you are using must not come any closer than 10 feet from the lines. Add 4 inches of distance for every 10 kV over 50 kV. The preferred setback distance from overhead wires during cleanup is 10 feet.

Good work habits soon become second nature. Treat electricity with the respect it deserves, and it will serve you efficiently and safely.

Example 18.3
What is the preferred setback distance from overhead wires during cleanup?

    A. 5 feet

    B. 7 feet

    C. 10 feet

    D. 12 feet

Answer: C

18.6 General Safety Requirements

Part 1926, Subpart K, Standard Number 1926.403 states the following.

18.6.1 Examination, Installation, and Use of Equipment

Examination. The employer shall ensure that electrical equipment is free from recognized hazards that are likely to cause death or serious physical harm to employees. Safety of equipment shall be determined on the basis of the following considerations:

- Suitability for installation and use in conformity with the provisions of this subpart. Suitability of equipment for an identified purpose may be evidenced by listing, labeling, or certification for that identified purpose.
- Mechanical strength and durability, including, for parts designed to enclose and protect other equipment, the adequacy of the protection thus provided.
- Electrical insulation.
- Heating effects under conditions of use.
- Arcing effects.
- Classification by type, size, voltage, current capacity, specific use.
- Other factors which contribute to the practical safeguarding of employees using or likely to come in contact with the equipment.

Installation and Use. Listed, labeled, or certified equipment shall be installed and used in accordance with instructions included in the listing, labeling, or certification.

18.6.2 Interrupting Rating

Equipment intended to break current shall have an interrupting rating at system voltage sufficient for the current that must be interrupted.

18.6.3 Mounting and Cooling of Equipment

Mounting. Electric equipment shall be firmly secured to the surface on which it is mounted. Wooden plugs driven into holes in masonry, concrete, plaster, or similar materials shall not be used.

Cooling. Electrical equipment which depends upon the natural circulation of air and convection principles for cooling of exposed surfaces shall be installed so that room air flow over such surfaces is not prevented by walls or by adjacent installed equipment. For equipment designed for floor mounting, clearance between top surfaces and adjacent surfaces shall be provided to dissipate rising warm air. Electrical equipment provided with ventilating openings shall be installed so that walls or other obstructions do not prevent the free circulation of air through the equipment.

18.6.4 Splices

Conductors shall be spliced or joined with splicing devices designed for the use or by brazing, welding, or soldering with a fusible metal or alloy. Soldered splices shall first be so spliced or joined as to be mechanically and electrically secure without solder and then soldered. All splices and joints and the free ends of conductors shall be covered with an insulation equivalent to that of the conductors or with an insulating device designed for the purpose.

18.6.5 Arcing Parts

Parts of electric equipment which in ordinary operation produce arcs, sparks, flames, or molten metal shall be enclosed or separated and isolated from all combustible materials.

18.6.6 Marking

Electrical equipment shall not be used unless the manufacturer's name, trademark, or other descriptive marking by which the organization responsible for the product may be identified is placed on the equipment and unless other markings are provided giving voltage, current, wattage, or other ratings as necessary. The marking shall be of sufficient durability to withstand the environment involved.

18.6.7 Identification of Disconnecting Means and Circuits

Each disconnecting means required by this subpart for motors and appliances shall be legibly marked to indicate its purpose, unless located and arranged so the purpose is evident. Each service, feeder, and branch circuit, at its disconnecting means or overcurrent device, shall be legibly marked to indicate its purpose, unless located and arranged so the purpose is evident. These markings shall be of sufficient durability to withstand the environment involved.

Readers are referred to the OSHA, Part 1926, Subpart K, to know some other electrical aspects such as Wiring design and protection, Wiring methods, components, and equipment for general use, Specific purpose equipment and installations, Hazardous (classified) locations, Special systems, Safety-related work practices, Lockout and tagging of circuits, Safety-related maintenance and environmental considerations, Maintenance of equipment, Environmental deterioration of equipment, Safety requirements for special equipment, Batteries and battery charging, and so on.

Example 18.4
Electrical equipment shall not be used unless the:

    A. Manufacturer's name is written.

    B. Manufacturer's trademark is written.

    C. Markings are provided giving voltage, current, wattage, etc.

    D. Markings are of sufficient durability to withstand the environment.

    E. All of the above.

Answer: *E*

18.7 Summary

The primary hazards of electricity and its use are shock, burns, arc-blast, explosions, and fires. Accidents and injuries in working with electricity are caused by one or a combination of the following factors such as unsafe equipment and/or installation, unsafe workplace caused by environmental factors, unsafe work practices, etc. Some protective methods to control electrical hazards include:

- Insulation
- Electrical protective devices
- Guarding
- Grounding
- PPE

The employer shall ensure that electrical equipment is free from recognized hazards that are likely to cause death or serious physical harm to employees. Electric equipment shall be firmly secured to the surface on which it is mounted. Electrical equipment shall not be used unless the manufacturer's name, trademark, or other descriptive marking by which the organization responsible for the product can be identified.

18.8 Multiple-Choice Questions

18.1 Workers who may work on or near exposed live (energized) parts:

A. Are unqualified workers
B. Are qualified workers
C. Don't need training
D. Don't need to know proper clearance distance

18.2 If you aren't trained to work on or near exposed live (energized) parts:

A. You are unqualified worker.
B. You are qualified worker.
C. You may not use any electrically powered equipment.
D. You do not need to know the specific safety practices for your job.

18.3 The flow of the electrons through a conductor:

A. Causes static
B. Is the generating source
C. Makes up an electrical current
D. Describes how the switch operates

18.4 A complete circuit is made up of:

A. A load
B. A conductor
C. A source of electricity
D. All of the above

18.5 You will get an electrical shock if:

A. You are wearing gloves.
B. You are an unqualified worker.
C. A part of your body completes an electrical circuit.
D. A GFCI is installed.

18.6 When electromotive force is exerted:

A. Electrons flow through a conductor.
B. An arc-blast occurs.

C. There is a file.
D. Workers must wear gloves.

18.7 The term "amperes" is used to describe:

A. The brightness of a light bulb
B. The number of ohms
C. The amount of current in a circuit
D. The poles of a battery

18.8 If a person has small burn mark on his skin after receiving an electric shock:

A. These surface burns are his only injury.
B. He can have internal injuries, such as never or muscle damage.
C. He was not really shocked.
D. The current was not large enough to cause any injury.

18.9 Fires can result from:

A. Using defective or damaged electrical equipment
B. High resistance connections
C. Improperly spliced or connected wires
D. All of the above

18. 10 Circuit breakers and GFCIs are examples of:

A. Insulation
B. Guarding
C. Electrical protective devices
D. Personal protective equipment

18.11 Electrical protective devices are designed to automatically _____ a circuit, if excess current from overload or ground-fault is detected.

A. Open
B. Bypass
C. Close
D. Circulate

18.9   Practice Problems

18.1 List some hazards of electricity.
18.2 List some causes of electrical accidents.
18.3 List some preventive measures of electrical accidents.
18.4 List some safe work practices for handling electricity.
18.5 List some general safety regulations for examination, installation, and use of equipment.
18.6 List some general safety regulations for mounting and cooling of equipment.
18.7 List some general safety regulations for identification of disconnecting means and circuits.

18.10   Critical Thinking and Discussion Topic

While working in a steel building construction, it is found that an electrician is essential to perform some electrical operation. However, no certified electrician is available at that moment. One machine operator said he is competent in electrical operation but he has no certification. As a site supervisor, what will you do in this situation?

(Photo courtesy of 123RF.)

CHAPTER **19**

Scaffolds

(Photo courtesy of 123RF.)

19.1   General

Scaffolding, also known as scaffolding or staging, is a temporary structure used to support the work crew, as shown in Fig. 19.1, and materials to assist in the construction, maintenance, and repair of buildings, bridges, and all other manmade structures. Scaffolds are widely used on-site to gain access to heights and areas that would otherwise be difficult to reach. Unsafe scaffolding may result in death or serious injury. Scaffolding is also used in adapted forms for formwork and shoring, grandstand seats, concert venues, access/view towers, exhibition stands, ski ramps, half pipes, and art projects.

Example 19.1
What is a scaffold?

    A. A permanent elevated platform used to support workers and/or materials.

    B. A temporary elevated platform used to support workers and/or materials.

    C. A temporary platform used to support a structure.

    D. A permanent platform used to support a structure.

Answer: *B*

19.2   Types of Scaffolds

19.2.1   Suspended Scaffolds

These are types of scaffolds that are suspended from a fixed overhead position (often placed at the top of a building, but it can be any fixed elevated structure). The various types of suspended scaffolds are discussed next.

Two-Point (Swing Stage).   These are one of the most common types of suspended scaffolds. You see them frequently on construction sites and on skyscrapers and other high-rises where they are used by window washers and exterior building workers. The scaffold is hung from the top of a building by two ropes that are secured at either end of the platform, usually with stirrups to provide for vertical movement.

Single-Point Adjustable (Spider Scaffolds).   This scaffold is smaller than the two-point, as it usually accommodates only one person at a time. It is secured to the top of the building using one rope from a fixed overhead support. As with the two-point, it has a mechanism to allow for vertical movement (up and down). Again, it is often used by window washers.

Suspended Platform.   This scaffold consists of horizontal and parallel ropes that are used to support a platform on which personnel and/or materials are placed. This is hung from vertical ropes suspended from a structure. These are often used in buildings still under construction. This type of scaffold is also called catenary scaffold.

Suspended Scaffold Platform.   These consist of at least one platform (they may be more, but they rarely exceed a single platform) that is supported by more than two ropes (usually four ropes are used, one secured at each corner of the platform) from an overhead structure and fined with a mechanism to allow for vertical movement. They are commonly used as chimney hoists in the cleaning and repair of chimney and/or other ventilation shafts.

Interior Hung.   This scaffold consists of two parallel, horizontal beams, upon which a board or plank is placed. It is then suspended by ropes (sometimes up to four, one at each corner of the platform) from a roof or ceiling. This is primarily used indoors, hence the name, because it must be hung from roof structures.

Needle Beam.   In this case, the platform (usually a simple wooden board) is supposed by two needle beams. The beams are hung from ropes. Usually one end of the platform is supported on the edge of a permanent structure.

Multilevel.   A multilevel scaffold is a series of platforms that are arranged vertically, one on top of the other. Visually, it looks like two or more two-point scaffolds (or multipoint scaffolds) on top of one another. The platforms are placed within the same framework, secured by a single stirrup from the same fixed overhead.

Float (Ship).   The float (or ship) scaffold consists of a platform supported by two bearers and is hung from a fixed overhead support with ropes.

Example 19.2
How is a float scaffold supported?

    A. By positive brackets between two ladders

    B. By two bearers hung from an overhead support

    C. By four ropes, one in each corner

    D. By tubes and couplers

Answer: B

19.2.2   Supported Scaffolds

These are scaffolds that consist of one or more platforms elevated on poles and beams which are placed upon a solid ground. The various types of supported scaffolds are discussed next.

Frame or Fabricated.   This is one of the most common types of scaffolds used. They are easy to build, economical, and practical. They range in size from small setups for one- or two-story residential houses to large setups used on commercial projects; however, it is possible to make them stand several stories high if the need arises. They consist of a single or multiple platforms elevated on tubes.

Mobile.   Mobile scaffolds are visually similar to frame scaffolds, but with one main difference: they are mounted on wheels or casters. They are commonly used by painters, plasterers, electricians, plumbers, and others who need to change position frequently, hence the wheels.

Pump Jack Scaffold.   This scaffold consists of a platform that is elevated on a pair of vertical poles and fitted with moveable brackets. The brackets allow the platform to be moved up and down using a jack (similar to those used to change a car tire). They are attractive because of their ability to be raised and lowered somewhat quickly and easily and because they are relatively inexpensive. A pump jack scaffold must be fitted with two positive gripping mechanisms to prevent slippage.

Ladder Jack.   This is a very simple scaffold. In this case, the platform is placed in brackets that are attached to a pair of ladders. They are mostly used for light applications. They are popular mostly due to their practicality, ease, and economy.

Tube and Coupler.   In this case, tubes (which look like pipes) are placed together in a lattice-like framework using couplers (joints) to hold them together. The lattice-like pattern into which they are arranged adds balance and strength to the framework. Platforms can be placed at the top of the framework and at multiple levels. Tube and coupler scaffolds are often used to carry heavy weights, and they can be built to a height of multiple stories while being able to be erected to conform to the shape of a building. A prime advantage lies in that they can be built in several directions and combinations that cater to any structure.

Poles. This type of scaffold is also known as Wood Pole, because every pan of the scaffold, from the platform on down, is built using wooden poles and boards. There are two types: single pole, which are supported on one side (the interior side) by a fixed structure, such as a building; and two-pole, which are supported by two sets of wooden poles on either side of the platform. They are not commonly used due to the fact that they must be built from scratch, cannot be reused, and are difficult to erect and use in a safe and compliant manner.

Example 19.3
Name the scaffolds.

   A. The scaffold is constructed using vertical parallel ropes or wires.

   B. The scaffold accommodates only one person at a time.

   C. The scaffold is most commonly used as chimney hoists to clean chimney.

   D. The scaffold involves a series of platforms that are arranged on top of the other.

   E. One end of the platform is supported on the edge of a permanent structure.

   F. The scaffold is mounted on casters and is commonly used by painters or plasterers.

   G. The scaffold's brackets allow the platform to be moved up and down effortlessly.

   H. Mostly used for light applications, this simple scaffold involves a platform placed in brackets.

   I. The lattice-like pattern of this scaffold adds balance and strength to the framework.

   J. This scaffold is not commonly used now as it must be built from scratch and cannot be reused.

Answers:
   A. *Catenary*

   B. *Single-point adjustable (spiders)*

   C. *Multipoint adjustable*

   D. *Multilevel*

   E. *Needle beam*

   F. *Mobile*

   G. *Pump jack*

   H. *Ladder jack*

   I. *Tube and coupler*

   J. *Pole*

Each type is made from several components which often include:

- A base jack or plate which is a load-bearing base for the scaffold
- The standard, the upright component with connector joints
- The ledger, a horizontal brace
- The transom, a horizontal cross-section load-bearing component which holds the batten, board, or decking unit
- Brace diagonal and/or cross-section bracing component
- Batten or board-decking component used to make the working platform
- Coupler, a fitting used to join components together
- Scaffold tie used to tie in the scaffold to structures
- Brackets used to extend the width of working platforms

Figure 19.1   A scaffold at a construction site.
(Photo courtesy of Daria Sannikova from Pexels.)

Specialized components used to assist in their use as a temporary structure often include heavy-duty bearing transitions, ladders, or stairwells for the entry and exit of the scaffold, beam ladder/unit types used to span obstacles, and rubbish chutes used to remove undesirable materials from the scaffold or construction project.

19.3   General Safety Requirements

Some major specifications by the 1926.451 are as follows:

- Each scaffold and scaffold component shall be capable of supporting, without failure, its own weight and at least four times the maximum intended load applied or transmitted to it.
- Direct connections to roofs and floors, and counterweights used to balance adjustable suspension scaffolds, shall be capable of resisting at least four times the tipping moment imposed by the scaffold operating at the rated load of the hoist, or 1.5 (minimum) times the tipping moment imposed by the scaffold operating at the stall load of the hoist, whichever is greater.
- Each suspension rope, including connecting hardware, used on nonadjustable suspension scaffolds shall be capable of supporting, without failure, at least six times the maximum intended load applied or transmitted to that rope.
- Each suspension rope, including connecting hardware, used on adjustable suspension scaffolds shall be capable of supporting, without failure, at least six times the maximum intended load applied or transmitted to that rope with the scaffold operating at either the rated load of the hoist, or two (minimum) times the stall load of the hoist, whichever is greater.
- The stall load of any scaffold hoist shall not exceed three times its rated load.
- Scaffolds shall be designed by a qualified person and shall be constructed and loaded in accordance with that design.
- Scaffolds shall not be moved horizontally while employees are on them unless required outrigger frames are installed on both sides of the scaffold.
- Catenary scaffolds have a maximum weight load of 500 pounds, so no more than two people should be on the scaffold at any one time.
- Scaffolds must contain safe, well-constructed guardrails that consist of three rails: top, mid, and, when necessary, a toe board, or employees must be provided with appropriate personal fall arrest systems; guardrails must be sufficiently strong and smooth according to OSHA guidelines.
- The midrails of guardrails must be capable of withstanding 150 pounds of force applied at any point, in any direction.
- Employers must also prohibit climbing on a scaffold's cross braces, and if there are no guardrails, provide personal fall arrest systems for workers.
- To avoid suspension trauma, when suspended, the worker must limit restriction points and move their arms and legs to maintain circulation.
- Toe board or mesh fencing can be installed on scaffolding to protect workers below the scaffolding from falling hazards.

Example 19.4
Scaffolds shall not be moved horizontally while employees are on them unless:

A. Required outrigger frames are installed on both sides of the scaffold.

B. The surface on which the scaffold is being moved is within 1 degree of level.

C. The scaffold being moved is near holes and obstructions.

D. An employee is present on the scaffold that extends outward beyond the wheels.

Answer: A

Each platform on all working levels of scaffolds shall be fully planked or decked between the front uprights and the guardrail supports as follows:

- Each platform unit (e.g., scaffold plank, fabricated plank, fabricated deck, or fabricated platform) shall be installed so that the space between adjacent units and the space between the platform and the uprights is no more than 1 inch (2.5 centimeters) wide, except where the employer can demonstrate that a wider space is necessary (e.g., to fit around uprights when side brackets are used to extend the width of the platform).
- Each scaffold platform and walkway shall be at least 18 inches (46 centimeter) wide.
- Each ladder jack scaffold, top plate bracket scaffold, roof bracket scaffold, and pump jack scaffold shall be at least 12 inches (30 centimeter) wide. There is no minimum width requirement for boatswains' chairs.
- The platform in a two-point scaffold should be no more than 36 inches wide.
- The maximum load for a single-point scaffold is 250 pounds.
- Roof bracket scaffolds be at least 12 inches wide is stayed until November 25, 1997 or until rulemaking regarding the minimum width of roof bracket scaffolds has been completed, whichever is later.
- Where scaffolds must be used in areas that the employer can demonstrate are so narrow that platforms and walkways cannot be at least 18 inches (46 centimeter) wide, such platforms and walkways shall be as wide as feasible, and employees on those platforms and walkways shall be protected from fall hazards by the use of guardrails and/or personal fall arrest systems.
- The front edge of all platforms shall not be more than 14 inches (36 centimeter) from the face of the work, unless guardrail systems are erected along the front edge and/ or personal fall arrest systems are used to protect employees from falling.
- The maximum distance from the face for outrigger scaffolds shall be 3 inches (8 centimeter).
- The maximum distance from the face for plastering and lathing operations shall be 18 inches (46 centimeter).
- Each end of a platform, unless cleated or otherwise restrained by hooks or equivalent means, shall extend over the centerline of its support at least 6 inches (15 centimeter).

- Each end of a platform 10 feet or less in length shall not extend over its support more than 12 inches (30 centimeter) unless the platform is designed and installed so that the cantilevered portion of the platform is able to support employees and/or materials without tipping, or has guardrails which block employee access to the cantilevered end.

- Each platform greater than 10 feet in length shall not extend over its support more than 18 inches (46 centimeter), unless it is designed and installed so that the cantilevered portion of the platform is able to support employees without tipping, or has guardrails which block employee access to the cantilevered end.

- The gaps between planks on the platforms of supported scaffolds should not exceed 1 inch.

- Scaffolds tend to be made mostly of metal. A minimum clearance to maintain from all electrical lines up to 50 kV is 10 feet.

Example 19.5

Falls are a leading cause of deaths on construction sites. On your jobsite, you have to work with different kinds of scaffolds. Do you know the correct requirements? Write down the correct values in the blanks.

A. The support ropes must be capable of handling up to weight of the scaffold and its load _____ times the intended maximum.

B. The platform for a two-point scaffold should be no more than _____ inches wide.

C. The maximum load for a single-point scaffold is _____ pounds.

D. Catenary scaffolds have a maximum weight load of _____ pounds, so no more than four people should be on the scaffold at any one time.

E. Scaffolds tend to be made mostly of metal. A minimum clearance to maintain from all electrical lines up to 50 kV is _____ feet.

F. A platform should be able to support its own weight in addition to _____ times the maximum allowed weight.

Answers:
A. 6
B. 36
C. 250
D. 500
E. 10
F. 4

19.4   Additional Requirements Applicable to Specific Types of Scaffolds

1926.452 covers the following specific types of scaffolds:

- Pole scaffolds
- Tube and coupler scaffolds

- Fabricated frame scaffolds (tubular welded frame scaffolds)
- Plasterers', decorators', and large area scaffolds
- Bricklayers square scaffolds (squares)
- Horse scaffolds
- Form scaffolds and carpenters' bracket scaffolds
- Roof bracket scaffolds
- Outrigger scaffolds
- Pump jack scaffolds
- Ladder jack scaffolds
- Window jack scaffolds
- Crawling boards (chicken ladders)
- Step, platform, and trestle ladder scaffolds
- Single-point adjustable suspension scaffolds
- Two-point adjustable suspension scaffolds (swing stages)
- Multipoint adjustable suspension scaffolds, stonesetters' multipoint adjustable suspension scaffolds, and masons' multipoint adjustable suspension scaffolds
- Catenary scaffolds
- Float (ship) scaffolds
- Interior hung scaffolds
- Needle beam scaffolds
- Multilevel suspended scaffolds
- Mobile scaffolds
- Repair bracket scaffolds
- Stilts

19.5   Aerial lifts

1926.453 covers the following specific types of scaffolds.

19.5.1   General

Aerial equipment may be made of metal, wood, fiberglass reinforced plastic (FRP), or other material; may be powered or manually operated; and are deemed to be aerial lifts whether or not they are capable of rotating about a substantially vertical axis. Aerial lifts include the following types of vehicle-mounted aerial devices used to elevate personnel to jobsites above ground:

- Extensible boom platforms
- Aerial ladders
- Articulating boom platforms

- Vertical towers
- A combination of any such devices

Aerial lifts may be field modified for uses other than those intended by the manufacturer, provided the modification has been certified in writing by the manufacturer or by any other equivalent entity, such as a nationally recognized testing laboratory, to be in conformity with all applicable provisions of ANSI A92.2-1969 and this section and to be at least as safe as the equipment was before modification.

19.5.2   Some Specific Requirements

For ladder trucks and tower trucks, aerial ladders shall be secured in the lower traveling position by the locking device on top of the truck cab, and the manually operated device at the base of the ladder before the truck is moved for highway travel.

For extensible and articulating boom platforms:

- Lift controls shall be tested each day prior to use to determine that such controls are in safe working condition.
- Only authorized persons shall operate an aerial lift.
- Belting off to an adjacent pole, structure, or equipment while working from an aerial lift shall not be permitted.
- Employees shall always stand firmly on the floor of the basket, and shall not sit or climb on the edge of the basket or use planks, ladders, or other devices for a work position.
- A body belt shall be worn and a lanyard attached to the boom or basket when working from an aerial lift.
- Boom and basket load limits specified by the manufacturer shall not be exceeded.
- The brakes shall be set and when outriggers are used, they shall be positioned on pads or a solid surface. Wheel chocks shall be installed before using an aerial lift on an incline, provided they can be safely installed.
- An aerial lift truck shall not be moved when the boom is elevated in a working position with men in the basket, except for equipment which is specifically designed for this type of operation.
- Articulating boom and extensible boom platforms, primarily designed as personnel carriers, shall have both platform (upper) and lower controls. Upper controls shall be in or beside the platform within easy reach of the operator. Lower controls shall provide for overriding the upper controls. Controls shall be plainly marked as to their function. Lower level controls shall not be operated unless permission has been obtained from the employee in the lift, except in case of emergency.
- Climbers shall not be worn while performing work from an aerial lift.
- The insulated portion of an aerial lift shall not be altered in any manner that might reduce its insulating value.
- Before moving an aerial lift for travel, the boom(s) shall be inspected to see that it is properly cradled and outriggers are in stowed position except as provided in 1926.453.

Example 19.6

Which of the following statements about guardrails is correct?

   A. Guardrails must be capable of withstanding up to 200 pounds of force upon the top rail and up to 150 pounds of force upon the midrails applied at any point and from any direction.

   B. The top rail and midrails must be constructed in such a way that they overhang the scaffold.

   C. The top rail must be between 50 and 75 inches above the platform or walking level of the scaffold.

   D. There are no provisions for installing guardrails.

Answer: *A*

Example 19.7

Toe board or mesh fencing installed on scaffolding protects workers below the scaffolding from what hazards?

   A. Slips and falls

   B. Falling objects

   C. Accessing the scaffold

   D. Coming in contract with tools

Answer: *B*

Example 19.8

The gaps between planks on the platforms of supported scaffolds should not exceed _____.

   A. 4 inches

   B. 3 inches

   C. 2 inches

   D. 1 inch

Answer: *D*

Example 19.9

To avoid suspension trauma, when suspended, the worker must limit restriction points and move their arms and legs to maintain circulation.

   • True

   • False

Answer: *True*

Example 19.10
A pump jack scaffold must be fitted with two _____ gripping mechanisms to prevent slippage.

    A. Negative

    B. Neutral

    C. Positive

    D. Grounded

Answer: C

Example 19.11
For which of the following cases is the climbing of cross-braces permitted?

    A. When a worker is using a fall arrest system.

    B. When other workers are present.

    C. When there are no materials on the scaffold which may fall.

    D. It is strictly prohibited.

Answer: D

19.6   Summary

A scaffold refers to a temporary, elevated platform that is used to support workers and/or materials. There are several types of scaffolds. Suspended scaffolds are the type that are suspended by a wire rope from a fixed overhead position (usually placed at the top of a building). They can be broken down further into the subcategories that include: two-point (or swing stage) scaffolds; single-point adjustable (spider) scaffolds; catenary (using horizontal and parallel wire ropes); multipoint adjustable; interior-hung; float (ship); and more. Supported scaffolds are the type that consist of one or more platforms elevated on poles and beams, which are placed upon a solid ground. They can be broken down into frame or fabricated, mobile, pump jack, and more. Specialty scaffolds includes decorators, plasterers, and large-area scaffolds, among others.

There are safety requirements applicable to all scaffolds and additional standards applicable to specific types of scaffolds. Unless otherwise stated, the safety regulations for the two-point scaffold apply to all other types of suspension scaffolds. With suspended scaffolding, a primary factor to consider in terms of safety is the anchorage of the scaffold from the building (or other structure). A scaffold must not just be anchored properly and be strong and stable; the workers themselves must be trained in how to safely use and maneuver (themselves and their equipment) in the often small space. Employers can minimize the risk to employees by making sure all scaffolding construction, materials, and other elements are compatible.

19.7 Multiple-Choice Questions

19.1 Each scaffold and scaffold component shall be capable of supporting at least _____ times the maximum intended load applied or transmitted to it.

    A. 2
    B. 3
    C. 4
    D. 5

19.2 The stall load of any scaffold hoist shall not exceed _____ times its rated load.

    A. 1.5
    B. 2.0
    C. 3.0
    D. 4.0

19.3 Each suspension rope, including connecting hardware, used on nonadjustable suspension scaffolds shall be capable of supporting at least _____ times the maximum intended load applied to that rope.

    A. 2
    B. 3
    C. 5
    D. 6

19.4 Each scaffold platform and walkway shall be at least:

    A. 1.5 feet wide
    B. 2.0 feet wide
    C. 2.5 feet wide
    D. 3.5 feet wide

19.5 Each ladder jack scaffold, top plate bracket scaffold, roof bracket scaffold, and pump jack scaffold shall be at least:

    A. 1.0 feet wide
    B. 2.0 feet wide
    C. 2.5 feet wide
    D. 3.5 feet wide

19.6 The front edge of all platforms shall not be more than _____ inches from the face of the work.

    A. 12
    B. 14
    C. 16
    D. 18

19.7 The maximum distance from the face for outrigger scaffolds shall be:

    A. 3 inches
    B. 6 inches
    C. 9 inches
    D. 12 inches

19.8 The maximum distance from the face for plastering and lathing operations shall be:

    A. 6 inches
    B. 9 inches
    C. 12 inches
    D. 18 inches

19.9 Each end of a platform, unless cleated or otherwise restrained by hooks or equivalent means, shall extend over the centerline of its support at least:

    A. 6 inches
    B. 9 inches
    C. 12 inches
    D. 18 inches

19.10 Each platform greater than 10 feet in length shall not extend over its support more than:

    A. 6 inches
    B. 9 inches
    C. 12 inches
    D. 18 inches

19.11 The platform in a two-point scaffold should be no more than _____ inches wide.

    A. 18
    B. 24
    C. 36
    D. 48

19.8   Practice Problems

19.1 List the main types of scaffolding used worldwide.
19.2 List the components of scaffolding.
19.3 List major general safety regulations of scaffolding.
19.4 List the types of vehicle-mounted aerial devices used to elevate personnel to jobsites above ground.
19.5 List the safety regulations for extensible and articulating boom platforms of aerial lift.

19.9   Critical Thinking and Discussion Topic

Your company uses proper scaffolding as specified by OSHA. The proper scaffolding costs a lot of money and effort. Some other company does not use proper scaffolding and relies on the workers' skill. The accident rate is almost similar in your company and the other company. Even sometimes, accident rate is lower if the company relies on workers' skill as workers become careful while working with poor scaffolding. On the other hand, workers do not pay much attention to safety when there is proper scaffolding and accident happens. How do you see these two scenarios?

CHAPTER **20**

Fall Protection

(Photo courtesy of Pexels.)

20.1 General

Fall protection is discussed lightly in Chapter 5. The importance of fall protection is very high in construction. That is why, this topic is further discussed here quoting some safety regulations from OSHA.

20.2 Duty to Have Fall Protection

This section sets forth requirements for employers to provide fall protection systems. All fall protection required by this section shall conform to the criteria set forth in Part 1926 Subpart M Standard 1926.501. The employer shall determine if the walking/working surfaces on which its employees are to work have the strength and structural integrity to support employees safely. Employees shall be allowed to work on those surfaces only when the surfaces have the requisite strength and structural integrity.

20.2.1   Unprotected Sides and Edges

Each employee on a walking/working surface (horizontal and vertical surface) with an unprotected side or edge which is 6 feet (1.8 meter) or more above a lower level shall be protected from falling by the use of guardrail systems, safety net systems, or personal fall arrest systems.

20.2.2   Leading Edges

- Each employee who is constructing a leading edge 6 feet (1.8 meter) or more above lower levels shall be protected from falling by guardrail systems, safety net systems, or personal fall arrest systems. Exception: When the employer can demonstrate that it is infeasible or creates a greater hazard to use these systems, the employer shall develop and implement a fall protection plan.

- Each employee on a walking/working surface 6 feet (1.8 meter) or more above a lower level where leading edges are under construction, but who is not engaged in the leading edge work, shall be protected from falling by a guardrail system, safety net system, or personal fall arrest system. If a guardrail system is chosen to provide the fall protection, and a controlled access zone has already been established for leading edge work, the control line may be used in lieu of a guardrail along the edge that parallels the leading edge.

20.2.3   Hoist Areas

Each employee in a hoist area shall be protected from falling 6 feet (1.8 meter) or more to lower levels by guardrail systems or personal fall arrest systems. If guardrail systems, (or chain, gate, or guardrail) or portions thereof, are removed to facilitate the hoisting operation (e.g., during landing of materials), and an employee must lean through the access opening or out over the edge of the access opening (to receive or guide equipment and materials, for example), that employee shall be protected from fall hazards by a personal fall arrest system.

20.2.4   Holes

- Each employee on walking/working surfaces shall be protected from falling through holes (including skylights) more than 6 feet (1.8 meter) above lower levels, by personal fall arrest systems, covers, or guardrail systems erected around such holes.

- Each employee on a walking/working surface shall be protected from tripping in or stepping into or through holes (including skylights) by covers.

- Each employee on a walking/working surface shall be protected from objects falling through holes (including skylights) by covers.

20.2.5   Formwork and Reinforcing Steel

Each employee on the face of formwork or reinforcing steel shall be protected from falling 6 feet (1.8 meter) or more to lower levels by personal fall arrest systems, safety net systems, or positioning device systems.

20.2.6 Ramps, Runways, and Other Walkways

Each employee on ramps, runways, and other walkways shall be protected from falling 6 feet (1.8 meter) or more to lower levels by guardrail systems.

20.2.7 Excavations

- Each employee at the edge of an excavation 6 feet (1.8 meter) or more in depth shall be protected from falling by guardrail systems, fences, or barricades when the excavations are not readily seen because of plant growth or other visual barrier.
- Each employee at the edge of a well, pit, shaft, and similar excavation 6 feet (1.8 meter) or more in depth shall be protected from falling by guardrail systems, fences, barricades, or covers.

20.2.8 Dangerous Equipment

- Each employee less than 6 feet (1.8 meter) above dangerous equipment shall be protected from falling into or onto the dangerous equipment by guardrail systems or by equipment guards.
- Each employee 6 feet (1.8 meter) or more above dangerous equipment shall be protected from fall hazards by guardrail systems, personal fall arrest systems, or safety net systems.

20.2.9 Overhand Bricklaying and Related Work

- Each employee performing overhand bricklaying and related work 6 feet (1.8 meter) or more above lower levels shall be protected from falling by guardrail systems, safety net systems, personal fall arrest systems, or shall work in a controlled access zone.
- Each employee reaching more than 10 inches (25 centimeter) below the level of the walking/working surface on which they are working shall be protected from falling by a guardrail system, safety net system, or personal fall arrest system.

20.2.10 Roofing Work on Low-Slope Roofs

Each employee engaged in roofing activities on low-slope roofs, with unprotected sides and edges 6 feet (1.8 meter) or more above lower levels shall be protected from falling by guardrail systems, safety net systems, personal fall arrest systems, or a combination of warning line system and guardrail system, warning line system and safety net system, or warning line system and personal fall arrest system, or warning line system and safety monitoring system. Or, on roofs 50 feet (15.25 meter) or less in width, the use of a safety monitoring system alone (i.e., without the warning line system) is permitted.

20.2.11 Steep Roofs

Each employee on a steep roof with unprotected sides and edges 6 feet (1.8 meter) or more above lower levels shall be protected from falling by guardrail systems with toe boards, safety net systems, or personal fall arrest systems.

20.2.12 Precast Concrete Erection

Each employee engaged in the erection of precast concrete members (including, but not limited to, the erection of wall panels, columns, beams, and floor and roof "tees") and related operations such as grouting of precast concrete members, who is 6 feet (1.8 meter) or more above lower levels shall be protected from falling by guardrail systems, safety net systems, or personal fall arrest systems. When the employer can demonstrate that it is infeasible or creates a greater hazard to use these systems, the employer shall develop and implement a fall protection plan.

Note: There is a presumption that it is feasible and will not create a greater hazard to implement at least one of the above-listed fall protection systems.

20.2.13 Residential Construction

Each employee engaged in residential construction activities 6 feet (1.8 meter) or more above lower levels shall be protected by guardrail systems, safety net system, or personal fall arrest system. When the employer can demonstrate that it is infeasible or creates a greater hazard to use these systems, the employer shall develop and implement a fall protection plan.

20.2.14 Wall Openings

Each employee working on, at, above, or near wall openings (including those with chutes attached) where the outside bottom edge of the wall opening is 6 feet (1.8 meter) or more above lower levels and the inside bottom edge of the wall opening is less than 39 inches (1.0 meter) above the walking/working surface shall be protected from falling by the use of a guardrail system, a safety net system, or a personal fall arrest system.

20.2.15 Protection from Falling Objects

When an employee is exposed to falling objects, the employer shall have each employee wear a hard hat and shall implement one of the following measures:

- Erect toe boards, screens, or guardrail systems to prevent objects from falling from higher levels.

- Erect a canopy structure and keep potential fall objects far enough from the edge of the higher level so that those objects would not go over the edge if they were accidentally displaced.

- Barricade the area to which objects could fall, prohibit employees from entering the barricaded area, and keep objects that may fall far enough away from the edge of a higher level so that those objects would not go over the edge if they were accidentally displaced.

20.3 Fall Protection Systems Criteria and Practices

Employers shall provide and install all fall protection systems required by this subpart for an employee, and shall comply with all other pertinent requirements of this subpart before that employee begins the work that necessitates the fall protection. Some major safety regulations from Part 1926 Subpart M Standard 1926.502 are stated in the following text.

20.3.1 Guardrail Systems

Guardrail systems and their use shall comply with the following provisions:

- Top edge height of top rails, or equivalent guardrail system members, shall be 42 inches (1.1 meter) plus or minus 3 inches (8 centimeter) above the walking/ working level.

- Midrails, screens, mesh, intermediate vertical members, or equivalent intermediate structural members shall be installed between the top edge of the guardrail system and the walking/working surface when there is no wall or parapet wall at least 21 inches (53 centimeter) high.

- Midrails, when used, shall be installed at a height midway between the top edge of the guardrail system and the walking/working level.

- Screens and mesh, when used, shall extend from the top rail to the walking/ working level and along the entire opening between top rail supports.

- Intermediate members (such as balusters), when used between posts, shall be not more than 19 inches (48 centimeter) apart.

- Other structural members (such as additional midrails and architectural panels) shall be installed such that there are no openings in the guardrail system that are more than 19 inches (0.5 meter) wide.

- Guardrail systems shall be capable of withstanding, without failure, a force of at least 200 pounds (890 newton) applied within 2 inches (5.1 centimeter) of the top edge, in any outward or downward direction, at any point along the top edge.

- When the 200 pound (890 newton) test load is applied in a downward direction, the top edge of the guardrail shall not deflect to a height less than 39 inches (1.0 meter) above the walking/working level.

- Midrails, screens, mesh, intermediate vertical members, solid panels, and equivalent structural members shall be capable of withstanding, without failure, a force of at least 150 pounds (666 newton) applied in any downward or outward direction at any point along the midrail or other member.

- Guardrail systems shall be so surfaced as to prevent injury to an employee from punctures or lacerations, and to prevent snagging of clothing.

- The ends of all top rails and midrails shall not overhang the terminal posts, except where such overhang does not constitute a projection hazard.

- Steel banding and plastic banding shall not be used as top rails or midrails.

- Top rails and midrails shall be at least one-quarter inch (0.6 centimeter) nominal diameter or thickness to prevent cuts and lacerations. If wire rope is used for top rails, it shall be flagged at not more than 6-foot intervals with high-visibility material.

- When guardrail systems are used at hoisting areas, a chain, gate, or removable guardrail section shall be placed across the access opening between guardrail sections when hoisting operations are not taking place.

- When guardrail systems are used at holes, they shall be erected on all unprotected sides or edges of the hole.

- When guardrail systems are used around holes used for the passage of materials, the hole shall have not more than two sides provided with removable guardrail sections to allow the passage of materials. When the hole is not in use, it shall be closed over with a cover, or a guardrail system shall be provided along all unprotected sides or edges.

- When guardrail systems are used around holes which are used as points of access (such as ladderways), they shall be provided with a gate, or be so offset that a person cannot walk directly into the hole.

- Guardrail systems used on ramps and runways shall be erected along each unprotected side or edge.

Example 20.1
Top rails and midrails shall be at least _____ inch nominal diameter or thickness to prevent cuts and lacerations.

    A. One-quarter

    B. Half

    C. Three-quarter

    D. One

Answer: A

20.3.2 Safety Net Systems

Safety net systems and their use shall comply with the following provisions:

- Safety nets shall be installed as close as practicable under the walking/working surface on which employees are working, but in no case more than 30 feet (9.1 meter) below such level. When nets are used on bridges, the potential fall area from the walking/working surface to the net shall be unobstructed.

- Safety nets shall extend outward from the outermost projection of the work surface as shown in Table 20.1.

20.3.3 Personal Fall Arrest Systems

Personal fall arrest systems and their use shall comply with the following provisions:

- Connectors shall be drop forged, pressed or formed steel, or made of equivalent materials.

Vertical Distance from Working Level to Horizontal Plane of Net	Minimum Required Horizontal Distance of Outer Edge of Net from the Edge of the Working Surface
Up to 5 feet	8 feet
More than 5 feet up to 10 feet	10 feet
More than 10 feet	13 feet

TABLE 20.1 Distance Requirements of Safety Net

- Connectors shall have a corrosion-resistant finish, and all surfaces and edges shall be smooth to prevent damage to interfacing parts of the system.

- Dee-rings and snaphooks shall have a minimum tensile strength of 5000 pounds (22.2 kilonewton).

- Dee-rings and snaphooks shall be proof-tested to a minimum tensile load of 3600 pounds (16 kilonewton) without cracking, breaking, or taking permanent deformation.

- Snaphooks shall be sized to be compatible with the member to which they are connected to prevent unintentional disengagement of the snaphook by depression of the snaphook keeper by the connected member, or shall be a locking-type snaphook designed and used to prevent disengagement of the snaphook by the contact of the snaphook keeper by the connected member.

20.3.4 Positioning Device Systems

Positioning device systems and their use shall conform to the following provisions:

- Positioning devices shall be rigged such that an employee cannot free fall more than 2 feet (0.6 meter).

- Positioning devices shall be secured to an anchorage capable of supporting at least twice the potential impact load of an employee's fall or 3000 pounds (13.3 kilonewton), whichever is greater.

- Connectors shall be drop forged, pressed or formed steel, or made of equivalent materials.

- Connectors shall have a corrosion-resistant finish, and all surfaces and edges shall be smooth to prevent damage to interfacing parts of this system.

- Connecting assemblies shall have a minimum tensile strength of 5000 pounds (22.2 kilonewton).

- Dee-rings and snaphooks shall be proof-tested to a minimum tensile load of 3600 pounds (16 kilonewton) without cracking, breaking, or taking permanent deformation.

20.3.5 Warning Line Systems

Warning line systems and their use shall comply with the following provisions:

- The warning line shall be erected around all sides of the roof work area.

- When mechanical equipment is not being used, the warning line shall be erected not less than 6 feet (1.8 meter) from the roof edge.

- When mechanical equipment is being used, the warning line shall be erected not less than 6 feet (1.8 meter) from the roof edge which is parallel to the direction of mechanical equipment operation, and not less than 10 feet (3.1 meter) from the roof edge which is perpendicular to the direction of mechanical equipment operation.

- Points of access, materials handling areas, storage areas, and hoisting areas shall be connected to the work area by an access path formed by two warning lines.

- When the path to a point of access is not in use, a rope, wire, chain, or other barricade, equivalent in strength and height to the warning line shall be placed across the path at the point where the path intersects the warning line erected around the work area, or the path shall be offset such that a person cannot walk directly into the work area.

Warning lines shall consist of ropes, wires, or chains, and supporting stanchions erected as follows:

- The rope, wire, or chain shall be flagged at not more than 6-foot (1.8 meter) intervals with high-visibility material.
- The rope, wire, or chain shall be rigged and supported in such a way that its lowest point (including sag) is no less than 34 inches (0.9 meter) from the walking/working surface and its highest point is no more than 39 inches (1.0 meter) from the walking/working surface.
- After being erected, with the rope, wire, or chain attached, stanchions shall be capable of resisting, without tipping over, a force of at least 16 pounds (71 newton) applied horizontally against the stanchion, 30 inches (0.8 meter) above the walking/working surface, perpendicular to the warning line, and in the direction of the floor, roof, or platform edge.
- The rope, wire, or chain shall have a minimum tensile strength of 500 pounds (2.22 kilonewton), and after being attached to the stanchions, shall be capable of supporting, without breaking.
- The line shall be attached at each stanchion in such a way that pulling on one section of the line between stanchions will not result in slack being taken up in adjacent sections before the stanchion tips over.
- No employee shall be allowed in the area between a roof edge and a warning line unless the employee is performing roofing work in that area.
- Mechanical equipment on roofs shall be used or stored only in areas where employees are protected by a warning line system, guardrail system, or personal fall arrest system.

Example 20.2
When mechanical equipment is not being used, the warning line shall be erected not less than _____ feet from the roof edge.

    A. 3

    B. 6

    C. 9

    D. 12

Answer: B

20.3.6 Controlled Access Zones

Controlled access zones and their use shall conform to the following provisions:

- When used to control access to areas where leading edge and other operations are taking place, the controlled access zone shall be defined by a control line or by any other means that restricts access.

- When control lines are used, they shall be erected not less than 6 feet (1.8 meter) nor more than 25 feet (7.7 meter) from the unprotected or leading edge, except when erecting precast concrete members.

- When erecting precast concrete members, the control line shall be erected not less than 6 feet (1.8 meter) nor more than 60 feet (18 meter) or half the length of the member being erected, whichever is less, from the leading edge.

- The control line shall extend along the entire length of the unprotected or leading edge and shall be approximately parallel to the unprotected or leading edge.

- The control line shall be connected on each side to a guardrail system or wall.

- When used to control access to areas where overhand bricklaying and related work are taking place:

  - The controlled access zone shall be defined by a control line erected not less than 10 feet (3.1 meter) nor more than 15 feet (4.5 meter) from the working edge.

  - The control line shall extend for a distance sufficient for the controlled access zone to enclose all employees performing overhand bricklaying and related work at the working edge and shall be approximately parallel to the working edge.

  - Additional control lines shall be erected at each end to enclose the controlled access zone. Only employees engaged in overhand bricklaying or related work shall be permitted in the controlled access zone.

- Control lines shall consist of ropes, wires, tapes, or equivalent materials, and supporting stanchions as follows:

  - Each line shall be flagged or otherwise clearly marked at not more than 6-foot (1.8 meter) intervals with high-visibility material.

  - Each line shall be rigged and supported in such a way that its lowest point (including sag) is not less than 39 inches (1 meter) from the walking/working surface and its highest point is not more than 45 inches (1.3 meter) (50 inches [1.3 meter] when overhand bricklaying operations are being performed) from the walking/working surface.

  - Each line shall have a minimum breaking strength of 200 pounds (0.88 kilonewton).

- On floors and roofs where guardrail systems are not in place prior to the beginning of overhand bricklaying operations, controlled access zones shall be enlarged, as necessary, to enclose all points of access, material handling areas, and storage areas.

- On floors and roofs where guardrail systems are in place, but need to be removed to allow overhand bricklaying work or leading edge work to take place, only that portion of the guardrail necessary to accomplish that day's work shall be removed.

20.3.7 Safety Monitoring Systems

Safety monitoring systems and their use shall comply with the following provisions:

- The employer shall designate a competent person to monitor the safety of other employees and the employer shall ensure that the safety monitor complies with the following requirements:
  - The safety monitor shall be competent to recognize fall hazards.
  - The safety monitor shall warn the employee when it appears that the employee is unaware of a fall hazard or is acting in an unsafe manner.
  - The safety monitor shall be on the same walking/working surface and within visual sighting distance of the employee being monitored.
  - The safety monitor shall be close enough to communicate orally with the employee.
  - The safety monitor shall not have other responsibilities which could take the monitor's attention from the monitoring function.
- Mechanical equipment shall not be used or stored in areas where safety monitoring systems are being used to monitor employees engaged in roofing operations on low-slope roofs.
- No employee, other than an employee engaged in roofing work (on low-sloped roofs) or an employee covered by a fall protection plan, shall be allowed in an area where an employee is being protected by a safety monitoring system.
- Each employee working in a controlled access zone shall be directed to comply promptly with fall hazard warnings from safety monitors.

20.3.8 Covers

Covers for holes in floors, roofs, and other walking/working surfaces shall meet the following requirements:

- Covers located in roadways and vehicular aisles shall be capable of supporting, without failure, at least twice the maximum axle load of the largest vehicle expected to cross over the cover.
- All other covers shall be capable of supporting, without failure, at least twice the weight of employees, equipment, and materials that may be imposed on the cover at any one time.
- All covers shall be secured when installed so as to prevent accidental displacement by the wind, equipment, or employees.
- All covers shall be color coded or they shall be marked with the word "HOLE" or "COVER" to provide warning of the hazard.

20.3.9 Protection from Falling Objects

Falling object protection shall comply with the following provisions:

- Toe boards, when used as falling object protection, shall be erected along the edge of the overhead walking/working surface for a distance sufficient to protect employees below.
- Toe boards shall be capable of withstanding, without failure, a force of at least 50 pounds (222 newton) applied in any downward or outward direction at any point along the toe board.
- Toe boards shall be a minimum of 3 1/2 inches (9 centimeter) in vertical height from their top edge to the level of the walking/working surface. They shall have not more than 1/4 inch (0.6 centimeter) clearance above the walking/working surface. They shall be solid or have openings not over 1 inch (2.5 centimeter) in greatest dimension.
- Where tools, equipment, or materials are piled higher than the top edge of a toe board, paneling or screening shall be erected from the walking/working surface or toe board to the top of a guardrail system's top rail or midrail, for a distance sufficient to protect employees below.
- Guardrail systems, when used as falling object protection, shall have all openings small enough to prevent passage of potential falling objects.

During the performance of overhand bricklaying and related work:

- No materials or equipment except masonry and mortar shall be stored within 4 feet (1.2 meter) of the working edge.
- Excess mortar, broken or scattered masonry units, and all other materials and debris shall be kept clear from the work area by removal at regular intervals.

During the performance of roofing work:

- Materials and equipment shall not be stored within 6 feet (1.8 meter) of a roof edge unless guardrails are erected at the edge.
- Materials which are piled, grouped, or stacked near a roof edge shall be stable and self-supporting.
- Canopies, when used as falling object protection, shall be strong enough to prevent collapse and to prevent penetration by any objects which may fall onto the canopy.

20.4 Fall Protection Plan

The fall protection plan option is available only to employees engaged in leading edge work, precast concrete erection work, or residential construction work who can demonstrate that it is unfeasible or it creates a greater hazard to use conventional fall protection equipment. If used, the plan should be strictly enforced.

- A fall protection plan must be prepared by a qualified person and developed specifically for each site.

- The fall protection plan must be maintained up to date.
- Any changes to the plan must be approved by a qualified person.
- A copy of the plan with all approved changes must be maintained at the site.
- The fall protection plan shall document the reasons why the use of conventional fall protection systems (guardrail systems, personal fall arrest systems, or safety nets systems) is infeasible or why their use would create a greater hazard.

A fall protection plan must consist of the following elements:

- Statement of policy
- Fall protection systems to be used
- Implementation of plan
- Enforcement
- Accident investigation
- Changes to the plan

This option is available only to employees engaged in leading edge work, precast concrete erection work, or residential construction work who can demonstrate that it is infeasible or it creates a greater hazard to use conventional fall protection equipment. The fall protection plan must conform to the following provisions:

- The fall protection plan shall be prepared by a qualified person and developed specifically for the site where the leading edge work, precast concrete work, or residential construction work is being performed and the plan must be maintained up to date.
- Any changes to the fall protection plan shall be approved by a qualified person.
- A copy of the fall protection plan with all approved changes shall be maintained at the jobsite.
- The implementation of the fall protection plan shall be under the supervision of a competent person.
- The fall protection plan shall document the reasons why the use of conventional fall protection systems (guardrail systems, personal fall arrest systems, or safety nets systems) is infeasible or why their use would create a greater hazard.
- The fall protection plan shall include a written discussion of other measures that will be taken to reduce or eliminate the fall hazard for workers who cannot be provided with protection from the conventional fall protection systems. For example, the employer shall discuss the extent to which scaffolds, ladders, or vehicle-mounted work platforms can be used to provide a safer working surface and thereby reduce the hazard of falling.
- The fall protection plan shall identify each location where conventional fall protection methods cannot be used.
- Where no other alternative measure has been implemented, the employer shall implement a safety monitoring system.

- The fall protection plan must include a statement which provides the name or other method of identification for each employee who is designated to work in controlled access zones. No other employee may enter controlled access zones.

- In the event an employee falls, or some other related, serious incident occurs (e.g., a near miss), the employer shall investigate the circumstances of the fall or other incident to determine if the fall protection plan needs to be changed (e.g., new practices, procedures, or training) and shall implement those changes to prevent similar types of falls or incidents.

Example 20.3
A fall protection plan must be prepared by a qualified person and developed specifically for each site and any changes to the plan must be approved by a:

A. Qualified person

B. Competent person

C. Licensed engineer

D. Employer

Answer: *A*

20.5   Training Requirements

The employer shall provide a training program for each employee who might be exposed to fall hazards. The program shall enable each employee to recognize the hazards of falling and shall train each employee in the procedures to be followed in order to minimize these hazards. According to the Part 1926 Subpart M Standard 1926.503, the employer shall assure that each employee has been trained, as necessary, by a competent person qualified in the following areas:

- The nature of fall hazards in the work area

- The correct procedures for erecting, maintaining, disassembling, and inspecting the fall protection systems to be used

- The use and operation of guardrail systems, personal fall arrest systems, safety net systems, warning line systems, safety monitoring systems, controlled access zones, and other protection to be used

- The role of each employee in the safety monitoring system when this system is used

- The limitations on the use of mechanical equipment during the performance of roofing work on low-sloped roofs

- The correct procedures for the handling and storage of equipment and materials and the erection of overhead protection

- The role of employees in fall protection plans

20.6 Summary

The importance of fall protection is very high in construction. Each employee on a walking/working surface (horizontal and vertical surface) with an unprotected side or edge, leading edge or hoist area or formwork, etc., which is 6 feet or more above a lower level shall be protected from falling by the use of guardrail systems, safety net systems, or personal fall arrest systems. This is true for overhand bricklaying, roofing work, steep roofs, and similar works. Employers shall provide and install all fall protection systems required by this subpart for an employee, and shall comply with all other pertinent requirements of this subpart before that employee begins the work that necessitates the fall protection. A Fall Protection Plan must be prepared by a qualified person and developed specifically for each site. The fall protection plan shall document the reasons why the use of conventional fall protection systems (guardrail systems, personal fall arrest systems, or safety nets systems) is infeasible or why their use would create a greater hazard.

20.7 Multiple-Choice Questions

20.1 Each employee on a walking/working surface with an unprotected side which is _____ feet or more above a lower level shall be protected from falling.

   A. 2
   B. 4
   C. 6
   D. 8

20.2 Each employee who is constructing a leading edge _____ feet or more above lower levels shall be protected from falling.

   A. 8
   B. 6
   C. 4
   D. 2

20.3 Each employee on the face of formwork or reinforcing steel shall be protected from falling _____ feet or more to lower levels by personal fall arrest systems, safety net systems, or positioning device systems.

   A. 2
   B. 4
   C. 6
   D. 8

20.4 Each employee reaching more than _____ inches below the level of the walking/working surface on which they are working shall be protected from falling.

   A. 6
   B. 10
   C. 14
   D. 18

20.5 On roofs _____ feet or less in width, the use of a safety monitoring system alone (i.e., without the warning line system) is permitted.

A. 30
B. 40
C. 50
D. 60

20.6 Each employee on a steep roof with unprotected sides and edges _____ feet or more above lower levels shall be protected from falling.

A. 1.5
B. 3.0
C. 4.5
D. 6.0

20.7 Each employee engaged in residential construction activities _____ feet or more above lower levels shall be protected by guardrail.

A. 1.5
B. 3.0
C. 4.5
D. 6.0

20.8 Top edge height of top rails, or equivalent guardrail system members, shall be _____ inches plus or minus _____ inches above the walking/working level.

A. 42, 3 respectively
B. 24, 3 respectively
C. 36, 3 respectively
D. 42, 6 respectively

20.9 Intermediate members (such as balusters) used in guardrails, when used between posts, shall be not more than:

A. 12 inches apart
B. 19 inches apart
C. 14 inches apart
D. 30 inches apart

20.8 Practice Problems

20.1 Mention the safety guidelines of fall protection for unprotected sides and edges.

20.2 Mention the safety guidelines of fall protection for leading edges.

20.3 Mention the safety guidelines of fall protection for hoist areas.

20.4 Mention the safety guidelines of fall protection for excavations areas.

20.5 Mention the safety guidelines of fall protection for roofing work on low-slope roofs.

20.6 Mention the safety guidelines of fall protection for steep roofs.

20.7 Mention the safety guidelines of fall protection for precast concrete erection.

20.8 Mention the safety guidelines of fall protection for protection from falling objects.

20.9   Mention the design criteria for the guardrail systems.

20.10   Mention the design criteria for safety net systems.

20.11   Mention the design criteria for personal fall arrest systems.

20.12   Mention the design criteria for warning line systems.

20.13   Mention the design criteria for controlled access zones.

20.14   Mention the safety guidelines for safety monitoring systems.

20.15   List the elements of fall protection plan.

20.16   List the qualifications of the trainer to train for fall protection.

20.9   Critical Thinking and Discussion Topic

The fall protection is very often too costly and ridiculous while working. Buying life-lines for all employees is also a lot of investment with no solid returns. Even, the cost of lifelines may be more than the cost of accident. Therefore, OSHA's standard for lifelines is too much for a construction company. What is your opinion about it?

CHAPTER 21

Helicopters, Hoists, Elevators, and Conveyors

(Photo courtesy of Daria Sannikova from Pexels.)

21.1 General

The construction and manufacturing industries require many types of specialized equipment, including cranes, derricks, hoists, elevators, and conveyors. These types of equipment are used to move large and heavy loads, providing a critical link between construction design and project management. While each of these has its own specific purpose, each also comes with its own set of safety risks and potential hazards.

21.2 Helicopters

Some major specifications by the OSHA Part 1926, Subpart N, Standard 1926.551 are as follows:

Helicopter regulations. Helicopter cranes shall be expected to comply with any applicable regulations of the Federal Aviation Administration.

Briefing. Prior to each day's operation a briefing shall be conducted. This briefing shall set forth the plan of operation for the pilot and ground personnel.

Slings and tag lines. Load shall be properly slung. Tag lines shall be of a length that will not permit their being drawn up into rotors. Pressed sleeve, swedged eyes, or equivalent means shall be used for all freely suspended loads to prevent hand splices from spinning open or cable clamps from loosening.

Cargo hooks. All electrically operated cargo hooks shall have the electrical activating device so designed and installed as to prevent inadvertent operation. In addition, these cargo hooks shall be equipped with an emergency mechanical control for releasing the load. The hooks shall be tested prior to each day's operation to determine that the release functions properly, both electrically and mechanically.

Personal protective equipment. Personal protective equipment for employees receiving the load shall consist of complete eye protection and hard hats secured by chinstraps. Loose-fitting clothing likely to flap in the downwash, and thus be snagged on hoist line, shall not be worn.

Loose gear and objects. Every practical precaution shall be taken to provide for the protection of the employees from flying objects in the rotor downwash. All loose gear within 100 feet of the place of lifting the load, depositing the load, and all other areas susceptible to rotor downwash shall be secured or removed.

Housekeeping. Good housekeeping shall be maintained in all helicopter loading and unloading areas.

Operator responsibility. The helicopter operator shall be responsible for size, weight, and manner in which loads are connected to the helicopter. If, for any reason, the helicopter operator believes the lift cannot be made safely, the lift shall not be made.

Hooking and unhooking loads. When employees are required to perform work under hovering craft, a safe means of access shall be provided for employees to reach the hoist line hook and engage or disengage cargo slings. Employees shall not perform work under hovering craft except when necessary to hook or unhook loads.

Static charge.   Static charge on the suspended load shall be dissipated with a grounding device before ground personnel touch the suspended load, or protective rubber gloves shall be worn by all ground personnel touching the suspended load.

Weight limitation.   The weight of an external load shall not exceed the manufacturer's rating.

Ground lines.   Hoist wires or other gear, except for pulling lines or conductors that are allowed to "pay out" from a container or roll off a reel, shall not be attached to any fixed ground structure, or allowed to foul on any fixed structure.

Visibility.   When visibility is reduced by dust or other conditions, ground personnel shall exercise special caution to keep clear of main and stabilizing rotors. Precautions shall also be taken by the employer to eliminate as far as practical reduced visibility.

Approach distance.   No unauthorized person shall be allowed to approach within 50 feet of the helicopter when the rotor blades are turning.

Approaching helicopter.   Whenever approaching or leaving a helicopter with blades rotating, all employees shall remain in full view of the pilot and keep in a crouched position. Employees shall avoid the area from the cockpit or cabin rearward unless authorized by the helicopter operator to work there.

Personnel.   Sufficient ground personnel shall be provided when required for safe helicopter loading and unloading operations.

Communications.   There shall be constant reliable communication between the pilot and a designated employee of the ground crew who acts as a signalman during the period of loading and unloading. This signalman shall be distinctly recognizable from other ground personnel.

Fires.   Open fires shall not be permitted in an area that could result in such fires being spread by the rotor downwash.

21.3   Material and Personnel Hoists

The employer shall comply with the manufacturer's specifications and limitations applicable to the operation of all hoists and elevators. Where manufacturer's specifications are not available, the limitations assigned to the equipment shall be based on the determinations of a professional engineer competent in the field. Rated load capacities, recommended operating speeds, and special hazard warnings or instructions shall be posted on cars and platforms.

21.3.1   Material Hoists

Some major specifications by the OSHA Part 1926, Subpart N, Standard 1926.552 are as follows:

- Operating rules shall be established and posted at the operator's station of the hoist. Such rules shall include signal system and allowable line speed for various loads. Rules and notices shall be posted on the car frame or crosshead in a conspicuous location, including the statement "No Riders Allowed."

- No person shall be allowed to ride on material hoists except for the purposes of inspection and maintenance.

- All entrances of the hoistways shall be protected by substantial gates or bars which shall guard the full width of the landing entrance. All hoistway entrance bars and gates shall be painted with diagonal contrasting colors, such as black and yellow stripes.

- Bars shall be not less than 2- by 4-inch wooden bars or the equivalent, located 2 feet from the hoistway line. Bars shall be located not less than 36 inches or more than 42 inches above the floor.

21.3.2   Personnel Hoists

Some major specifications by the OSHA Part 1926, Subpart N, Standard 1926.552 are as follows:

- Hoist towers outside the structure shall be enclosed for the full height on the side or sides used for entrance and exit to the structure. At the lowest landing, the enclosure on the sides not used for exit or entrance to the structure shall be enclosed to a height of at least 10 feet. Other sides of the tower adjacent to floors or scaffold platforms shall be enclosed to a height of 10 feet above the level of such floors or scaffolds.

- Towers inside of structures shall be enclosed on all four sides throughout the full height.

- Towers shall be anchored to the structure at intervals not exceeding 25 feet. In addition to tie-ins, a series of guys shall be installed. Where tie-ins are not practical the tower shall be anchored by means of guys made of wire rope at least one-half inch in diameter, securely fastened to anchorage to ensure stability.

- Hoistway doors or gates shall be not less than 6 feet 6 inches high and shall be provided with mechanical locks which cannot be operated from the landing side, and shall be accessible only to persons on the car (Fig. 21.1).

Example 21.1
For personnel hoists, towers inside of structures shall be enclosed on all _____ sides throughout the full height.

    A. 2

    B. 3

    C. 4

    D. No need to enclose

Answer: C

Figure 21.1   A personal hoist for construction works.
(Photo courtesy of ANSI.)

21.4   Base-Mounted Drum Hoists

Some major specifications by the OSHA Part 1926, Subpart N, Standard 1926.553 are as follows:

- Exposed moving parts such as gears, projecting screws, setscrews, chain, cables, chain sprockets, and reciprocating or rotating parts, which constitute a hazard, shall be guarded.
- All controls used during the normal operation cycle shall be located within easy reach of the operator's station.
- Electric motor operated hoists shall be provided with:
  - A device to disconnect all motors from the line upon power failure and not permit any motor to be restarted until the controller handle is brought to the "off" position.
  - Where applicable, an overspeed preventive device.
  - A means whereby remotely operated hoists stop when any control is ineffective.
- All base-mounted drum hoists in use shall meet the applicable requirements for design, construction, installation, testing, inspection, maintenance, and operations, as prescribed by the manufacturer.

21.5   Overhead Hoists

Some major specifications by the OSHA Part 1926, Subpart N, Standard 1926.554 are as follows:

- The safe working load of the overhead hoist, as determined by the manufacturer, shall be indicated on the hoist, and this safe working load shall not be exceeded.

- The supporting structure to which the hoist is attached shall have a safe working load equal to that of the hoist.

- The support shall be arranged so as to provide for free movement of the hoist and shall not restrict the hoist from lining itself up with the load.

- The hoist shall be installed only in locations that will permit the operator to stand clear of the load at all times.

- Air hoists shall be connected to an air supply of sufficient capacity and pressure to safely operate the hoist. All air hoses supplying air shall be positively connected to prevent their becoming disconnected during use.

- All overhead hoists in use shall meet the applicable requirements for construction, design, installation, testing, inspection, maintenance, and operation, as prescribed by the manufacturer.

21.6   Conveyors

Some major specifications by the OSHA Part 1926, Subpart N, Standard 1926.555 are as follows:

- Means for stopping the motor or engine shall be provided at the operator's station. Conveyor systems shall be equipped with an audible warning signal to be sounded immediately before starting up the conveyor.

- If the operator's station is at a remote point, similar provisions for stopping the motor or engine shall be provided at the motor or engine location.

- Emergency stop switches shall be arranged so that the conveyor cannot be started again until the actuating stop switch has been reset to running or "on" position.

- Screw conveyors shall be guarded to prevent employee contact with turning flights.

- Where a conveyor passes over work areas, aisles, or thoroughfares, suitable guards shall be provided to protect employees required to work below the conveyors.

- All crossovers, aisles, and passageways shall be conspicuously marked by suitable signs, as required by Subpart G of Part 1926.

- Conveyors shall be locked out or otherwise rendered inoperable, and tagged out with a "Do Not Operate" tag during repairs and when operation is hazardous to employees performing maintenance work (Fig. 21.2).

Figure 21.2   A stone handling conveyor.
(Photo courtesy of H. S. Engineers.)

21.7   Summary

Many types of specialized equipment, including cranes, derricks, hoists, elevators, and conveyors, are very common in construction industry. The employer should comply with the manufacturer's specifications and limitations applicable to these equipment. Where manufacturer's specifications are not available, the limitations assigned to the equipment shall be based on the determinations of a professional engineer competent in the field. The safe working load of the overhead hoist, as determined by the manufacturer, shall be indicated on the hoist, and this safe working load shall not be exceeded.

21.8   Multiple-Choice Questions

21.1   Helicopter cranes shall be expected to comply with any applicable regulations of the:
   A.   Federal Aviation Administration
   B.   AASHTO
   C.   OSHA only
   D.   Manual on Uniform Traffic Control Devices

21.2 All electrically operated cargo hooks in helicopters shall have the electrical activating device so designed and installed as to prevent:

A. Inadvertent operation
B. Falling apart
C. Electric shock
D. Mechanical failure

21.3 While operating helicopters, personal protective equipment for employees receiving the load shall consist of complete eye protection and hard hats secured by:

A. Chinstraps
B. Nylon cable
C. Galvanized iron wire
D. Synthetic cable

21.4 Good housekeeping shall be maintained in all helicopter:

A. Loading areas
B. Unloading areas
C. Loading and unloading areas
D. All over the helicopter areas

21.5 No unauthorized person shall be allowed to approach within _____ feet of the helicopter when the rotor blades are turning.

A. 20
B. 30
C. 40
D. 50

21.6 For material hoists, bars shall be not less than _____ wooden bars or the equivalent, located 2 feet from the hoistway line.

A. 2- by 2-inch
B. 2- by 4-inch
C. 2- by 6-inch
D. 4- by 4-inch

21.7 For material hoists, bars shall be located not less than _____ inches nor more than _____ inches above the floor.

A. 36, 42 respectively
B. 24, 42 respectively
C. 42, 42 respectively
D. 36, 48 respectively

21.8 For personnel hoists, hoistway doors or gates shall be not less than:

A. 3.5 feet high
B. 5.5 feet high
C. 6.5 feet high
D. 8.5 feet high

21.9 For personnel hoists, towers shall be anchored to the structure at intervals not exceeding:

A. 10 feet
B. 15 feet
C. 25 feet
D. 35 feet

21.9   Practice Problems

21.1   List some safety regulations for helicopters used in construction.

21.2   List some safety regulations for material hoists used in construction.

21.3   List some safety regulations for personnel hoists used in construction.

21.4   List some safety regulations for base-mounted drum hoists used in construction.

21.5   List some safety regulations for overhead hoists used in construction.

21.6   List some safety regulations for conveyors used in construction.

21.10   Critical Thinking and Discussion Topic

The new forklift operator at your jobsite concerns you. He has proper training but he looks very clumsy to you. You called your boss over phone and got a reply that he did pretty well in his training and he will do well here as well. You were assured of his work quality as well. As a site supervisor, what will you do in this situation?

(Photo courtesy of Tom Fisk from Pexels.)

CHAPTER 22

Mechanized Equipment, Motor Vehicles, and Marine Operations

(Photo courtesy by Weir ESCO from Pexels.)

22.1 General

The majority of fatalities that occur in road construction work zones in the United States involve a worker being struck by a piece of construction equipment or other vehicle. A worker in this industry is just as likely to be struck by a piece of construction equipment inside the work zone as by passing traffic. Highway and street construction workers are at risk of fatal and serious nonfatal injury when working in the vicinity of passing motorists, construction vehicles, and equipment. According to U.S. Department of Health and Human Services (DHHS), National Institute of Occupational Safety and Health (NIOSH) Publication No. 2001-128, each year more than 100 workers are killed and over 20,000 are injured in the highway and street construction industry. Vehicles and equipment operating in and around the work zone are involved in over half of the worker fatalities in the construction industry.

A motor vehicle, also known as motorized vehicle or automotive vehicle, is a self-propelled vehicle, commonly wheeled, that does not operate on rails and is used for the transportation of people or cargo (Figure 22.1). Mechanical equipment means equipment, devices, and accessories, the use of which relates to water supply, drainage, heating, ventilating, air conditioning, and similar purposes. All vehicles must be checked at the start of each shift to ensure that parts, equipment, and accessories are in safe operating condition and are free of apparent damage that could cause failure while in use. These components include:

- Service brakes, including trailer brake connections
- Parking system (hand brakes)

FIGURE 22.1 A motor vehicle and a construction equipment.
(Photo courtesy of Ivan from Pexels.)

- Emergency stopping system (brakes)
- Tires
- Horn
- Steering mechanism
- Coupling devices
- Seat belts
- Operating controls
- Safety devices

All defects must be corrected before the vehicles are placed in service.

22.2 Safety Regulations for Mechanized Equipment

Some major specifications by the Part 1926 Subpart O Standard 1926.600 are as follows:

- All equipment left unattended at night, adjacent to a highway in normal use, or adjacent to construction areas where work is in progress, shall have appropriate lights or reflectors, or barricades equipped with appropriate lights or reflectors, to identify the location of the equipment.
- A safety tire rack, cage, or equivalent protection shall be provided and used when inflating, mounting, or dismounting tires installed on split rims, or rims equipped with locking rings or similar devices.
- Heavy machinery, equipment, or parts thereof, which are suspended or held aloft by use of slings, hoists, or jacks shall be substantially blocked or cribbed to prevent falling or shifting before employees are permitted to work under or between them. Bulldozer and scraper blades, end-loader buckets, dump bodies, and similar equipment shall be either fully lowered or blocked when being repaired or when not in use. All controls shall be in a neutral position, with the motors stopped and brakes set, unless work being performed requires otherwise.
- Whenever the equipment is parked, the parking brake shall be set. Equipment parked on inclines shall have the wheels chocked and the parking brake set.
- All cab glass shall be safety glass, or equivalent, that introduces no visible distortion affecting the safe operation of any machine covered by this subpart.
- All equipment covered by this subpart shall comply with the following requirements when working or being moved in the vicinity of power lines or energized transmitters, except where electrical distribution and transmission lines have been de-energized and visibly grounded at point of work or where insulating barriers, not a part of or an attachment to the equipment or machinery, have been erected to prevent physical contact with the lines:
  - For lines rated 50 kilovolt or below, minimum clearance between the lines and any part of the crane or load shall be 10 feet.
  - For lines rated over 50 kilovolt, minimum clearance between the lines and any part of the crane or load shall be 10 feet plus 0.4 inch for each 1 kilovolt over 50 kilovolt, or twice the length of the line insulator, but never less than 10 feet.

- In transit with no load and boom lowered, the equipment clearance shall be a minimum of 4 feet for voltages less than 50 kilovolt, and 10 feet for voltages over 50 kilovolt, up to and including 345 kilovolt, and 16 feet for voltages up to and including 750 kilovolt.

- A person shall be designated to observe clearance of the equipment and give timely warning for all operations where it is difficult for the operator to maintain the desired clearance by visual means.

- Cage-type boom guards, insulating links, or proximity warning devices may be used on cranes, but the use of such devices shall not alter the requirements of any other regulation of this part even if such device is required by law or regulation.

- Any overhead wire shall be considered to be an energized line unless and until the person owning such line or the electrical utility authorities indicate that it is not an energized line and it has been visibly grounded.

Example 22.1

All equipment left unattended at night, adjacent to a highway in normal use, or adjacent to construction areas where work is in progress, shall have appropriate:

A. Lights

B. Reflectors

C. Barricades equipped with appropriate lights

D. Barricades equipped with appropriate reflectors

E. All of the above four items

F. Any one of the above four items

Answer: F

22.3   Safety Regulations for Motor Vehicles

Some major specifications by the Part 1926 Subpart O Standard 1926.601 are as follows:

- All vehicles shall have a service brake system, an emergency brake system, and a parking brake system. These systems may use common components, and shall be maintained in operable condition.

- Whenever visibility conditions warrant additional light, all vehicles, or combinations of vehicles, in use shall be equipped with at least two headlights and two taillights in operable condition.

- All vehicles, or combination of vehicles, shall have brake lights in operable condition regardless of light conditions.

- All vehicles shall be equipped with an adequate audible warning device at the operator's station and in an operable condition.

- No employer shall use any motor vehicle equipment having an obstructed view to the rear unless:

  - The vehicle has a reverse signal alarm audible above the surrounding noise level.

  - The vehicle is backed up only when an observer signals that it is safe to do so.

- All vehicles with cabs shall be equipped with windshields and powered wipers. Cracked and broken glass shall be replaced. Vehicles operating in areas or under conditions that cause fogging or frosting of the windshields shall be equipped with operable defogging or defrosting devices.

- All haulage vehicles, whose pay load is loaded by means of cranes, power shovels, loaders, or similar equipment, shall have a cab shield and/or canopy adequate to protect the operator from shifting or falling materials.

- Tools and material shall be secured to prevent movement when transported in the same compartment with employees.

- Vehicles used to transport employees shall have seats firmly secured and adequate for the number of employees to be carried.

- Trucks with dump bodies shall be equipped with positive means of support, permanently attached, and capable of being locked in position to prevent accidental lowering of the body while maintenance or inspection work is being done.

- Operating levers controlling hoisting or dumping devices on haulage bodies shall be equipped with a latch or other device which will prevent accidental starting or tripping of the mechanism.

- Trip handles for tailgates of dump trucks shall be so arranged that, in dumping, the operator will be in the clear.

- Mud flaps may be used in lieu of fenders whenever motor vehicle equipment is not designed for fenders.

- All vehicles in use shall be checked at the beginning of each shift to assure that the following parts, equipment, and accessories are in safe operating condition and free of apparent damage that could cause failure while in use: service brakes, including trailer brake connections; parking system (hand brake); emergency stopping system (brakes); tires; horn; steering mechanism; coupling devices; seat belts; operating controls; and safety devices. All defects shall be corrected before the vehicle is placed in service. These requirements also apply to equipment such as lights, reflectors, windshield wipers, defrosters, fire extinguishers, etc., where such equipment is necessary.

Example 22.2
Equipped with an adequate audible warning device at the operator's station in a vehicle is:

  A. A must

  B. Optional but recommended

  C. Optional and rewardable

  D. No such specification exists

Answer: *A*
All vehicles shall be equipped with an adequate audible warning device at the operator's station and in an operable condition.

22.4 Safety Regulations for Material Handling Equipment

Some major specifications by the Part 1926 Subpart O Standard 1926.602 are provided in the following text.

22.4.1 Earthmoving Equipment

Seat belts. Seat belts shall be provided on all equipment covered by this section and shall meet the requirements of the Society of Automotive Engineers, J386-1969, Seat Belts for Construction Equipment. Seat belts for agricultural and light industrial tractors shall meet the seat belt requirements of Society of Automotive Engineers J333a-1970, Operator Protection for Agricultural and Light Industrial Tractors. Seat belts need not be provided for equipment which is designed only for standup operation. Seat belts need not be provided for equipment which does not have roll-over protective structure (ROPS) or adequate canopy protection.

Access roadways and grades. No employer shall move or cause to be moved construction equipment or vehicles upon any access roadway or grade unless the access roadway or grade is constructed and maintained to accommodate safely the movement of the equipment and vehicles involved. Every emergency access ramp and berm used by an employer shall be constructed to restrain and control runaway vehicles.

Fenders. Pneumatic-tired earth-moving haulage equipment (trucks, scrapers, tractors, and trailing units) whose maximum speed exceeds 15 miles per hour shall be equipped with fenders on all wheels. An employer may, of course, at any time seek to show that the uncovered wheels present no hazard to personnel from flying materials.

Audible alarms. All bidirectional machines, such as rollers, compacters, front-end loaders, bulldozers, and similar equipment, shall be equipped with a horn, distinguishable from the surrounding noise level, which shall be operated as needed when the machine is moving in either direction. The horn shall be maintained in an operative condition. No employer shall permit earthmoving or compacting equipment which has an obstructed view to the rear to be used in reverse gear unless the equipment has in operation a reverse signal alarm distinguishable from the surrounding noise level or an employee signals that it is safe to do so.

Scissor points. Scissor points on all front-end loaders, which constitute a hazard to the operator during normal operation, shall be guarded.

22.4.2 Excavating and Other Equipment

Tractors shall have seat belts as required for the operators when seated in the normal seating arrangement for tractor operation, even though back-hoes, breakers, or other similar attachments are used on these machines for excavating or other work.

22.4.3 Lifting and Hauling Equipment

Industrial trucks shall meet the requirements of 1926.600 and the following:

- Lift trucks, stackers, etc., shall have the rated capacity clearly posted on the vehicle so as to be clearly visible to the operator. When auxiliary removable counterweights are provided by the manufacturer, corresponding alternate rated capacities also shall be clearly shown on the vehicle. These ratings shall not be exceeded.

- No modifications or additions which affect the capacity or safe operation of the equipment shall be made without the manufacturer's written approval. If such modifications or changes are made, the capacity, operation, and maintenance instruction plates, tags, or decals shall be changed accordingly. In no case shall the original safety factor of the equipment be reduced.

- If a load is lifted by two or more trucks working in unison, the proportion of the total load carried by any one truck shall not exceed its capacity.

- Steering or spinner knobs shall not be attached to the steering wheel unless the steering mechanism is of a type that prevents road reactions from causing the steering handwheel to spin. The steering knob shall be mounted within the periphery of the wheel.

- Unauthorized personnel shall not be permitted to ride on powered industrial trucks. A safe place to ride shall be provided where riding of trucks is authorized.

- Whenever a truck is equipped with vertical only, or vertical and horizontal controls elevatable with the lifting carriage or forks for lifting personnel, the following additional precautions shall be taken for the protection of personnel being elevated.

  - Use of a safety platform firmly secured to the lifting carriage and/or forks.

  - Means shall be provided whereby personnel on the platform can shut off power to the truck.

  - Such protection from falling objects as indicated necessary by the operating conditions shall be provided.

Example 22.3
Scissor points on all front-end loaders, which constitute a hazard to the operator during normal operation, shall be:

    A. Guarded

    B. Removed

    C. Lighted

    D. Provided

Answer: *A*

22.5 Safety Regulations for Pile-Driving Equipment

Some major specifications by the Part 1926 Subpart O Standard 1926.603 are as follows:

- Boilers and piping systems which are a part of, or used with, pile-driving equipment shall meet the applicable requirements of the American Society of Mechanical Engineers, Power Boilers (section I).

- All pressure vessels which are a part of, or used with, pile-driving equipment shall meet the applicable requirements of the American Society of Mechanical Engineers, Pressure Vessels (section VIII).

- Overhead protection, which will not obscure the vision of the operator and which meets the requirements of Subpart N of the Part 1926, shall be provided. Protection shall be the equivalent of 2-inch planking or other solid material of equivalent strength.

- Stop blocks shall be provided for the leads to prevent the hammer from being raised against the head block.

- A blocking device, capable of safely supporting the weight of the hammer, shall be provided for placement in the leads under the hammer at all times while employees are working under the hammer.

- Guards shall be provided across the top of the head block to prevent the cable from jumping out of the sheaves.

- When the leads must be inclined in the driving of batter piles, provisions shall be made to stabilize the leads.

- Fixed leads shall be provided with ladder, and adequate rings, or similar attachment points, so that the loft worker may engage his safety belt lanyard to the leads. If the leads are provided with loft platforms(s), such platform(s) shall be protected by standard guardrails.

- Steam hose leading to a steam hammer or jet pipe shall be securely attached to the hammer with an adequate length of at least 1/4-inch diameter chain or cable to prevent whipping in the event the joint at the hammer is broken. Air hammer hoses shall be provided with the same protection as required for steam lines.

- Safety chains, or equivalent means, shall be provided for each hose connection to prevent the line from thrashing around in case the coupling becomes disconnected.

- Steam line controls shall consist of two shutoff valves, one of which shall be a quick-acting lever type within easy reach of the hammer operator.

- Guys, outriggers, thrustouts, or counterbalances shall be provided as necessary to maintain stability of pile driver rigs.

22.6   Safety Regulations for Site Clearing

Some major specifications by the Part 1926 Subpart O Standard 1926.604 are as follows:

- Employees engaged in site clearing shall be protected from hazards of irritant and toxic plants and suitably instructed in the first aid treatment available.

- All equipment used in site clearing operations shall be equipped with rollover guards meeting the requirements. In addition, rider-operated equipment shall be equipped with an overhead and rear canopy guard meeting the following requirements:

  - The overhead covering on this canopy structure shall be of not less than 1/8-inch steel plate or 1/4-inch woven wire mesh with openings no greater than 1 inch, or equivalent.

- The opening in the rear of the canopy structure shall be covered with not less than 1/4-inch woven wire mesh with openings no greater than 1 inch.

Example 22.4

Employees engaged in site clearing shall be protected from hazards of:

  A. Irritant plants

  B. Toxic plants

  C. None of the above

  D. Both of the above

Answer: D

22.7   Safety Regulations for Marine Operations and Equipment

Some major specifications by the Part 1926 Subpart O Standard 1926.605 are provided in the following text.

22.7.1   Access to Barges

- Ramps for access of vehicles to or between barges shall be of adequate strength, provided with side boards, well maintained, and properly secured.
- Unless employees can step safely to or from the wharf, float, barge, or river towboat, either a ramp, or a safe walkway, shall be provided.
- Jacob's ladders shall be of the double rung or flat tread type. They shall be well maintained and properly secured.
- A Jacob's ladder shall either hang without slack from its lashings or be pulled up entirely.
- When the upper end of the means of access rests on or is flush with the top of the bulwark, substantial steps properly secured and equipped with at least one substantial hand rail approximately 33 inches in height shall be provided between the top of the bulwark and the deck.
- Obstructions shall not be laid on or across the gangway.
- The means of access shall be adequately illuminated for its full length.
- Unless the structure makes it impossible, the means of access shall be so located that the load will not pass over employees.

22.7.2   Working Surfaces of Barges

- Employees shall not be permitted to walk along the sides of covered lighters or barges with coamings more than 5 feet high, unless there is a 3-foot clear walkway, or a grab rail, or a taut handline is provided.
- Employees shall not be permitted to pass fore and aft, over, or around deckloads, unless there is a safe passage.

- Employees shall not be permitted to walk over deckloads from rail to coaming unless there is a safe passage. If it is necessary to stand at the outboard or inboard edge of the deckload where less than 24 inches of bulwark, rail, coaming, or other protection exists, all employees shall be provided with a suitable means of protection against falling from the deckload.

22.7.3   First-Aid and Lifesaving Equipment

- Provisions for rendering first aid and medical assistance shall be in accordance with Subpart D of this part.
- The employer shall ensure that there is in the vicinity of each barge in use at least one U.S. Coast Guard-approved 30-inch liftering with not less than 90 feet of line attached, and at least one portable or permanent ladder which will reach the top of the apron to the surface of the water. If the above equipment is not available at the pier, the employer shall furnish it during the time that he is working the barge.
- Employees walking or working on the unguarded decks of barges shall be protected with U.S. Coast Guard-approved work vests or buoyant vests.

Example 22.5

Which features do OSHA standards require for motor vehicles? Select all that apply.

A. All vehicles must have adequate audible warning devices at the operator's station.

B. All vehicles must have a parking brake system that works.

C. All vehicles must have a service brake system and an emergency brake system that work.

D. Whenever visibility conditions warrant additional light, all vehicles must be equipped with at least two working taillights.

E. Whenever visibility is poor, all vehicles must be equipped with at least two working brake lights.

F. The braking systems must be kept in operating condition and must have no components in common.

G. Whenever additional light is needed, all vehicles must have at least two working headlights unless supplemental lights are used onsite.

Answers: A, B, C, and D

22.8   Summary

All vehicles must be checked at the start of each shift to ensure that parts, equipment, and accessories are in safe operating condition and are free of apparent damage that could cause failure while in use. All defects must be corrected before the vehicles are placed in service. Vehicles used to transport employees must have seats firmly secured and adequate for the number of employees to be carried. Tools and materials must also be secured to prevent movement when transported in the same compartment with employees. All haulage vehicles with pay loads loaded by cranes, power shovels, loaders, or similar equipment must have a cab shield and/or canopy adequate to protect the operator from shifting or falling materials. No employer shall use any motor vehicle equipment having an obstructed view to the rear unless:

- The vehicle has a reverse signal alarm audible above the surrounding noise levels.
- The vehicle is backed up only when the observer signals that it is safe to do so.

All vehicles must be equipped with adequate audible warning devices at the operator's station. These devices must be kept operational, and whenever visibility conditions warrant additional light, all vehicles or combinations of vehicles in use, must be equipped with at least two operable headlights and taillights. All vehicles must have a service brake system, an emergency brake system, and a parking brake system. These systems can utilize common components, and they always must be maintained in operable condition.

22.9 Multiple-Choice Questions

22.1 Which pneumatic-tired earth-moving haulage equipment shall be equipped with fenders on all wheels?

A. Whose maximum speed exceeds 5 miles per hour
B. Whose maximum speed exceeds 10 miles per hour
C. Whose maximum speed exceeds 15 miles per hour
D. Whose maximum speed exceeds 20 miles per hour

22.2 Employees shall not be permitted to walk along the sides of covered lighters or barges with coamings more than:

A. 5 feet high
B. 7 feet high
C. 8 feet high
D. 8.5 feet high

22.3 All haulage vehicles with pay loads loaded by cranes, power shovels, loaders, or similar equipment must have a cab shield and/or canopy adequate to protect the operator from shifting or falling materials.

- True
- False

22.4 Whenever visibility conditions warrant additional light, all vehicles or combinations of vehicles in use must be equipped with at least _____ operable headlights and taillights.

A. 1
B. 2
C. 3
D. 4

22.5 All vehicles must be equipped with adequate warning devices at the operator's station. These devices must be kept operational.

- True
- False

22.6 All vehicles with cabs must be equipped with _____ and power wipers.

    A. Hoods
    B. Radios
    C. Windshields
    D. Floor mats

22.7 Trucks with dump bodies must be equipped with _____ means of support.

    A. Negative
    B. Competent
    C. Weak
    D. Positive

22.8 All vehicles must have:

    A. A service brake system
    B. An emergency brake system
    C. A parking brake system
    D. All of the above

22.10 Practice Problems

22.1 List the safety regulations for mechanized equipment.

22.2 List the safety regulations for motor vehicles.

22.3 List the safety regulations for material handling equipment.

22.4 List the safety regulations for pile-driving equipment.

22.5 List the safety regulations for site clearing.

22.6 List the safety regulations for marine operations and equipment.

22.11 Critical Thinking and Discussion Topic

The forklift at your jobsite concerns you. It came from the repair and inspection shop recently and was certified as good as new. You called your boss over phone and got a reply that it should work perfect. As a site supervisor, what will you do in this situation?

CHAPTER 23

Excavations

(Photo courtesy by Paula from Pexels.)

23.1 General

Excavation (Fig. 23.1) means any man-made cut, cavity, trench, or depression in an earth surface, formed by earth removal using tools, machinery, or explosives. The removed soil at an excavation site is called spoil. Cave-ins pose the greatest risk and are much more likely than other excavation-related accidents to result in worker fatalities. Both trench boxes and shoring serve to protect workers from cave-ins. Other potential hazards include falls, falling loads, water accumulations, hazardous atmospheres, fire, and incidents involving mobile equipment. Trench collapses cause dozens of fatalities and hundreds of injuries each year. Therefore, OSHA recommends a competent person must be designated by the employer and have knowledge related to soil classification, protective systems, and safety standards related to excavation.

FIGURE 23.1   Excavation in highway in Colorado.
(Photo courtesy of Armando Perez.)

Example 23.1
A _____ person must be designated by the employer and have knowledge related to soil classification, protective systems, and safety standards related to excavation.

    A. Clarified

    B. Competent

    C. Senior

    D. Tired

Answer: B

23.2   Specific Excavation Requirements

Some major specifications by the OSHA 1926.651 are as follows:

Surface encumbrances.   All surface encumbrances that are located so as to create a hazard to employees shall be removed or supported, as necessary, to safeguard employees. Excavations 20 feet and greater in depth must have a protective system that is planned and designed by a professional engineer. OSHA standards prohibit excessive water accumulations or any water accumulation where properly monitored water removal equipment is not in place when workers are in the trench.

23.2.1 Underground Installations

- The estimated location of utility installations, such as sewer, telephone, fuel, electric, water lines, or any other underground installations that reasonably may be expected to be encountered during excavation work, shall be determined prior to opening an excavation.

- Utility companies or owners shall be contacted within established or customary local response times, advised of the proposed work, and asked to establish the location of the utility underground installations prior to the start of actual excavation. When utility companies or owners cannot respond to a request to locate underground utility installations within 24 hours (unless a longer period is required by state or local law), or cannot establish the exact location of these installations, the employer may proceed, provided the employer does so with caution, and provided detection equipment or other acceptable means to locate utility installations are used.

- When excavation operations approach the estimated location of underground installations, the exact location of the installations shall be determined by safe and acceptable means.

- While the excavation is open, underground installations shall be protected, supported, or removed as necessary to safeguard employees.

Example 23.2
Excavations _____ feet and greater in depth must have a protective system that is planned and designed by a professional engineer.

    A. 5

    B. 10

    C. 15

    D. 20

Answer: D

23.2.2 Access and Egress

Structural ramps.   Structural ramps that are used solely by employees as a means of access or egress from excavations shall be designed by a competent person. Structural ramps used for access or egress of equipment shall be designed by a competent person qualified in structural design, and shall be constructed in accordance with the design.

- Ramps and runways constructed of two or more structural members shall have the structural members connected together to prevent displacement.

- Structural members used for ramps and runways shall be of uniform thickness.

- Cleats or other appropriate means used to connect runway structural members shall be attached to the bottom of the runway or shall be attached in a manner to prevent tripping.
- Structural ramps used in lieu of steps shall be provided with cleats or other surface treatments on the top surface to prevent slipping.

Means of egress from trench excavations. A stairway, ladder, ramp, or other safe means of egress shall be located in trench excavations that are 4 feet (1.22 meter) or more in depth so as to require no more than 25 feet (7.62 meter) of lateral travel for employees.

23.2.3   Exposure to Vehicular Traffic

Employees exposed to public vehicular traffic shall be provided with, and shall wear, warning vests or other suitable garments marked with or made of reflectorized or high-visibility material.

23.2.4   Exposure to Falling Loads

No employee shall be permitted underneath loads handled by lifting or digging equipment. Employees shall be required to stand away from any vehicle being loaded or unloaded to avoid being struck by any spillage or falling materials. Operators may remain in the cabs of vehicles being loaded or unloaded when the vehicles are equipped, to provide adequate protection for the operator during loading and unloading operations.

23.2.5   Warning System for Mobile Equipment

When mobile equipment is operated adjacent to an excavation, or when such equipment is required to approach the edge of an excavation, and the operator does not have a clear and direct view of the edge of the excavation, a warning system shall be utilized such as barricades, hand or mechanical signals, or stop logs. If possible, the grade should be away from the excavation.

23.2.6   Hazardous Atmospheres

- Where oxygen deficiency (atmospheres containing less than 19.5% oxygen) or a hazardous atmosphere exists or could reasonably be expected to exist, such as in excavations in landfill areas or excavations in areas where hazardous substances are stored nearby, the atmospheres in the excavation shall be tested before employees enter excavations greater than 4 feet (1.22 meter) in depth.
- Adequate precautions shall be taken to prevent employee exposure to atmospheres containing less than 19.5% oxygen and other hazardous atmospheres. These precautions include providing proper respiratory protection or ventilation.
- Adequate precaution shall be taken, such as providing ventilation, to prevent employee exposure to an atmosphere containing a concentration of a flammable gas in excess of 20% of the lower flammable limit of the gas.
- When controls are used that are intended to reduce the level of atmospheric contaminants to acceptable levels, testing shall be conducted as often as necessary to ensure that the atmosphere remains safe.

23.2.7 Protection from Hazards Associated with Water Accumulation

Employees shall not work in excavations in which there is accumulated water, or in excavations in which water is accumulating, unless adequate precautions have been taken to protect employees against the hazards posed by water accumulation. The precautions necessary to protect employees adequately vary with each situation, but could include special support or shield systems to protect from cave-ins, water removal to control the level of accumulating water, or use of a safety harness and lifeline. If water is controlled or prevented from accumulating by the use of water removal equipment, the water removal equipment and operations shall be monitored by a competent person to ensure proper operation. If excavation work interrupts the natural drainage of surface water (such as streams), diversion ditches, dikes, or other suitable means shall be used to prevent surface water from entering the excavation and to provide adequate drainage of the area adjacent to the excavation. Excavations subject to runoff from heavy rains will require an inspection by a competent person.

23.2.8 Stability of Adjacent Structures

Where the stability of adjoining buildings, walls, or other structures is endangered by excavation operations, support systems such as shoring, bracing, or underpinning shall be provided to ensure the stability of such structures for the protection of employees. Excavation below the level of the base or footing of any foundation or retaining wall that could be reasonably expected to pose a hazard to employees shall not be permitted except when:

- A support system, such as underpinning, is provided to ensure the safety of employees and the stability of the structure.
- The excavation is in stable rock.
- A registered professional engineer has approved the determination that the structure is sufficiently removed from the excavation so as to be unaffected by the excavation activity; or
- A registered professional engineer has approved the determination that such excavation work will not pose a hazard to employees.

Sidewalks, pavements, and appurtenant structure shall not be undermined unless a support system or another method of protection is provided to protect employees from the possible collapse of such structures.

23.2.9 Protection of Employees from Loose Rock or Soil

Adequate protection shall be provided to protect employees from loose rock or soil that could pose a hazard by falling or rolling from an excavation face. Such protection shall consist of scaling to remove loose material; installation of protective barricades at intervals as necessary on the face to stop and contain falling material; or other means that provide equivalent protection.

Employees shall be protected from excavated or other materials or equipment that could pose a hazard by falling or rolling into excavations. Protection shall be provided by placing and keeping such materials or equipment at least 2 feet (0.61 meter) from the edge of excavations, or by the use of retaining devices that are sufficient to prevent materials or equipment from falling or rolling into excavations, or by a combination of both if necessary.

23.2.10   Inspections

Daily inspections of excavations, the adjacent areas, and protective systems shall be made by a competent person for evidence of a situation that could result in possible cave-ins, indications of failure of protective systems, hazardous atmospheres, or other hazardous conditions. An inspection shall be conducted by the competent person prior to the start of work and as needed throughout the shift. Inspections shall also be made after every rainstorm or other hazard-increasing occurrence. These inspections are only required when employee exposure can be reasonably anticipated. Where the competent person finds evidence of a situation that could result in a possible cave-in, indications of failure of protective systems, hazardous atmospheres, or other hazardous conditions, exposed employees shall be removed from the hazardous area until the necessary precautions have been taken to ensure their safety.

23.3   Requirements for Protective Systems

Some major specifications by the OSHA 1926.652 are provided in the following text.

23.3.1   Protection of Employees in Excavations

Each employee in an excavation shall be protected from cave-ins by an adequate protective system except when:

- Excavations are made entirely in stable rock.
- Excavations are less than 5 feet (1.52 meter) in depth and examination of the ground by a competent person provides no indication of a potential cave-in.

Protective systems shall have the capacity to resist without failure all loads that are intended or could reasonably be expected to be applied or transmitted to the system.

Example 23.3
Excavations are less than _____ feet in depth, and examination of the ground by a competent person provides no indication of a potential cave-in.

   A. 1
   B. 3
   C. 5
   D. 9

Answer: C

23.3.2   Design of Sloping and Benching Systems

The slopes and configurations of sloping and benching systems shall be selected and constructed by the employer or his designee and shall be in accordance with the requirements of as follows:

Option (1)—Allowable configurations and slopes

Option (2)—Determination of slopes and configurations

Option (3)—Designs using other tabulated data

Option (4)—Design by a registered professional engineer

23.3.3 Materials and Equipment

- Materials and equipment used for protective systems shall be free from damage or defects that might impair their proper function.

- Manufactured materials and equipment used for protective systems shall be used and maintained in a manner that is consistent with the recommendations of the manufacturer, and in a manner that will prevent employee exposure to hazards.

- When material or equipment that is used for protective systems is damaged, a competent person shall examine the material or equipment and evaluate its suitability for continued use. If the competent person cannot assure the material or equipment is able to support the intended loads or is otherwise suitable for safe use, then such material or equipment shall be removed from service, and shall be evaluated and approved by a registered professional engineer before being returned to service.

23.3.4 Installation and Removal of Support

- Members of support systems shall be securely connected together to prevent sliding, falling, kickouts, or other predictable failure.

- Support systems shall be installed and removed in a manner that protects employees from cave-ins, structural collapses, or from being struck by members of the support system.

- Individual members of support systems shall not be subjected to loads exceeding those which those members were designed to withstand.

- Before temporary removal of individual members begins, additional precautions shall be taken to ensure the safety of employees, such as installing other structural members to carry the loads imposed on the support system.

- Removal shall begin at, and progress from, the bottom of the excavation. Members shall be released slowly so as to note any indication of possible failure of the remaining members of the structure or possible cave-in of the sides of the excavation.

- Backfilling shall progress together with the removal of support systems from excavations.

Excavation of material to a level no greater than 2 feet (0.61 meter) below the bottom of the members of a support system shall be permitted, but only if the system is designed to resist the forces calculated for the full depth of the trench, and there are no indications while the trench is open of a possible loss of soil from behind or below the bottom of the support system.

23.3.5 Sloping and Benching Systems

Employees shall not be permitted to work on the faces of sloped or benched excavations at levels above other employees except when employees at the lower levels are adequately protected from the hazard of falling, rolling, or sliding material or equipment.

23.3.6 Shield Systems

- Shield systems shall not be subjected to loads exceeding those which the system was designed to withstand.

- Shields shall be installed in a manner to restrict lateral or other hazardous movement of the shield in the event of the application of sudden lateral loads.

- Employees shall be protected from the hazard of cave-ins when entering or exiting the areas protected by shields.

- Employees shall not be allowed in shields when shields are being installed, removed, or moved vertically.

23.3.7 Additional Requirement for Shield Systems Used in Trench Excavations

Excavations of earth material to a level not greater than 2 feet (0.61 meter) below the bottom of a shield shall be permitted, but only if the shield is designed to resist the forces calculated for the full depth of the trench, and there are no indications while the trench is open of a possible loss of soil from behind or below the bottom of the shield.

Example 23.4

You're a supervisor on a crew digging a 12-foot-deep trench. Which of the following practices are safe? Select all that apply.

A. Place spoils within 1 foot of an excavation's edge.

B. Spoils are placed so that water is diverted away from the excavation.

C. While workers are in the trench, a competent person monitors the excavation project until it is completed.

D. Lots of water accumulates when workers are in the trench. There's no water removal equipment available.

E. Slightly damaged excavation equipment is being used because it still works.

F. No support or shield systems have been installed.

G. Spoils have been placed 4 feet away from the trench.

H. Workers are using water removal equipment to keep the water level in the trench low.

Answers: *B, C, G, and H*

23.4 Soil Classification

Part 1926, Subpart P App A provides specifications of soil classification. Soil classification system means, for the purpose of this subpart, a method of categorizing soil and rock deposits in a hierarchy of Stable Rock, Type A, Type B, and Type C, in decreasing order of stability. The categories are determined based on an analysis of the properties and performance characteristics of the deposits and the environmental conditions of exposure. Stable rock means natural solid mineral matter that can be excavated with vertical sides and remain intact while exposed. Submerged soil means soil which is underwater or is free seeping. Unconfined compressive strength means the load per unit area at which a soil will fail in compression. It can be determined by laboratory testing, or estimated in the field using a pocket penetrometer, by thumb penetration tests, and other methods.

Type A means cohesive soils with an unconfined compressive strength of 1.5 ton per square foot (tsf) (144 kilopascal) or greater. Examples of cohesive soils are: clay, silty clay, sandy clay, clay loam, and, in some cases, silty clay loam and sandy clay loam. Cemented soils such as caliche and hardpan are also considered Type A. Great effort is required to penetrate the soil with thumb.

However, no soil is Type A if:

   (i) The soil is fissured.

  (ii) The soil is subject to vibration from heavy traffic, pile driving, or similar effects.

 (iii) The soil has been previously disturbed.

 (iv) The soil is part of a sloped, layered system where the layers dip into the excavation on a slope of four horizontal to one vertical (4H:1V) or greater.

  (v) The material is subject to other factors that would require it to be classified as a less stable material.

Type B means:

   (i) Cohesive soil with an unconfined compressive strength greater than 0.5 tsf (48 kilopascal) but less than 1.5 tsf (144 kilopascal).

  (ii) Granular cohesionless soils including: angular gravel (similar to crushed rock), silt, silt loam, sandy loam, and, in some cases, silty clay loam and sandy clay loam.

 (iii) Previously disturbed soils except those which would otherwise be classified as Type C soil.

 (iv) Soil that meets the unconfined compressive strength or cementation requirements for Type A, but is fissured or subject to vibration.

  (v) Dry rock that is not stable.

 (vi) Material that is part of a sloped, layered system where the layers dip into the excavation on a slope less steep than four horizontal to one vertical (4H:1V), but only if the material would otherwise be classified as Type B.

Type C means:

   (i) Cohesive soil with an unconfined compressive strength of 0.5 tsf (48 kilopascal) or less.

  (ii) Granular soils including gravel, sand, and loamy sand.

 (iii) Submerged soil or soil from which water is freely seeping.

 (iv) Submerged rock that is not stable.

    (v) Material in a sloped, layered system where the layers dip into the excavation on a slope of four horizontal to one vertical (4H:1V) or steeper.

    (vi) Soil that can be penetrated very easily with thumb.

Part 1926, Subpart P also provides specifications for the following:

- App A—Soil Classification
- App B—Sloping and Benching
- App C—Timber Shoring for Trenches
- App D—Aluminum Hydraulic Shoring for Trenches
- App E—Alternatives to Timber Shoring
- App F—Selection of Protective Systems

These are not discussed here as these are beyond the scope of this book.

Example 23.5
Match the characteristics or examples of soils listed in Table 23.1 to the correct soil type by OSHA.

	Type A soil	Type B soil	Type C soil
Cemented soils	X		
Sand			X
Clay	X		
Sandy loam		X	
Silt		x	
Great effort to penetrate with thumb	X		
Easily penetrable with thumb			X
Granular soils lacking cohesion		X	
Submerged soil			X

TABLE 23.1 Matching Characteristics or Examples of Soils to the Correct Soil Type

23.5 Summary

Excavating is considered one of the most hazardous operations in the field of construction. Excavations must be immediately filled back to their normal state after work is completed. Trench boxes are usually made of aluminum or steel, and may be used in combination with sloping and benching. Trench boxes must be inspected regularly, maintained properly, and used under the supervision of a competent person. Once the excavation has been cleared, workers should carefully remove the protective system, starting from the bottom to up. Visual analysis is conducted to determine qualitative information regarding the excavation site in general, the soil adjacent to the excavation, the soil forming the sides of the open excavation, and the soil taken as samples from excavated material.

23.6   Multiple-Choice Questions

23.1 If utility companies or owners are contacted to establish the location of the utility underground installations, how much time is allowed to them to locate the underground utility?

  A. 1 day
  B. 2 days
  C. 3 days
  D. 4 days

23.2 Which of the following works must be performed prior to opening an excavation?

  A. Determine the density of the soil.
  B. Identify the location of underground utilities.
  C. Identify the groundwater table.
  D. Identify the agricultural value of the soil.

23.3 A stairway, ladder, ramp, or other safe means of egress shall be located in trench excavations that are _____ or more in depth.

  A. 2 feet
  B. 4 feet
  C. 6 feet
  D. 8 feet

23.4 Each employee in an excavation is required to be protected from cave-ins by an adequate protective system when:

  A. Excavations are made entirely in stable rock.
  B. Excavations are less than 5 feet in depth.
  C. Examination of the ground by a competent person provides no indication of a potential cave-in.
  D. Any other cases not listed.

23.5 Protection for falling or rolling into excavations must be provided by placing and keeping such materials or equipment at least _____ feet from the edge of excavations.

  A. 0.5
  B. 1.0
  C. 1.5
  D. 2.0

23.6 Type C soil is that which has unconfined compressive strength of 0.5 tsf.

  • True
  • False

23.7 OSHA standards _____ excessive water accumulations or any water accumulation where properly monitored water removal equipment is not in place when workers are in the trench.

  A. Welcomes
  B. Permits
  C. Allows
  D. Prohibits

23.8 The removed soil at an excavation site is called _____.

    A. Spilled

    B. Soil

    C. Spoil

    D. Space

23.9 What are some dangers of excavations?

    A. Lack of oxygen

    B. Fire

    C. Water accumulation

    D. All of the above

23.10 Both trench boxes and shoring serve to protect workers from cave-ins.

    • True

    • False

23.11 The atmospheres in the excavation shall be tested before employees enter excavations greater than _____ feet.

    A. 1

    B. 3

    C. 4

    D. 7

23.7 Practice Problems

23.1 List some general safety regulations for underground installations.

23.2 List some general safety regulations for hazardous atmosphere for excavations.

23.3 List some general safety regulations for stability of adjacent structures during excavation.

23.4 List some general safety regulations for protection of employees in excavations.

23.5 List some general safety regulations for installation and removal of support.

23.6 List some general safety regulations for shield systems.

23.8 Critical Thinking and Discussion Topic

While working in an excavation site, you installed the excavation support as recommended by OSHA. However, suddenly a small rain started and it looked that the rain will not affect the slope stability. One of the workers said that the rain may affect the slope stability. You are scheduled to finish the job today. As a site supervisor, what will you do in this situation?

CHAPTER 24

Concrete and Masonry Construction

(Photo courtesy by Yury Kim from Pexels.)

24.1   General

Concrete work (Fig. 24.1a) involves the following basic processes: preparation of the concrete mixture; delivery of the mixture to the construction site; feeding, distribution, and compaction of the mixture in the formwork (molds); curing of the concrete while it is hardening; and quality control of the concrete work.

- **Preparation.** Preparation of the concrete mix is usually carried out at concrete plants or in movable concrete-mixing units. Also used for this purpose are stock (sectional-assembly) plants, whose equipment, transportable on railroad flatcars or trailer trucks, may be set up in large units.

- **Delivery.** Delivery of concrete mix to the construction site is usually carried out by truck transport. In addition to dump trucks, concrete trucks are used, which are specially equipped for hauling concrete mix; for long distances, concrete-mixer trucks are utilized.

- **Feeding.** The feeding of concrete mix is carried out by belt conveyors, concrete pumps, concrete hoists, pneumatic blowers, and vibration chutes. Feeding and distributing concrete mix for the laying of concrete footings under structural components and under equipment to be used in industrial buildings are also carried out by mobile concrete placement units, equipped with directable belt conveyors.

- **Compaction.** Compaction of concrete mix is an extremely important process in concrete work. It ensures that the mix will compactly fill all the interstices between the reinforcement rods as well as between the reinforcement rods and the formwork in order to achieve the required strength, impermeability to water, and frost resistance of the concrete.

- **Curing.** Curing of cement consists of establishing a temperature-humidity system required for the hardening of a compacted concrete mix and in the protection of the concrete from jarring vibrations, impacts, and so on.

- **Quality control.** Quality control of the concrete work includes making concrete samples at the work site, maintaining them under conditions close to those of production, and testing the samples for strength.

The masonry work (Fig. 24.1b) is the construction work carried out during the erection of stonework on buildings and structures made of natural and artificial masonry materials. The masonry work is a complex of processes which includes, in addition to the basic processes (the laying of bricks or other stones in mortar, the delivery and laying of stones, and the smoothing of mortars), related auxiliary processes (the erection of scaffolding and trestles and the preparation of materials at the construction site).

(a) Concrete pouring   (b) Masonry wall construction

FIGURE 24.1   Concrete and masonry construction.
(Photo courtesy of 123RF.)

Head-and-face gear is required for workers applying cement, sand, and water mixture through a pneumatic hose.

Example 24.1
What PPE is required for workers applying cement, sand, and water mixture through a pneumatic hose?

    A. Hard hat

    B. Steel-toes shoes

    C. Head-and-face gear

    D. Respirator

Answer: C

24.2 Requirements for Equipment and Tools

Some general regulations for concrete and masonry equipment and tools based on the Part 1926, Subpart Q, Standard Number 1926.702 are listed as follows.

Bulk cement storage. Bulk storage bins, containers, and silos shall be equipped with conical or tapered bottoms; and mechanical or pneumatic means of starting the flow of material. No employee shall be permitted to enter storage facilities unless the ejection system has been shut down, locked out, and tagged to indicate that the ejection system is not to be operated.

Concrete mixers. Concrete mixers with one cubic yard (8 m³) or larger loading skips shall be equipped with a mechanical device to clear the skip of materials; and guardrails installed on each side of the skip.

Power concrete trowels. Powered and rotating-type concrete troweling machines that are manually guided shall be equipped with a control switch that will automatically shut off the power whenever the hands of the operator are removed from the equipment handles.

Concrete buggies. Concrete buggy handles shall not extend beyond the wheels on either side of the buggy.

Concrete pumping systems. Concrete pumping systems using discharge pipes shall be provided with pipe supports designed for 100% overload. Compressed air hoses used on concrete pumping system shall be provided with positive fail-safe joint connectors to prevent separation of sections when pressurized.

Concrete buckets. Concrete buckets equipped with hydraulic or pneumatic gates shall have positive safety latches or similar safety devices installed to prevent premature or accidental dumping. Concrete buckets shall be designed to prevent concrete from hanging up on top and the sides.

Tremies. Sections of tremies and similar concrete conveyances shall be secured with wire rope (or equivalent materials) in addition to the regular couplings or connections.

Bull floats.   Bull float handles are used where they might contact energized electrical conductors. These shall be constructed of nonconductive material or insulated with a nonconductive sheath whose electrical and mechanical characteristics provide protection equivalent to that of a handle constructed of nonconductive material.

Masonry saws.   Masonry saw shall be guarded with a semicircular enclosure over the blade. A method for retaining blade fragments shall be incorporated in the design of the semicircular enclosure.

Lockout/Tagout procedures.   No employee shall be permitted to perform maintenance or repair activity on equipment (such as compressors mixers, screens, or pumps used for concrete and masonry construction activities) where the inadvertent operation of the equipment could occur and cause injury, unless all potentially hazardous energy sources have been locked out and tagged. Tags shall read Do Not Start or similar language to indicate that the equipment is not to be operated.

Example 24.2
Storage bins and silos must be equipped with _____ bottoms.

    A. Flat

    B. Round

    C. Conical

    D. Square

Answer: C

24.3   Requirements for Cast-in-Place Concrete

Part 1926, Subpart Q, Standard Number 1926.703 provides Requirements for cast-in-place concrete. Some major items are stated here.

24.3.1   General Requirements for Formwork

- Formwork shall be designed, fabricated, erected, supported, braced, and maintained so that it will be capable of supporting without failure all vertical and lateral loads that may reasonably be anticipated to be applied to the formwork, which is designed, fabricated, erected, supported, braced, and maintained.
- Drawings or plans, including all revisions, for the jack layout, formwork (including shoring equipment), working decks, and scaffolds, shall be available at the jobsite.

24.3.2   Shoring and Reshoring

- All shoring equipment (including equipment used in reshoring operations) shall be inspected prior to erection to determine that the equipment meets the requirements specified in the formwork drawings.
- Shoring equipment found to be damaged such that its strength is reduced to less than that required shall not be used for shoring.

- Erected shoring equipment shall be inspected immediately prior to, during, and immediately after concrete placement.
- Shoring equipment that is found to be damaged or weakened after erection, such that its strength is reduced to less than that required, shall be immediately reinforced.
- The sills for shoring shall be sound, rigid, and capable of carrying the maximum intended load.
- All base plates, shore heads, extension devices, and adjustment screws shall be in firm contact, and secured when necessary, with the foundation and the form.
- Eccentric loads on shore heads and similar members shall be prohibited unless these members have been designed for such loading.
- Whenever single post shores are used one on top of another (tiered), the employer shall comply with the following specific requirements in addition to the general requirements for formwork:
  - The design of the shoring shall be prepared by a qualified designer and the erected shoring shall be inspected by an engineer qualified in structural design.
  - The single post shores shall be vertically aligned.
  - The single post shores shall be spliced to prevent misalignment.
  - The single post shores shall be adequately braced in two mutually perpendicular directions at the splice level. Each tier shall also be diagonally braced in the same two directions.
- Adjustment of single post shores to raise formwork shall not be made after the placement of concrete.
- Reshoring shall be erected, as the original forms and shores are removed, whenever the concrete is required to support loads in excess of its capacity.

Example 24.3
Erected shoring equipment must be inspected in the following circumstances, except:

A. Immediately prior to concrete placement

B. During concrete placement

C. Immediately after concrete placement

D. A few days after concrete placement

Answer: D

24.3.3 Vertical Slip Forms

- The steel rods or pipes on which jacks climb or by which the forms are lifted shall be:
  - Specifically designed for that purpose; and
  - Adequately braced where not encased in concrete.

- Forms shall be designed to prevent excessive distortion of the structure during the jacking operation.
- All vertical slip forms shall be provided with scaffolds or work platforms where employees are required to work or pass.
- Jacks and vertical supports shall be positioned in such a manner that the loads do not exceed the rated capacity of the jacks.
- The jacks or other lifting devices shall be provided with mechanical dogs or other automatic holding devices to support the slip forms whenever failure of the power supply or lifting mechanism occurs.
- The form structure shall be maintained within all design tolerances specified for plumbness during the jacking operation.
- Jacking equipment being used to raise a 2-ton steel column must be capable of supporting at least 5 tons.
- The predetermined safe rate of lift shall not be exceeded.

24.3.4   Reinforcing Steel

- Reinforcing steel for walls, piers, columns, and similar vertical structures shall be adequately supported to prevent overturning and to prevent collapse.
- Employers shall take measures to prevent unrolled wire mesh from recoiling. Such measures may include, but are not limited to, securing each end of the roll or turning over the roll.

Example 24.4
Steel rods and pipes on which forms are lifted must meet all these requirements, except:

A. They must be encased in concrete.

B. They must be adequately braced.

C. They must be specifically designed for that purpose.

D. They must be physically attached to the forms.

Answer: D

Example 24.5
Jacking equipment being used to raise a 2-ton steel column must be capable of supporting at least _____ tons.

A. 3

B. 5

C. 8

D. 10

Answer: B

24.3.5   Removal of Formwork

- Forms and shores (except those used for slabs on grade and slip forms) shall not be removed until the employer determines that the concrete has gained sufficient strength to support its weight and superimposed loads. Such determination shall be based on compliance with one of the following:

  - The plans and specifications stipulate conditions for removal of forms and shores, and such conditions have been followed.

  - The concrete has been properly tested with an appropriate ASTM standard test method designed to indicate the concrete compressive strength, and the test results indicate that the concrete has gained sufficient strength to support its weight and superimposed loads.

- Reshoring shall not be removed until the concrete being supported has attained adequate strength to support its weight and all loads in place upon it.

Example 24.6
Who must determine when forms and shores may be removed?

  A. The workers

  B. The employer

  C. The design engineer

  D. A competent person

Answer: *B*

24.4   Requirements for Precast Concrete

Part 1926, Subpart Q, Standard Number 1926.704 states that:

- Precast concrete wall units, structural framing, and tilt-up wall panels shall be adequately supported to prevent overturning and to prevent collapse until permanent connections are completed.

- Lifting inserts which are embedded or otherwise attached to tilt-up precast concrete members shall be capable of supporting at least two times the maximum intended load applied or transmitted to them.

- Lifting inserts which are embedded or otherwise attached to precast concrete members, other than the tilt-up members, shall be capable of supporting at least four times the maximum intended load applied or transmitted to them.

- Lifting hardware shall be capable of supporting at least five times the maximum intended load applied or transmitted to the lifting hardware.

- No employee shall be permitted under precast concrete members being lifted or tilted into position except those employees required for the erection of those members.

24.5   Requirements for Lift-Slab Construction Operations

Part 1926, Subpart Q, Standard Number 1926.705 states that:

- Lift-slab operations shall be designed and planned by a registered professional engineer who has experience in lift-slab construction. Such plans and designs shall be implemented by the employer and shall include detailed instructions and sketches indicating the prescribed method of erection. These plans and designs shall also include provisions for ensuring lateral stability of the building/structure during construction.

- Jacks/lifting units shall be marked to indicate their rated capacity as established by the manufacturer.

- Jacks/lifting units shall not be loaded beyond their rated capacity as established by the manufacturer.

- Jacking equipment shall be capable of supporting at least two and one-half times the load being lifted during jacking operations and the equipment shall not be overloaded. For the purpose of this provision, jacking equipment includes any load-bearing component which is used to carry out the lifting operation(s). Such equipment includes, but is not limited, to the following: threaded rods, lifting attachments, lifting nuts, hook-up collars, T-caps, shearheads, columns, and footings.

- Jacks/lifting units shall be designed and installed so that they will neither lift nor continue to lift when they are loaded in excess of their rated capacity.

- Jacks/lifting units shall have a safety device installed which will cause the jacks/lifting units to support the load in any position in the event any jack lifting unit malfunctions or loses its lifting ability.

- Jacking operations shall be synchronized in such a manner to ensure even and uniform lifting of the slab. During lifting, all points at which the slab is supported shall be kept within 1/2 inch of that needed to maintain the slab in a level position.

- If leveling is automatically controlled, a device shall be installed that will stop the operation when the 1/2 inch tolerance is exceeded or where there is a malfunction in the jacking (lifting) system.

- If leveling is maintained by manual controls, such controls shall be located in a central location and attended by a competent person while lifting is in progress.

- The maximum number of manually controlled jacks/lifting units on one slab shall be limited to a number that will permit the operator to maintain the slab level within specified tolerances of this section, but in no case shall that number exceed 14.

- All welding on temporary and permanent connections shall be performed by a certified welder, familiar with the welding requirements specified in the plans and specifications for the lift-slab operation.

- Load transfer from jacks/lifting units to building columns shall not be executed until the welds on the column shear plates (weld blocks) are cooled to air temperature.

- Jacks/lifting units shall be positively secured to building columns so that they do not become dislodged or dislocated.

- Equipment shall be designed and installed so that the lifting rods cannot slip out of position or the employer shall institute other measures, such as the use of locking or blocking devices, which will provide positive connection between the lifting rods and attachments and will prevent components from disengaging during lifting operations.

Example 24.7

Who may design and plan lift-slab operations?

    A. The employer

    B. A competent person

    C. A professional engineer

    D. A trained worker

Answer: C

24.6 Requirements for Masonry Construction

A limited access zone shall be established whenever a masonry wall is being constructed. Some major regulations based on the Part 1926, Subpart Q, Standard Number 1926.706 for masonry construction are as follows:

- The limited access zone shall be established prior to the start of construction of the wall.

- The limited access zone shall be equal to the height of the wall to be constructed plus 4 feet, and shall run the entire length of the wall.

- The limited access zone shall be established on the side of the wall which will be unscaffolded.

- The limited access zone shall be restricted to entry by employees actively engaged in constructing the wall. No other employee shall be permitted to enter the zone.

- The limited access zone shall remain in place until the wall is adequately supported to prevent overturning and to prevent collapse unless the height of wall is over 8 feet, in which case, the limited access zone shall remain in place.

All masonry walls over 8 feet in height shall be adequately braced to prevent overturning and to prevent collapse unless the wall is adequately supported so that it will not overturn or collapse. The bracing shall remain in place until permanent supporting elements of the structure are in place.

Example 24.8
Limited access zones around masonry wall construction must meet all these requirements, except:

    A. Equal to the height of the wall plus 4 feet

    B. Set up on the side of the wall that is scaffolded

    C. Run the entire length of the wall

    D. Restricted to workers actively engaged in building the wall

Answer: D

24.7 Summary

Employees must not be permitted to ride on or work under concrete buckets while the buckets are being elevated or lowered into position. This is just one example of the measures employees on the worksite must take in order to prevent potential injury or death due to negligent behavior. An example of an engineered protective measure is found in powered and rotating-type concrete troweling machines that are manually guided and equipped with a control switch that will automatically shut off the power whenever the hands of the operator are removed from the equipment handles.

Jacking equipment must be capable of supporting at least two and one-half times the load being lifted during jacking operations, and the equipment must not be overloaded. Lifting inserts that are embedded or otherwise attached to tilt-up wall panels must be capable of supporting at least two times the maximum intended load applied or transmitted to them. Lifting inserts for other precast members, excluding tilt-up members, must be capable of supporting four times the load. Lifting hardware members must be capable of supporting five times the maximum intended load applied to the lifting hardware.

Erected shoring equipment must be inspected immediately before, during, and after concrete placement. All base plates, shore heads, extension devices, and adjustment screws must be in firm contact, and secured when necessary, with the form and foundation. Shoring equipment that is found to be damaged or weakened after erection must be reinforced immediately.

24.8 Multiple-Choice Questions

24.1 Which size of concrete loading skips shall be equipped with a mechanical device to clear the skip of materials?

    A. One cubic yard or larger loading
    B. Two cubic yard or larger loading
    C. Three cubic yard or larger loading
    D. Four cubic yard or larger loading

24.2 Concrete pumping systems using discharge pipes shall be provided with pipe supports designed for:

    A. 50% overload

    B. 75% overload

    C. 100% overload

    D. 150% overload

24.3 In addition to the regular couplings or connections, sections of tremies and similar concrete conveyances shall be secured with:

    A. Wire rope

    B. Bolting

    C. Welding

    D. Heavy-duty slip-critical bolts

24.4 Jacks/lifting units shall not be loaded beyond their:

    A. Rated capacity

    B. 0.75 times of the rated capacity

    C. 0.90 times of the rated capacity

    D. 1.2 times of the rated capacity

24.5 When shall a limited access zone be established?

    A. Whenever a masonry wall is being constructed

    B. Whenever a concrete wall is being constructed

    C. Whenever a masonry foundation is being constructed

    D. Whenever a concrete paving is being done

24.6 All masonry walls over _____ in height shall be adequately braced to prevent overturning and to prevent collapse.

    A. 4 feet

    B. 6 feet

    C. 8 feet

    D. 10 feet

24.7 What is the maximum number of manually controlled jacks/lifting units on one slab?

    A. 8

    B. 10

    C. 12

    D. 14

24.8 For precast concrete, lifting hardware shall be capable of supporting at least _____ times the maximum intended load applied transmitted to the lifting hardware.

    A. 2

    B. 4

    C. 5

    D. 6

24.9 Lifting inserts which are embedded or otherwise attached to tilt-up precast concrete members shall be capable of supporting at least _____ times the maximum intended load applied or transmitted to them.

A. 2
B. 4
C. 5
D. 6

24.10. Lifting inserts which are embedded or otherwise attached to precast concrete members shall be capable of supporting at least _____ times the maximum intended load applied or transmitted to them.

A. 2
B. 4
C. 5
D. 6

24.11 When embedded in concrete which one of these materials strengthens it?

A. Reinforcing steel
B. Gravel
C. Cement
D. Asbestos fibers

24.9 Practice Problems

24.1 Discuss the different steps of concrete construction.

24.2 List some general requirements for concrete and masonry equipment and tools.

24.3 List some general requirements for formwork of cast-in-place concrete.

24.4 List some general requirements for shoring and reshoring of cast-in-place concrete.

24.5 List some general requirements for vertical slip forms of cast-in-place concrete.

24.6 List some general requirements for removal of formwork of cast-in-place concrete.

24.7 List some general requirements for precast concrete.

24.8 List some general requirements for lift-slab construction operations.

24.9 List some general requirements for masonry construction.

24.10 Critical Thinking and Discussion Topic

In a masonry wall construction, you installed all the safety features recommended by OSHA. However, the mortar prepared had little more water and became so weak. The material engineer assured that it would meet the strength requirement. However, you suspect that it might collapse if you build more than 2 feet of wall. However, the wall must be finished today as it falls in the critical path. As a site supervisor, what will you do in this situation?

CHAPTER 25

Steel Erection

(Photo by the author.)

25.1 General

The erection of structural steel work consists of assembling steel components into on-site frames. Processes involve lifting and positioning the components, then connecting them together. Usually this is achieved by bolting, but sometimes site welding is used.

Steel erection essentially consists of four main tasks:

1. Establishing that the foundations are suitable and safe for an erection to commence.

2. Lifting and placing components into position, generally using cranes but sometimes by jacking. To secure components in place, bolted connections will be made, but will not yet be fully tightened. Bracings may similarly not be fully secured.

3. Aligning the structure, principally by checking that column bases are lined, and level and columns are plumb. Packing in beam-to-column connections may need to be changed to allow column plumb to be adjusted.

4. Bolting-up, which means completing all the bolted connections to secure and impart rigidity to the frame.

Steel erection is one of the top 10 most dangerous occupations according to BLS fatality data year after year. Steel erection work includes structures of heavy-duty high-rise, metal buildings, and even signs. The erection of steel is often the skeleton of bridges, office buildings, commercial, retail, and industrial buildings.

25.2   Hazards in Steel Erections

The hazards associated with the handling, transport, and erection of steel structures vary at each location, depending on the type of material, the different sizes of steel structures, and other working conditions. A job hazard analysis (JHA) is required for all of these activities as the risk levels are very high. Steel structure erection is an essential activity, with a historically significant incidence of injuries and illness accidents. Second, these steel structure erections can be overlooked when preparing safety plans. The erection activity ranges from the mobilization and installation of equipment, materials, and the facility till demobilization of the same. In keeping with the same view, this article focuses on the major hazards and controls involved exclusively during the erection of the steel structure. The following list shows hazards that have been identified for steel structure erection.

Hazards for housekeeping:

- Injuries like cut, fall, slip, and strain, sprain of body due to handling, arranging, lifting, or shifting of scaffold components
- Injury due to fall and trip, equipment failure
- Exposure to hazardous chemicals, materials, fumes gases, etc.

Hazards for materials handling/movement:

- Injury caused by falls or being struck by steel structures
- Electric shock or electrocution caused by coming in contact with live electricity wire or touching steel structural materials
- Injury due to overload, drop, trap, trike, or fall of structural members
- Injuries caused on being crushed by swing loads or struck by extended parts of steel structures; injuries such as sprains, back damage, back strain due to fall of

steel structures, unexpected trip overs while moving or climbing on the outside of steel structures

- Injury caused by damaged lifting tools tackles, defective rigging, or equipment failure
- Injury due to inadequate visibility and no communication between crane operator and signaler

Hazards for cutting, grinding, and welding:

- Injury due to fall of person and/or materials from height
- Injury due to spark and fires while welding

Hazards for forklift or crane operations:

- Fall or strike injury due to faulty operation of fork lift, hydra, or crane
- Injury to the workmen due to load shifting or load bouncing; prevent overturning or overloading of forklift, hydra, or crane
- Body injury due to fall or overturn of crane, and crane striking or hitting people

25.3 General Safety Measures in Steel Erections

The following control measures can be followed to provide a safer working environment for steel structure erection.

Safety measures for hazards for housekeeping:

- Provide adequate manpower for loading/unloading. Adequate supervision is must.
- Provide required tools, tackles, and lifting equipment.
- Use hosting facilities by crane or hydra.
- Prepare small size of each structural member for safe handling/lifting.
- Provide hardhats and safety glasses, work boots, full-length work pants, and shirts.
- The training for erection of steel structures should address some important hazards and control measures.
- Keep work areas clean and organized.
- Be aware of changes in elevations in and around jobsite, including pot holes.
- MSDS sheets for all materials shall be maintained on site. All chemicals shall be kept in the original container or marked appropriately.

Safety measures for hazards for materials handling/movement:

- Use designated trucking routes and a designated person to assist the trucks in entering and exiting the jobsite as required.
- Keep eye out for people or pedestrians on jobsite.
- Trailer carrying steel structures to be located on stable ground where it is to be unloaded prior to removing chains.

- Inspect load prior to removing chains.
- Use spotter to help guide forks to material.
- Use only trained forklift/crane operators with certification.
- Do not store steel unsafely.
- Barricade the area covering radius of crane under the load being lifted.
- Use tag line to control loads while load may swing.
- Only one person to give signals and ensure that all personnel are out of barricade area.
- Rigger to be free and away from rigging of steel structures prior to crane lifting each load. It includes hands, body, other body parts from surrounding materials, etc.
- Inspect rigging tools and tackles daily. Damaged/defective rigging tools to be destroyed to prevent use, and removed from the jobsite.
- Signalmen to be used wherever it is applicable.
- No structural members are to be left unsecured. No vertical steel member to be left standing unless secured in a minimum of two directions by horizontal members.

Safety measures for hazards for cutting, grinding, and welding:

- Use inspected and certified scaffold platform while working at height.
- Keep access ladders that are secured and extended 1.0 meter above the deck being accessed.
- While climbing on ladder, maintain tree point contact.
- Ensure that structural materials are clamped and secured firmly.
- Use required personal protective equipment.
- Keep unauthorized person out of area.
- Keep fuel container out of away from welding area.
- Keep fire extinguishers nearby to all welding operations, as well as additional fire extinguishers should be available.
- All workers must tie/anchor safety hook of safety belt.
- Transfer all loose materials like bolts, nuts, etc., by rope and bucket.
- Hardhats are required by all crew members always. Welders may wear brim backward during times of welding only.
- Barricade/flag the areas below welding operations or provide a fire watch to prevent burns to others from falling slag, etc.

Safety measures for hazards for forklift/crane operations:

- Any employee operating the forklift shall be certified and have proof of certification on forklift operation.

- No employee shall ride on the forks, on a load on the forks, or anywhere on the forklift or its load.

- Forklift operator must travel at a safe speed with loads on the lift to ensure that bouncing due to road condition/elevation (potholes, ruts, etc.) around the jobsite do not pose any hazard.

- Unbalanced loads with tendency to roll, etc., needs to be secured back to the mast of the forklift.

- All workmen are to stay clear of the sides and front of the forklift a distance equal to the full width of the load, in case the load shifts.

- The operator shall not pick objects exceeding the weight limits of the lift.

- The swing radius of the crane, including counterweights, shall be flagged off to prevent any employee from entering a zone where they could be struck by the crane.

- Ensure that the crane and all rigging materials are identified with their load capacity and possess a valid load test certificate and a thorough examination certificate issued by a competent person authorized by the government.

- Experienced signalman appointed for the task.

- In case of a heavy lift or if there is low headroom for the operation, it is treated as a critical lift and a lift plan is to be prepared.

- Ensure that lifting points are available on the load. If not they are to be fabricated, examined, and load tested by a competent person.

- Operation shall be in accordance with the site safe system of work, that is, permit to work.

- Barricade and sign erected to prohibit the entry of unauthorized personnel.

- The ground crews are wearing all basic PPE including hand gloves.

- Provide sufficient lighting for clear visibility.

25.4 Site Layout, Site-Specific Erection Plan, and Construction Sequence

Part 1926, Subpart R, Standard Number 1926.752 lists the following:

Commencement of steel erection. A steel erection contractor shall not erect steel unless it has received written notification that the concrete in the footings, piers, and walls or the mortar in the masonry piers and walls has attained, on the basis of an appropriate ASTM standard test method of field-cured samples, either 75% of the intended minimum compressive design strength or sufficient strength to support the loads imposed during steel erection.

Site layout. The controlling contractor shall ensure that the following are provided and maintained:

- Adequate access roads into and through the site for the safe delivery and movement of derricks, cranes, trucks, other necessary equipment, and the material to be erected and means and methods for pedestrian and vehicular control. Exception: this requirement does not apply to roads outside of the construction site.

- A firm, properly graded, drained area, readily accessible to the work with adequate space for the safe storage of materials and the safe operation of the erector's equipment.

Site-specific erection plan. Where employers elect, due to conditions specific to the site, to develop alternate means and methods that provide employee protection, a site-specific erection plan shall be developed by a qualified person and be available at the worksite.

Example 25.1

A steel erection contractor shall not erect steel unless it has received written notification that the concrete in the footings, piers, and walls or the mortar in the masonry piers and walls has attained:

A. 50% of the intended minimum compressive design strength

B. 65% of the intended minimum compressive design strength

C. 85% of the intended minimum compressive design strength

D. Sufficient strength to support the loads imposed during steel erection

Answer: D

25.5   Hoisting and Rigging

Major safety regulations based on the Part 1926, Subpart R, Standard Number 1926.753 for hoisting and rigging are as follows.

25.5.1   Preshift Visual Inspection of Cranes

- Cranes being used in steel erection activities shall be visually inspected prior to each shift by a competent person; the inspection shall include observation for deficiencies during operation. At a minimum this inspection shall include the following:
  - All control mechanisms for maladjustments
  - Control and drive mechanism for excessive wear of components and contamination by lubricants, water, or other foreign matter
  - Safety devices, including but not limited to boom angle indicators, boom stops, boom kickout devices, anti-two block devices, and load moment indicators where required
  - Air, hydraulic, and other pressurized lines for deterioration or leakage, particularly those which flex in normal operation
  - Hooks and latches for deformation, chemical damage, cracks, or wear

- Wire rope reeving for compliance with hoisting equipment manufacturer's specifications
- Electrical apparatus for malfunctioning, signs of excessive deterioration, dirt, or moisture accumulation
- Hydraulic system for proper fluid level
- Tires for proper inflation and condition
- Ground conditions around the hoisting equipment for proper support, including ground settling under and around outriggers, groundwater accumulation, or similar conditions
- The hoisting equipment for level position
- The hoisting equipment for level position after each move and setup

- If any deficiency is identified, an immediate determination shall be made by the competent person as to whether the deficiency constitutes a hazard.
- If the deficiency is determined to constitute a hazard, the hoisting equipment shall be removed from service until the deficiency has been corrected.
- The operator shall be responsible for those operations under the operator's direct control. Whenever there is any doubt regarding safety, the operator shall have the authority to stop and refuse to handle loads until safety has been assured.

Example 25.2

If any deficiency in a hoisting equipment is identified, what is the first work needed to be done?

A. The hoisting equipment shall be removed from service until the deficiency has been corrected.

B. An immediate correction is to be made.

C. An immediate determination shall be made by the competent person as to whether the deficiency constitutes a hazard.

D. An immediate request is to be sent to the employer for the possible actions.

Answer: C

Example 25.3

Whenever there is any doubt as to the safety of a hoisting equipment, who should have the authority to stop and refuse to handle loads until safety has been assured?

A. The employer

B. The operator

C. The surveyor

D. The OSHA inspector

Answer: B

25.5.2 Working under Loads

Routes for suspended loads shall be preplanned to ensure that no employee is required to work directly below a suspended load except for:

- Employees engaged in the initial connection of the steel
- Employees necessary for the hooking or unhooking of the load

When working under suspended loads, the following criteria shall be met:

- Materials being hoisted shall be rigged to prevent unintentional displacement.
- Hooks with self-closing safety latches or their equivalent shall be used to prevent components from slipping out of the hook.
- All loads shall be rigged by a qualified rigger.

25.5.3 Multiple Lift Rigging Procedure

A multiple lift shall only be performed if the following criteria are met:

- A multiple lift rigging assembly is used.
- A maximum of five members are hoisted per lift.
- Only beams and similar structural members are lifted.
- All employees engaged in the multiple lift have been trained.
- No crane is permitted to be used for a multiple lift where such use is contrary to the manufacturer's specifications and limitations.

Components of the multiple lift rigging assembly shall be specifically designed and assembled with a maximum capacity for total assembly and for each individual attachment point. This capacity, certified by the manufacturer or a qualified rigger, shall be based on the manufacturer's specifications with a 5 to 1 safety factor for all components.

25.6 Structural Steel Assembly

Major safety regulations based on the Part 1926, Subpart R, Standard Number 1926.754 for structural steel assembly are as follows.

25.6.1 Regulations for Multistory Structures

The following additional requirements shall apply for multistory structures:

- The permanent floors shall be installed as the erection of structural members progresses, and there shall be not more than eight stories between the erection floor and the uppermost permanent floor, except where the structural integrity is maintained as a result of the design.
- At no time shall there be more than four floors or 48 feet (14.6 meter), whichever is less, of unfinished bolting or welding above the foundation or uppermost permanently secured floor, except where the structural integrity is maintained as a result of the design.

- A fully planked or decked floor or nets shall be maintained within two stories or 30 feet (9.1 meter), whichever is less, directly under any erection work being performed.

25.6.2 Working with Shear Connectors

Tripping hazards. Shear connectors (such as headed steel studs, steel bars, or steel lugs), reinforcing bars, deformed anchors, or threaded studs shall not be attached to the top flanges of beams, joists, or beam attachments so that they project vertically from or horizontally across the top flange of the member until after the metal decking, or other walking/working surface, has been installed.

Installation of shear connectors. When shear connectors are used in construction of composite floors, roofs, and bridge decks, employees shall lay out and install the shear connectors after the metal decking has been installed, using the metal decking as a working platform. Shear connectors shall not be installed from within a controlled decking zone (CDZ).

25.6.3 Plumbing-Up

When deemed necessary by a competent person, plumbing-up equipment shall be installed in. When used, plumbing-up equipment shall be in place and properly installed before the structure is loaded with construction material such as loads of joists, bundles of decking, or bundles of bridging. Plumbing-up equipment shall be removed only with the approval of a competent person.

25.6.4 Metal Decking

Hoisting, landing, and placing of metal decking bundles. Bundle packaging and strapping shall not be used for hoisting unless specifically designed for that purpose. If loose items such as dunnage, flashing, or other materials are placed on the top of metal decking bundles to be hoisted, such items shall be secured to the bundles. Metal decking bundles shall be landed on framing members so that enough support is provided to allow the bundles to be unbanded without dislodging the bundles from the supports. At the end of the shift or when environmental or jobsite conditions require, metal decking shall be secured against displacement.

Roof and floor holes and openings. Metal decking at roof and floor holes and openings shall be installed as follows:

- Framed metal deck openings shall have structural members turned down to allow continuous deck installation except where not allowed by structural design constraints or constructability.

- Roof and floor holes and openings shall be decked over. Where large size, configuration or other structural design does not allow openings to be decked over (such as elevator shafts, stair wells, etc.), employees shall be protected.

- Metal decking holes and openings shall not be cut until immediately prior to being permanently filled with the equipment or structure needed or intended to fulfill its specific use shall be immediately covered.

Covering roof and floor openings.

- Covers for roof and floor openings shall be capable of supporting, without failure, twice the weight of the employees, equipment, and materials that may be imposed on the cover at any one time.
- All covers shall be secured when installed to prevent accidental displacement by the wind, equipment, or employees.
- All covers shall be painted with high-visibility paint or shall be marked with the word "HOLE" or "COVER" to provide warning of the hazard.
- Smoke dome or skylight fixtures that have been installed are not considered covers for the purpose of this section unless they meet the strength requirements.

Decking gaps around columns.   Wire mesh, exterior plywood, or equivalent shall be installed around columns where planks or metal decking do not fit tightly. The materials used must be of sufficient strength to provide fall protection for personnel and prevent objects from falling through.

Installation of metal decking.   Metal decking shall be laid tightly and immediately secured upon placement to prevent accidental movement or displacement. During initial placement, metal decking panels shall be placed to ensure full support by structural members.

Derrick floors.   A derrick floor shall be fully decked and/or planked and the steel member connections completed to support the intended floor loading. Temporary loads placed on a derrick floor shall be distributed over the underlying support members so as to prevent local overloading of the deck material.

25.7   Column Anchorage

Major safety regulations based on the Part 1926, Subpart R, Standard Number 1926.755 for column anchorage are as follows.

25.7.1   General Requirements for Erection Stability

- All columns shall be anchored by a minimum of four anchor rods (anchor bolts).
- Each column anchor rod (anchor bolt) assembly, including the column-to-base plate weld and the column foundation, shall be designed to resist a minimum eccentric gravity load of 300 pounds (136.2 kilograms) located 18 inches (0.46 meter) from the extreme outer face of the column in each direction at the top of the column shaft.
- Columns shall be set on level finished floors, pregrouted leveling plates, leveling nuts, or shim packs which are adequate to transfer the construction loads.
- All columns shall be evaluated by a competent person to determine whether guying or bracing is needed; if guying or bracing is needed, it shall be installed.

25.7.2   Repair, Replacement, or Field Modification of Anchor Rods (Anchor Bolts)

Anchor rods (anchor bolts) shall not be repaired, replaced, or field-modified without the approval of the project structural engineer of record. Prior to the erection of a column, the controlling contractor shall provide written notification to the steel erector if there has been any repair, replacement, or modification of the anchor rods (anchor bolts) of that column.

25.8   Beams and Columns

Major safety regulations based on the Part 1926, Subpart R, Standard Number 1926.756 for beams and columns erections are as follows.

General.   During the final placing of solid web structural members, the load shall not be released from the hoisting line until the members are secured with at least two bolts per connection, of the same size and strength as shown in the erection drawings, drawn up wrench-tight or the equivalent as specified by the project structural engineer of record. A competent person shall determine if more than two bolts are necessary to ensure the stability of cantilevered members; if additional bolts are needed, they shall be installed.

Diagonal bracing.   Solid web structural members used as diagonal bracing shall be secured by at least one bolt per connection drawn up wrench-tight or the equivalent as specified by the project structural engineer of record.

Double connections at columns and/or at beam webs over a column.   When two structural members on opposite sides of a column web, or a beam web over a column, are connected sharing common connection holes, at least one bolt with its wrench-tight nut shall remain connected to the first member unless a shop-attached or field-attached seat or equivalent connection device is supplied with the member to secure the first member and prevent the column from being displaced.

If a seat or equivalent device is used, the seat (or device) shall be designed to support the load during the double connection process. It shall be adequately bolted or welded to both a supporting member and the first member before the nuts on the shared bolts are removed to make the double connection.

Column splices.   Each column splice shall be designed to resist a minimum eccentric gravity load of 300 pounds (136.2 kilogram) located 18 inches (0.46 meter) from the extreme outer face of the column in each direction at the top of the column shaft.

Perimeter columns.   Perimeter columns shall not be erected unless the perimeter columns extend a minimum of 48 inches (1.2 meter) above the finished floor to permit installation of perimeter safety cables prior to erection of the next tier, except where constructability does not allow.

25.9   Open Web Steel Joists

Where steel joists are used and columns are not framed in at least two directions with solid web structural steel members, a steel joist shall be field-bolted at the column to provide lateral stability to the column during erection. For the installation of this joist:

- A vertical stabilizer plate shall be provided on each column for steel joists. The plate shall be a minimum of 6 inch by 6 inch (152 millimeter by 152 millimeter) and shall extend at least 3 inches (76 millimeter) below the bottom chord of the joist with a 13/16 inch (21 millimeter) hole to provide an attachment point for guying or plumbing cables.
- The bottom chords of steel joists at columns shall be stabilized to prevent rotation during erection.
- Hoisting cables shall not be released until the seat at each end of the steel joist is field-bolted, and each end of the bottom chord is restrained by the column stabilizer plate.

Where constructability does not allow a steel joist to be installed at the column:

- An alternate means of stabilizing joists shall be installed on both sides near the column and shall:
  - Provide stability
  - Be designed by a qualified person
  - Be shop installed
  - Be included in the erection drawings
- Hoisting cables shall not be released until the seat at each end of the steel joist is field-bolted and the joist is stabilized.
- Where steel joists at or near columns span 60 feet (18.3 meter) or less, the joist shall be designed with sufficient strength to allow one employee to release the hoisting cable without the need for erection bridging.
- Where steel joists at or near columns span more than 60 feet (18.3 meter), the joists shall be set in tandem with all bridging installed unless an alternative method of erection, which provides equivalent stability to the steel joist, is designed by a qualified person and is included in the site-specific erection plan.
- A steel joist or steel joist girder shall not be placed on any support structure unless such structure is stabilized.
- When steel joist(s) are landed on a structure, they shall be secured to prevent unintentional displacement prior to installation.
- No modification that affects the strength of a steel joist or steel joist girder shall be made without the approval of the project structural engineer of record.

Some other areas where OSHA 1926.757 provided specifications are as follows:

- Field-bolted joists
- Attachment of steel joists and steel joist girders
- Erection of steel joists
- Erection bridging
- Landing and placing loads

25.10   Systems-Engineered Metal Buildings

Major safety regulations based on the Part 1926, Subpart R, Standard Number 1926.758 for systems-engineered metal buildings are as follows:

- Each structural column shall be anchored by a minimum of four anchor rods (anchor bolts).

- Rigid frames shall have 50% of their bolts or the number of bolts specified by the manufacturer (whichever is greater) installed and tightened on both sides of the web adjacent to each flange before the hoisting equipment is released.

- Construction loads shall not be placed on any structural steel framework unless such framework is safely bolted, welded, or otherwise adequately secured.

- In girt and eave strut-to-frame connections, when girts or eave struts share common connection holes, at least one bolt with its wrench-tight nut shall remain connected to the first member unless a manufacturer-supplied, field-attached seat or similar connection device is present to secure the first member so that the girt or eave strut is always secured against displacement.

- Both ends of all steel joists or cold-formed joists shall be fully bolted and/or welded to the support structure before:
  - Releasing the hoisting cables
  - Allowing an employee on the joists
  - Allowing any construction loads on the joists

- Purlins and girts shall not be used as an anchorage point for a fall arrest system unless written approval is obtained from a qualified person.

- Purlins may only be used as a walking/working surface when installing safety systems, after all permanent bridging has been installed and fall protection is provided.

- Construction loads may be placed only within a zone that is within 8 feet (2.5 meter) of the centerline of the primary support member.

25.11   Fall Protection

Part 1926, Subpart R, Standard Number 1926.760 states that each employee engaged in a steel erection activity who is on a walking/working surface with an unprotected side or edge more than 15 feet (4.6 meter) above a lower level shall be protected from fall hazards by guardrail systems, safety net systems, personal fall arrest systems, positioning device systems, or fall restraint systems.

Perimeter safety cables.   On multistory structures, perimeter safety cables shall be installed at the final interior and exterior perimeters of the floors as soon as the metal decking has been installed.

Connectors.   Each connector shall:

- Be protected from fall hazards of more than two stories or 30 feet (9.1 meter) above a lower level, whichever is less.

- Be provided, at heights over 15 and up to 30 feet above a lower level, with a personal fall arrest system, positioning device system, or fall restraint system and wear the equipment necessary to be able to be tied off; or be provided with other means of protection from fall hazards.

Controlled decking zone (CDZ).   A CDZ may be established in that area of the structure over 15 and up to 30 feet above a lower level where metal decking is initially being installed and forms the leading edge of a work area. In each CDZ, the following shall apply:

- Each employee working at the leading edge in a CDZ shall be protected from fall hazards of more than two stories or 30 feet (9.1 meter), whichever is less.

- Access to a CDZ shall be limited to only those employees engaged in leading edge work.

- The boundaries of a CDZ shall be designated and clearly marked. The CDZ shall not be more than 90 feet (27.4 meter) wide and 90 (27.4 meter) feet deep from any leading edge. The CDZ shall be marked by the use of control lines or the equivalent.

- Each employee working in a CDZ shall have completed CDZ training.

- Unsecured decking in a CDZ shall not exceed 3000 square feet (914.4 square meter).

- Safety deck attachments shall be performed in the CDZ from the leading edge back to the control line and shall have at least two attachments for each metal decking panel.

- Final deck attachments and installation of shear connectors shall not be performed in the CDZ.

Criteria for fall protection equipment.   Guardrail systems, safety net systems, personal fall arrest systems, positioning device systems and their components shall conform to the criteria in OSHA 1926.502. Fall arrest system components shall be used in fall restraint systems and shall conform to the criteria. Either body belts or body harnesses shall be used in fall restraint systems.

Custody of fall protection.   Fall protection provided by the steel erector shall remain in the area where steel erection activity has been completed, to be used by other trades, only if the controlling contractor or its authorized representative:

- Has directed the steel erector to leave the fall protection in place

- Has inspected and accepted control and responsibility of the fall protection prior to authorizing persons other than steel erectors to work in the area

Example 25.4
Each employee working at the leading edge in a controlled decking zone (CDZ) shall be protected from fall hazards of more than:

    A. Two stories

    B. One story

    C. 15 feet

    D. 25 feet

Answer: *A*
Each employee working at the leading edge in a CDZ shall be protected from fall hazards of more than two stories or 30 feet (9.1 meter), whichever is less.

25.12 Summary

Steel erection work includes structures of heavy-duty, high-rise, metal buildings, and even signs. The erection of steel is often the skeleton of bridges, office buildings, and commercial, retail, and industrial buildings. The hazards associated with the handling, transport, and erection of steel structures vary at each location, depending on the type of material, the different sizes of steel structures, and other working conditions. Proper safety specifications must be followed to avoid hazards while steel erection.

25.13 Multiple-Choice Questions

25.1 A steel erection contractor shall not erect steel unless the concrete in the footings, piers, and walls or the mortar in the masonry piers and walls has attained:

    A.   25% of the intended minimum compressive design strength
    B.   50% of the intended minimum compressive design strength
    C.   75% of the intended minimum compressive design strength
    D.   100% of the intended minimum compressive design strength

25.2 What is the maximum number of floors regulated of unfinished bolting or welding above the foundation or uppermost permanently secured floor?

    A.   One floor
    B.   Two floors
    C.   Three floors
    D.   Four floors

25.3 What is the maximum height for which a fully planked or decked floor or nets shall be maintained directly under any erection work being performed?

    A.   10 feet
    B.   20 feet
    C.   30 feet
    D.   40 feet

25.4 Each column anchor rod (anchor bolt) assembly, including the column-to-base plate weld and the column foundation, shall be designed to resist a minimum eccentric gravity load of:

A. 100 pounds
B. 200 pounds
C. 300 pounds
D. 400 pounds

25.5 All columns shall be anchored by a minimum of:

A. 3 anchor rods
B. 4 anchor rods
C. 5 anchor rods
D. 6 anchor rods

25.6 Each column splice shall be designed to resist a minimum eccentric gravity load of:

A. 100 pounds
B. 200 pounds
C. 300 pounds
D. 400 pounds

25.7 A vertical stabilizer plate shall be provided on each column for steel joists. The plate shall be a minimum of:

A. 2 inch-square
B. 4 inch-square
C. 6 inch-square
D. 8 inch-square

25.8 What is the percent of bolts to be installed and tightened on both sides of the web adjacent to each flange in rigid frame before the hoisting equipment is released?

A. 25
B. 50
C. 75
D. 100

25.9 Construction loads may be placed only within a zone that is within _____ of the centerline of the primary support member.

A. 2 feet
B. 4 feet
C. 6 feet
D. 8 feet

25.14 Practice Problems

25.1 List the tasks associated with steel erection.

25.2 List the hazards associated with steel erection.

25.3 List the general safety measures to be adopted to avoid the hazards associated with steel erection.

25.4 List the safety regulations for site layout of steel erection process.

25.5 List the safety regulations for the preshift visual inspection of cranes.

25.6 List the safety regulations associated with multiple lift rigging procedure.

25.7 List safety regulations associated with installation of shear connectors on composite floors, roofs, and bridge decks.

25.8 List safety regulations associated with column anchorage steel erection.

25.15   Critical Thinking and Discussion Topic

Steel erection works require careful attention. The properties of steel depend on the temperature a lot. Do you think any difference in safety practices for steel erection considering day versus night, cold versus hot temperature?

(Photo courtesy of 123RF.)

CHAPTER 26

Underground Construction, Caissons, Cofferdams, and Compressed Air

(Photo courtesy of 123RF.)

26.1 General

Underground construction means underground tunnels, shafts, chambers, and passageways. It also includes cut-and-cover excavations which are both physically connected to ongoing underground construction operations. A caisson is a watertight retaining structure used to work on the foundations of a bridge pier, for the construction of a concrete dam, or for the repair of ships. Caissons are constructed in such a way that the water can be pumped out, keeping the work environment dry. A cofferdam is a structure that retains water and soil that allows the enclosed area to be pumped out and excavated dry. Cofferdams are commonly used for construction of bridge piers and other support structures built within water.

26.2 Underground Construction

26.2.1 General Safety

Part 1926, Subpart S, Standard Number 1926.800 states that the employer shall provide and maintain safe means of access and egress to all work stations. The employer shall provide access and egress in such a manner that employees are protected from being struck by excavators, haulage machines, trains, and other mobile equipment. The employer shall control access to all openings to prevent unauthorized entry underground. Unused chutes, manways, or other openings shall be tightly covered, bulkheaded, or fenced off, and shall be posted with warning signs indicating "Keep Out" or similar language. Completed or unused sections of the underground facility shall be barricaded.

26.2.2 Check-In/Check-Out

The employer shall maintain a check-in/check-out procedure that will ensure that aboveground personnel can determine an accurate count of the number of persons underground in the event of an emergency. However, this procedure is not required when the construction of underground facilities designed for human occupancy has been sufficiently completed so that the permanent environmental controls are effective, and when the remaining construction activity will not cause any environmental hazard or structural failure within the facilities.

Example 26.1

The _____ shall maintain a check-in/check-out procedure that will ensure that aboveground personnel can determine an accurate count of the number of persons underground.

    A. Site supervisor

    B. Competent person

    C. Professional engineer

    D. Employer

Answer: D

26.2.3 Safety Specifications

All employees shall be instructed in the recognition and avoidance of hazards associated with underground construction activities including, where appropriate, the following subjects:

- Air monitoring
- Ventilation

- Illumination
- Communications
- Flood control
- Mechanical equipment
- Personal protective equipment
- Explosives
- Fire prevention and protection and
- Emergency procedures, including evacuation plans and check-in/check-out systems

26.3   Safety Regulations for Caissons

Part 1926, Subpart S, Standard Number 1926.801 states that:

- Wherever, in caisson work in which compressed air is used, and the working chamber is less than 11 feet in length, and when such caissons are at any time suspended or hung while work is in progress so that the bottom of the excavation is more than 9 feet below the deck of the working chamber, a shield shall be erected therein for the protection of the employees.
- Shafts shall be subjected to a hydrostatic or air-pressure test, at which pressure they shall be tight. The shaft shall be stamped on the outside shell about 12 inches from each flange to show the pressure to which they have been subjected.
- Whenever a shaft is used, it shall be provided, where space permits, with a safe, proper, and suitable staircase for its entire length, including landing platforms, not more than 20 feet apart. Where this is impracticable, suitable ladders shall be installed with landing platforms located about 20 feet apart to break the climb.
- All caissons having a diameter or side greater than 10 feet shall be provided with a man lock and shaft for the exclusive use of employees.
- In addition to the gauge in the locks, an accurate gauge shall be maintained on the outer and inner side of each bulkhead. These gauges shall be accessible at all times and kept in accurate working order.

Example 26.2
Whenever a shaft is used, it shall be provided, where space permits, with a safe, proper, and suitable staircase for its entire length, including landing platforms, not more than:

A. 5 feet apart

B. 10 feet apart

C. 20 feet apart

D. 25 feet apart

Answer: C

26.4 Safety Regulations for Cofferdams

Part 1926, Subpart S, Standard Number 1926.802 states that:

- If overtopping of the cofferdam by high waters is possible, means shall be provided for controlled flooding of the work area.

- Warning signals for evacuation of employees in case of emergency shall be developed and posted.

- Cofferdam walkways, bridges, or ramps with at least two means of rapid exit shall be provided with guardrails.

- Cofferdams located close to navigable shipping channels shall be protected from vessels in transit, where possible.

26.5 Compressed Air

Part 1926, Subpart S, Standard Number 1926.803 provides specifications in the following areas.

26.5.1 Medical Attendance, Examination, and Regulations

- There shall be retained one or more licensed physicians familiar with and experienced in the physical requirements and the medical aspects of compressed air work and the treatment of decompression illness. He shall be available at all times while work is in progress in order to provide medical supervision of employees employed in compressed air work. He shall himself be physically qualified and be willing to enter a pressurized environment.

- No employee shall be permitted to enter a compressed air environment until he has been examined by the physician and reported by him to be physically qualified to engage in such work.

- In the event an employee is absent from work for 10 days, or is absent due to sickness or injury, he shall not resume work until he is reexamined by the physician, and his physical condition reported, as provided in this paragraph, to be such as to permit him to work in compressed air.

- After an employee has been employed continuously in compressed air for a period designated by the physician, but not to exceed 1 year, he shall be reexamined by the physician to determine if he is still physically qualified to engage in compressed air work.

- Such physician shall at all times keep a complete and full record of examinations made by him. The physician shall also keep an accurate record of any decompression illness or other illness or injury incapacitating any employee for work, and of all loss of life that occurs in the operation of a tunnel, caisson, or other compartment in which compressed air is used.

- Records shall be available for the inspection of the Secretary or his representatives, and a copy thereof shall be forwarded to OSHA within 48 hours following the occurrence of the accident, death, injury, or decompression illness. It shall state as fully as possible the cause of said death or decompression illness, and the place where the injured or sick employee was taken, and such other relative information as may be required by the Secretary.

- A fully equipped first-aid station shall be provided at each tunnel project regardless of the number of persons employed. An ambulance or transportation suitable for a litter case shall be at each project.

- Where tunnels are being excavated from portals more than 5 road miles apart, a first-aid station and transportation facilities shall be provided at each portal.

- A medical lock shall be established and maintained in immediate working order whenever air pressure in the working chamber is increased above the normal atmosphere.

26.5.2 Telephone and Signal Communication

Effective and reliable means of communication, such as bells, whistles, or telephones, shall be maintained, at all times between all the following locations:

- The working chamber face
- The working chamber side of the man lock near the door
- The interior of the man lock
- Lock attendant's station
- The compressor plant
- The first-aid station
- The emergency lock (if one is required)
- The special decompression chamber (if one is required)

26.5.3 Signs and Records

Any code of signals used shall be conspicuously posted near workplace entrances and such other locations as may be necessary to bring them to the attention of all employees concerned. For each 8-hour shift, a record of employees employed under air pressure shall be kept by an employee who shall remain outside the lock near the entrance. This record shall show the period each employee spends in the air chamber and the time taken from decompression. A copy shall be submitted to the appointed physician after each shift.

26.5.4 Compression

- Every employee going under air pressure for the first time shall be instructed on how to avoid excessive discomfort.

- During the compression of employees, the pressure shall not be increased to more than 3 pounds per square inch gauge within the first minute. The pressure shall be held at 3 pounds per square inch gauge and again at 7 pounds per square inch gauge sufficiently long to determine if any employee is experiencing discomfort.

- After the first minute the pressure shall be raised uniformly and at a rate not to exceed 10 pounds per square inch per minute.

- If any employee complains of discomfort, the pressure shall be held to determine if the symptoms are relieved. If, after 5 minutes the discomfort does not disappear, the lock attendant shall gradually reduce the pressure until the employee signals that the discomfort has ceased. If he does not indicate that the discomfort has disappeared, the lock attendant shall reduce the pressure to atmospheric and the employee shall be released from the lock.

- No employee shall be subjected to pressure exceeding 50 pounds per square inch except in emergency.

Example 26.3
For normal operation, no employee shall be subjected to pressure exceeding:

    A. 20 pounds per square inch

    B. 40 pounds per square inch

    C. 50 pounds per square inch

    D. 75 pounds per square inch

Answer: C

26.5.5 Decompression

In the event it is necessary for an employee to be in compressed air more than once in a 24-hour period, the appointed physician shall be responsible for the establishment of methods and procedures of decompression applicable to repetitive exposures. If decanting is necessary, the appointed physician shall establish procedures before any employee is permitted to be decompressed by decanting methods. The period of time that the employees spend at atmospheric pressure between the decompression following the shift and recompression shall not exceed 5 minutes.

26.5.6 Man Locks

- Except in emergency, no employees employed in compressed air shall be permitted to pass from the working chamber to atmospheric pressure until after decompression.

- The lock attendant in charge of a man lock shall be under the direct supervision of the appointed physician. He shall be stationed at the lock controls on the free air side during the period of compression and decompression and shall remain at the lock control station whenever there are men in the working chamber or in the man lock.

- Except where air pressure in the working chamber is below 12 pounds per square inch gauge, each man lock shall be equipped with automatic controls which, through taped programs, cams, or similar apparatus, shall automatically regulate decompressions. It shall also be equipped with manual controls to permit the lock attendant to override the automatic mechanism in the event of an emergency.

26.5.7   Special Decompression Chambers

The headroom in the special decompression chamber shall not be less than a minimum 7 feet and the cubical content shall provide at least 50 cubic feet of airspace for each employee. For each occupant, there shall be provided 4 square feet of free walking area and 3 square feet of seating space, exclusive of area required for lavatory and toilet facilities. The rated capacity shall be based on the stated minimum space per employee and shall be posted at the chamber entrance. The posted capacity shall not be exceeded, except in case of emergency. Each special decompression chamber shall be equipped with the following:

- A clock or clocks suitably placed so that the attendant and the chamber occupants can readily ascertain the time

- Pressure gauges which will indicate to the attendants and to the chamber occupants the pressure in the chamber

- Valves to enable the attendant to control the supply and discharge of compressed air into and from the chamber

- Valves and pipes, in connection with the air supply and exhaust, arranged so that the chamber pressure can be controlled from within and without

- Effective means of oral intercommunication between the attendant, occupants of the chamber, and the air compressor plant

- An observation port at the entrance to permit observation of the chamber occupants

26.5.8   Compressor Plant and Air Supply

- At all times there shall be a thoroughly experienced, competent, and reliable person on duty at the air control valves as a gauge tender who shall regulate the pressure in the working areas. During tunneling operations, one gauge tender may regulate the pressure in not more than two headings: Provided that the gauge and controls are all in one location. In caisson work, there shall be a gauge tender for each caisson.

- The low air compressor plant shall be of sufficient capacity to not only permit the work to be done safely, but shall also provide a margin to meet emergencies and repairs.

- Low air compressor units shall have at least two independent and separate sources of power supply and each shall be capable of operating the entire low air plant and its accessory systems.

- The capacity, arrangement, and number of compressors shall be sufficient to maintain the necessary pressure without overloading the equipment and to assure maintenance of such pressure in the working chamber during periods of breakdown, repair, or emergency.

- Switching from one independent source of power supply to the other shall be done periodically to ensure the workability of the apparatus in an emergency.

- Duplicate low-pressure air feedlines and regulating valves shall be provided between the source of air supply and a point beyond the locks with one of the lines extending to within 100 feet of the working face.

- All high- and low-pressure air supply lines shall be equipped with check valves.
- Low-pressure air shall be regulated automatically. In addition, manually operated valves shall be provided for emergency conditions.
- The air intakes for all air compressors shall be located at a place where fumes, exhaust, gases, and other air contaminants will be at a minimum.
- Gauges indicating the pressure in the working chamber shall be installed in the compressor building, the lock attendant's station, and at the employer's field office.

26.5.9   Ventilation and Air Quality

- Exhaust valves and exhaust pipes shall be provided and operated so that the working chamber shall be well ventilated, and there shall be no pockets of dead air. Outlets may be required at intermediate points along the main low-pressure air supply line to the heading to eliminate such pockets of dead air. Ventilating air shall not be less than 30 cubic feet per minute.
- The air in the workplace shall be analyzed by the employer not less than once each shift, and records of such tests shall be kept on file at the place where the work is in progress. The test results shall be within the threshold limit values, for hazardous gases, and within 10% of the lower explosive limit of flammable gases. If these limits are not met, immediate action to correct the situation shall be taken by the employer.
- The temperature of all working chambers which are subjected to air pressure shall, by means of after-coolers or other suitable devices, be maintained at a temperature not to exceed 85°F.
- Forced ventilation shall be provided during decompression. During the entire decompression period, forced ventilation through chemical or mechanical air purifying devices that will ensure a source of fresh air shall be provided.
- Whenever heat-producing machines (moles, shields) are used in compressed air tunnel operations, a positive means of removing the heat buildup at the heading shall be provided.

26.5.10   Electricity

- All lighting in compressed air chambers shall be by electricity exclusively, and two independent electric-lighting systems with independent sources of supply shall be used. The emergency source shall be arranged to become automatically operative in the event of failure of the regularly used source.
- The minimum intensity of light on any walkway, ladder, stairway, or working level shall not be less than 10 foot-candles, and in all workplaces the lighting shall at all times be such as to enable employees to see clearly.
- All electrical equipment and wiring for light and power circuits shall comply with the requirements for use in damp, hazardous, high temperature, and compressed air environments.
- External parts of lighting fixtures and all other electrical equipment, when within 8 feet of the floor, shall be constructed of noncombustible, nonabsorptive, insulating materials, except that metal may be used if it is effectively grounded.

- Portable lamps shall be equipped with noncombustible, nonabsorptive, insulating sockets, approved handles, basket guards, and approved cords.
- The use of worn or defective portable and pendant conductors is prohibited.

26.5.11 Sanitation

- Sanitary, heated, lighted, and ventilated dressing rooms and drying rooms shall be provided for all employees engaged in compressed air work. Such rooms shall contain suitable benches and lockers. Bathing accommodations (showers at the ratio of one to 10 employees per shift), equipped with running hot and cold water, and suitable and adequate toilet accommodations, shall be provided. One toilet for each 15 employees, or fractional part thereof, shall be provided.
- When the toilet bowl is shut by a cover, there should be an air space so that the bowl or bucket does not implode when pressure is increased.
- All parts of caissons and other working compartments shall be kept in a sanitary condition.

Example 26.4

In an underground construction area, there are 40 employees working. How many toilets are to be provided at a minimum level?

    A. 2

    B. 3

    C. 4

    D. 5

Answer: B

One toilet for each 15 employees shall be provided.

26.5.12 Fire Prevention and Protection

- Firefighting equipment shall be available at all times and shall be maintained in working condition.
- While welding or flame-cutting is being done in compressed air, a firewatch with a fire hose or approved extinguisher shall stand by until such operation is completed.
- Shafts and caissons containing flammable material of any kind, either above or below ground, shall be provided with a waterline and a fire hose connected thereto, so arranged that all points of the shaft or caisson are within reach of the hose stream.
- Fire hose shall be at least 1.5 inches in nominal diameter; the water pressure shall at all times be adequate for efficient operation of the type of nozzle used; and the water supply shall be such as to ensure an uninterrupted flow. Fire hose, when not in use, shall be located or guarded to prevent injury thereto.
- The power house, compressor house, and all buildings housing ventilating equipment shall be provided with at least one hose connection in the water line,

with a fire hose connected thereto. A fire hose shall be maintained within reach of structures of wood over or near shafts.

- Tunnels shall be provided with a 2-inch minimum diameter water line extending into the working chamber and to within 100 feet of the working face. Such line shall have hose outlets with 100 feet of fire hose attached and maintained as follows: One at the working face, one immediately inside of the bulkhead of the working chamber, and one immediately outside such bulkhead. In addition, hose outlets shall be provided at 200-foot intervals throughout the length of the tunnel, and 100 feet of fire hose shall be attached to the outlet nearest to any location where flammable material is being kept or stored or where any flame is being used.

- In addition to fire hose protection required by this subpart, on every floor of every building not under compressed air, but used in connection with the compressed air work, there shall be provided at least one approved fire extinguisher of the proper type for the hazard involved. At least two approved fire extinguishers shall be provided in the working chamber as follows: One at the working face and one immediately inside the bulkhead (pressure side). Extinguishers in the working chamber shall use water as the primary extinguishing agent and shall not use any extinguishing agent which could be harmful to the employees in the working chamber. The fire extinguisher shall be protected from damage.

- Highly combustible materials shall not be used or stored in the working chamber. Wood, paper, and similar combustible material shall not be used in the working chamber in quantities which could cause a fire hazard. The compressor building shall be constructed of noncombustible material.

- Man locks shall be equipped with a manual-type fire extinguisher system that can be activated inside the man lock and also by the outside lock attendant. In addition, a fire hose and portable fire extinguisher shall be provided inside and outside the man lock. The portable fire extinguisher shall be the dry chemical type.

- Equipment, fixtures, and furniture in man locks and special decompression chambers shall be constructed of noncombustible materials. Bedding, etc., shall be chemically treated so as to be fire resistant.

- Head frames shall be constructed of structural steel or open frame work fireproofed timber. Head houses and other temporary surface buildings or structures within 100 feet of the shaft, caisson, or tunnel opening shall be built of fire-resistant materials.

- No oil, gasoline, or other combustible material shall be stored within 100 feet of any shaft, caisson, or tunnel opening, except that oils may be stored in suitable tanks in isolated fireproof buildings, provided such buildings are not less than 50 feet from any shaft, caisson, or tunnel opening, or any building directly connected thereto.

- Positive means shall be taken to prevent leaking flammable liquids from flowing into the areas specifically mentioned in the preceding paragraph.

- All explosives used in connection with compressed air work shall be selected, stored, transported, and used as specified.

26.5.13   Bulkheads and Safety Screens

- Intermediate bulkheads with locks, or intermediate safety screens or both, are required where there is the danger of rapid flooding.

- In tunnels 16 feet or more in diameter, hanging walkways shall be provided from the face to the man lock as high in the tunnel as practicable, with at least 6 feet of head room. Walkways shall be constructed of noncombustible material. Standard railings shall be securely installed throughout the length of all walkways on open sides. Where walkways are ramped under safety screens, the walkway surface shall be skidproofed by cleats or by equivalent means.

- Bulkheads used to contain compressed air shall be tested, where practicable, to prove their ability to resist the highest air pressure which may be expected to be used.

Example 26.5

Hanging walkways shall be provided from the face to the man lock as high in the tunnel as practicable, with at least 6 feet of head room, for tunnel diameter of:

A. 5 feet or less

B. 6–10 feet

C. 11–15 feet

D. 16 feet or more

Answer: D

26.6   Summary

Underground construction includes underground tunnels, shafts, chambers, passageways, cut-and-cover excavations, etc. The employer should provide access and egress in a safe way so that employees are protected from being struck by excavators, haulage machines, trains, and other mobile equipment. The employer shall control access to all openings to prevent unauthorized entry underground. All employees shall be instructed in the recognition and avoidance of hazards associated with underground construction activities. Any code of signals used shall be conspicuously posted near workplace entrances and such other locations as may be necessary to bring them to the attention of all employees concerned.

26.7   Multiple-Choice Questions

26.1 Underground construction means:

    A. Underground tunnels

    B. Shafts

    C. Chambers

    D. Passageways

    E. All of the above

26.2 Who shall control access to all openings to prevent unauthorized entry underground?

    A.   The employer
    B.   The employee
    C.   The workers
    D.   None of the above

26.3 Which of the following tasks can be performed to underground construction safety?

    A.   Unused chutes, manways, or other openings shall be fenced off.
    B.   Unused chutes, manways, or other openings shall be posted with warning signs.
    C.   Completed or unused sections of the underground facility shall be barricaded.
    D.   All of the above.

26.4 How can an aboveground personnel determine an accurate count of the number of persons underground in the event of an emergency?

    A.   Maintain a check-in/check-out procedure
    B.   Maintaining a camera
    C.   Keep a watchman to count
    D.   None of the above

26.5 All caissons having a diameter or side greater than _____ shall be provided with a man lock and shaft for the exclusive use of employees.

    A.   10 feet
    B.   20 feet
    C.   11 feet
    D.   12 inches

26.6 Cofferdam walkways, bridges, or ramps with at least _____ means of rapid exit shall be provided with guardrails.

    A.   One
    B.   Two
    C.   Three
    D.   Four

26.7 Records shall be available for the inspection of the Secretary or his representatives, and a copy thereof shall be forwarded to OSHA within _____ hours following the occurrence of the accident, death, injury, or decompression illness.

    A.   24
    B.   48
    C.   72
    D.   96

26.8 During the compression of employees, the pressure shall not be increased to more than _____ pounds per square inch gauge within the first minute.

    A.   1
    B.   2
    C.   3
    D.   4

26.9 The headroom in the special decompression chamber shall be not less than a minimum:

    A.   3 feet
    B.   5 feet
    C.   7 feet
    D.   9 feet

26.10 Duplicate low-pressure air feedlines and regulating valves shall be provided between the source of air supply and a point beyond the locks with one of the lines extending to within _____ feet of the working face.

A. 25
B. 50
C. 75
D. 100

26.11 For fire prevention and protection, fire hose shall be at least _____ inches in nominal diameter.

A. 1/2
B. 1-1/2
C. 2-1/2
D. 4-1/2

26.8   Practice Problems

26.1 Define cofferdam and caisson.
26.2 Mention some general safety regulations for underground construction.
26.3 What is the benefit of checking-in and checking-out for underground construction?
26.4 List the safety regulations for caissons.
26.5 List the safety regulations for cofferdams.
26.6 List the safety regulations for medical attendance, examination, and regulations for using compressed air.
26.7 List the safety regulations for compression of compressed air.
26.8 List the safety regulations for decompression of compressed air.
26.9 List the safety regulations for man locks of compressed air.
26.10 List the safety regulations for compressor plant and air supply of compressed air.
26.11 List the safety regulations for ventilation and air quality of compressed air.

26.9   Critical Thinking and Discussion Topic

What should be the difference in training of workers working in tunnels versus workers working in conventional residential building construction?

(Photo courtesy of 123RF.)

CHAPTER 27

Demolition

(Photo courtesy of 123RF.)

27.1 General

Demolition (Fig. 27.1) is the dismantling, razing, destroying, or wrecking by mechanical or explosive means of any building or structure, or any part thereof. Construction involves putting up a structure, whereas demolition involves pulling it down. Building age and the safety condition of the building are the most common reasons for demolition job. If a building no longer fulfills its purpose, it will be demolished to make way for the construction of new buildings. Demolition work involves many of the same

FIGURE 27.1 Demolition of a highway bridge.
(Photo courtesy of Armando Perez.)

hazards that surface during other construction work. Demolition, however, also entails additional hazards due to a variety of other factors.

27.2 Types of Demolition

There are different types of demolition. The major types are as follows:

Interior demolition. Interior demolition is the taking apart of interior portions of a structure while preserving the exterior, usually in preparation for a renovation project. This usually includes removal of walls, ceilings, pipes, etc.

Selective demolition. A selective demolition project involves the removal of specific interior or exterior portions of a building while protecting the remaining structure and nearby structures and areas.

Dismantling/Deconstruction. This method involves the careful dismantlement or deconstruction of a structure to preserve components for reuse, recycling, or refurbishment. Dismantling is generally more labor intensive than demolition.

Total demolition. Total demolition is self-explanatory. It is the demolition of an entire structure, and it can be achieved by a number of methods discussed in the next section.

Example 27.1
Deconstruction means:

    A. The demolition of an entire structure

    B. The removal of specific interior or exterior portions

    C. The taking apart of interior portions

    D. The careful dismantlement to preserve components for reuse

Answer: D

27.3   Methods of Demolition

Some demolition methods are as follows:

Hand demolition.   This type of demolition is also known as green demolition or demolition by hand. Deconstruction is the process of manually stripping and deconstructing the house piece by piece with the intent of salvaging as many of the materials inside the home as possible. Some power tools required for the hand demolition are reciprocating saw, circular saw, rotary/jackhammers, cordless drill, angle grinder, air cleaner, etc. Hand tools may include bars, pliers, nippers and snips, etc.

Mechanical demolition.   This type of demolition uses specialized mechanical equipment and tools. These include hydraulic excavators equipped with specialized attachments that can break concrete and steel, effectively "chewing" the structure apart. Smaller equipment like skid steer loaders and demolition robots are used for smaller tasks and interior and selective demolition.

Implosion.   A highly specialized type of demolition that employs the use of explosives to bring down high structures by undermining structural supports so that it collapses within its own footprint or along a predetermined path. Implosion is used in less than 1% of demolition projects.

Crane and wrecking ball.   One of the earliest methods of demolition, the wrecking ball is largely outmoded, replaced by excavators and other mechanical means that offer better precision, efficiency, and safety. Figure 27.2 shows a crawler crane with a heavy metal wrecking ball on a steel cable at demolition of tall building.

Factors affecting selection of demolition method depends on many factors such as:

- Type of structure such as tall building, agricultural silo, etc.
- Size of structure such as tall building, small building, etc.
- Major materials of structure such as steel, concrete, masonry, etc.
- Presence of any hazardous materials in the structure
- Available time

Figure **27.2** Crawler crane with a heavy metal wrecking ball on a steel cable at demolition of tall building.

- Location and surrounding of structure
- Existence of utilities
- Limitation of noise, dust, and vibrations in the area
- Skills of persons available
- Safety issue around the area
- Availability of equipment

27.4   Demolition Hazards

Demolition sites pose physical and health hazards to workers. These and other common hazards can, however, be avoided or controlled with proper planning.

27.4.1   Collapses

Changes in the original design of the structure, as well as the unknown strength or weakness of certain construction materials (e.g., post-tensioned concrete), may lead to the collapse of walls or floors during demolition. The engineering survey will identify areas where premature collapses may occur. These areas should then be mounted or braced prior to demolition to ensure a safe working environment. Any areas previously damaged by fire, flood, or other causes should also be shoreline or braced. Starting at the top of the structure and demolishing the exterior walls and floors while proceeding downward also reduces the chances of a collapse.

Example 27.2

The _____ survey will identify areas where premature collapses may occur.

   A. Engineering
   B. Geographical
   C. Visual
   D. Mechanical

Answer: *A*

27.4.2   Falls

Falls are another significant physical hazard frequently present on demolition sites. Wall openings should be protected to a height of approximately 42 inches by guardrails, and floor openings should be marked and covered with material capable of withstanding the loads to be imposed. Often fall protection is required (e.g., for workers in aerial lifts). Furthermore, stairs and ladders should be properly installed, inspected, and provided with proper lighting.

27.4.3   Hazardous Materials

Demolition work is often carried out on old structures which are more likely to include hazardous materials such as asbestos, lead, or heavy metals requiring special handling of material. If these materials are present (confirmation tests may be required), their removal should be factored into the work schedule and carried out before demolition work begins. Other hazardous materials, such as chemicals, gases, explosives, or flammable materials, should also be removed prior to the start of demolition work.

Lead dust is caused by the removal, grinding, or cutting of materials covered with lead-based paint or by the handling of metallic lead. Lead fumes may also be generated when a torch is used to cut tanks containing leaded petrol or other lead-containing products. Since lead is a toxic material that can cause serious illness, respirators should be worn if lead dust or fumes are present.

Silica can be found in many building materials, such as natural stone, brick, and concrete. Breaking, cutting, crushing, or grinding this material will produce dust containing crystalline silica. Excessive dust exposure can cause silicosis, a disease that causes lung problems. If it is not possible to control dust to an acceptable level by keeping the material wet or moist, the breathing equipment should be used.

Asbestos dust can be generated whenever materials containing asbestos are handled or removed. Typical asbestos-containing materials include sprayed asbestos coatings on steel columns, insulation materials, fire-resistant walls, asbestos cement sheets, and flooring materials. Breathing this dust can cause asbestosis and lung cancer. Asbestos coating or insulation should be removed by a certified asbestos removal worker before any demolition is started.

Gases and vapors are chemical hazards that may be present in buildings previously used for the manufacture or storage of chemicals. These vapors can be found in pre-existing tanks and pipes, from the burning of waste materials, and even from natural processes such as metal rust. The degree of hazard depends on the type, toxicity, and concentration of the gas present. Adequate ventilation and appropriate respiratory protection equipment must be provided when toxic chemicals are exposed.

27.4.4   Noise

Demolition operations often cause a noise level that exceeds the hearing protection requirement. These levels are caused by the use of explosives, machinery, or other tools and should be controlled by the use of engineering controls (e.g., equipment modification and/or the use of barriers), administrative controls (e.g., insulation or rotation of workers), and hearing protection. Sometimes a combination of controls and hearing protection is needed to ensure proper protection. High noise levels from equipment such as compressors and jackhammers often produce noise in excess of the maximum permissible levels. Long-term exposure to excessive noise may result in permanent hearing loss. To avoid this, always wear hearing protection when noise levels are high.

27.4.5   Dust

As many laborers can attest, demolishing structures produce a large amount of dust. Breaking down concrete and other building materials, such as drywall, releases silica dust into the air, and exposures to other types of dust are harmful to workers as well. These exposures can be significantly reduced through the use of engineering controls, such as wet methods. If exposures cannot be reduced below safe levels through engineering or administrative controls, the appropriate respirator should be worn during dust-generating tasks.

27.4.6   Confined Space

Confined space hazards include basements, reservoirs, and excavations. Hazards may include oxygen deficiency and/or the presence of toxic or flammable gasses such as carbon monoxide, methane, or hydrogen sulfide. Be sure that atmospheres in confined spaces have been tested and the levels determined to be safe before entering these areas.

27.4.7   Dermatitis or Skin Irritation

Dermatitis or irritation of the skin may result from contacting substances such as mineral oil, pitch, disinfectant, solvents, oils, acids, and alkalis. Exposure to epoxy, formaldehyde, nickel, cobalt, and chromium may cause allergic reactions in some people. If contact with any of these substances is likely, protective clothing and gloves with cotton linen should be worn to prevent contact with the skin.

27.4.8   Miscellaneous Hazards

Demolition hazards vary depending on the nature of the work. The major hazards are discussed above. Some other hazards include:

- Work at height
- Plant and machinery overturning or failure
- Contact with live overheads
- Contact with buried utilities and services
- Movement of vehicles due to vibration

- Smokes or possible fires from burning waste timber
- Electric shock and tearing down of powerlines
- The silting up of drainage system by dust
- The problems arising from split fuel oils
- Sharp objects, including glass and nails from the demolition, or syringes left by trespassers
- Manual handling required in most cases

27.5   Steps of Demolition

The following steps can be followed:

Step 1. Surveying such as checking the drawing, material content, surrounding, volume of demolition, etc.

Step 2. Removal of hazardous materials from the structure

Step 3. Preparation of plan

Step 4. Hazard/risk identifications

Step 5. Safety measures for the potential hazard/risk

Step 6. Execution of the plan

Step 7. Removal and reuse of the materials

27.6   Preparatory Operations

Part 1926, Subpart T, Standard Number 1926.850 provides some regulations for the preparatory demolition operation. Few major items are as follows:

- Prior to permitting employees to start demolition operations, an engineering survey shall be made, by a competent person, of the structure to determine the condition of the framing, floors, and walls, and possibility of unplanned collapse of any portion of the structure. Any adjacent structure where employees may be exposed shall also be similarly checked. The employer shall have in writing evidence that such a survey has been performed.
- When employees are required to work within a structure to be demolished which has been damaged by fire, flood, explosion, or other cause, the walls or floor shall be shored or braced.
- All electric, gas, water, steam, sewer, and other service lines shall be shut off, capped, or otherwise controlled, outside the building line before demolition work is started. In each case, any utility company which is involved shall be notified in advance.
- If it is necessary to maintain any power, water, or other utilities during demolition, such lines shall be temporarily relocated, as necessary, and protected.
- It shall also be determined if any type of hazardous chemicals, gases, explosives, flammable materials, or similarly dangerous substances have been used in any

pipes, tanks, or other equipment on the property. When the presence of any such substances is apparent or suspected, testing and purging shall be performed and the hazard eliminated before demolition is started.

- Where a hazard exists from fragmentation of glass, such hazards shall be removed.

- Where a hazard exists to employees falling through wall openings, the opening shall be protected to a height of approximately 42 inches.

- When debris is dropped through holes in the floor without the use of chutes, the area onto which the material is dropped shall be completely enclosed with barricades not less than 42 inches high and not less than 6 feet back from the projected edge of the opening above. Signs, warning of the hazard of falling materials, shall be posted at each level. Removal shall not be permitted in this lower area until debris handling ceases above.

- All floor openings, not used as material drops, shall be covered over with material substantial enough to support the weight of any load which may be imposed. Such material shall be properly secured to prevent its accidental movement.

- Except for the cutting of holes in floors for chutes, holes through which to drop materials, preparation of storage space, and similar necessary preparatory work, the demolition of exterior walls and floor construction shall begin at the top of the structure and proceed downward. Each story of exterior wall and floor construction shall be removed and dropped into the storage space before commencing the removal of exterior walls and floors in the story next below.

- Employee entrances to multistory structures being demolished shall be completely protected by sidewalk sheds or canopies, or both, providing protection from the face of the building for a minimum of 8 feet. All such canopies shall be at least 2 feet wider than the building entrances or openings (1 foot wider on each side thereof), and shall be capable of sustaining a load of 150 pounds per square foot.

Example 27.3

Prior to permitting employees to start demolition operations, an engineering survey shall be made by:

A. A competent person

B. A professional engineer

C. The owner

D. The city officer

Answer: A

27.7 Stairs, Passageways, and Ladders

Part 1926, Subpart T, Standard Number 1926.851 states that:

- Only those stairways, passageways, and ladders, designated as means of access to the structure of a building, shall be used. Other access ways shall be entirely closed at all times.

- All stairs, passageways, ladders, and incidental equipment thereto, which are covered by this section, shall be periodically inspected and maintained in a clean safe condition.

- In a multistory building, when a stairwell is being used, it shall be properly illuminated by either natural or artificial means, and completely and substantially covered over at a point not less than two floors below the floor on which work is being performed, and access to the floor where the work is in progress shall be through a properly lighted, protected, and separate passageway.

27.8   Chutes

Part 1926, Subpart T, Standard Number 1926.852 states that:

- No material shall be dropped to any point lying outside the exterior walls of the structure unless the area is effectively protected.

- All material chutes, or sections thereof, at an angle of more than 45 degrees from the horizontal, shall be entirely enclosed, except for openings equipped with closures at or about floor level for the insertion of materials. The openings shall not exceed 48 inches in height measured along the wall of the chute. At all stories below the top floor, such openings shall be kept closed when not in use.

- A substantial gate shall be installed in each chute at or near the discharge end. A competent employee shall be assigned to control the operation of the gate, and the backing and loading of trucks.

- When operations are not in progress, the area surrounding the discharge end of a chute shall be securely closed off.

- Any chute opening, into which workmen dump debris, shall be protected by a substantial guardrail approximately 42 inches above the floor or other surface on which the men stand to dump the material. Any space between the chute and the edge of openings in the floors through which it passes shall be solidly covered over.

- Where the material is dumped from mechanical equipment or wheelbarrows, a securely attached toe board or bumper, not less than 4 inches thick and 6 inches high, shall be provided at each chute opening.

- Chutes shall be designed and constructed of such strength as to eliminate failure due to impact of materials or debris loaded therein.

27.9   Removal of Materials through Floor Openings

Part 1926, Subpart T, Standard Number 1926.853 states that any openings cut in a floor for the disposal of materials shall be no larger in size than 25% of the aggregate of the total floor area, unless the lateral supports of the removed flooring remain in place. Floors weakened or otherwise made unsafe by demolition operations shall be shored to carry safely the intended imposed load from demolition operations.

27.10   Removal of Walls, Masonry Sections, and Chimneys

Part 1926, Subpart T, Standard Number 1926.854 states that:

- Masonry walls, or other sections of masonry, shall not be permitted to fall upon the floors of the building in such masses as to exceed the safe carrying capacities of the floors.
- No wall section, which is more than one story in height, shall be permitted to stand alone without lateral bracing, unless such wall was originally designed and constructed to stand without such lateral support, and is in a condition safe enough to be self-supporting. All walls shall be left in a stable condition at the end of each shift.
- Employees shall not be permitted to work on the top of a wall when weather conditions constitute a hazard.
- Structural or load-supporting members on any floor shall not be cut or removed until all stories above such a floor have been demolished and removed. This provision shall not prohibit the cutting of floor beams for the disposal of materials or for the installation of equipment.
- Floor openings within 10 feet of any wall being demolished shall be planked solid, except when employees are kept out of the area below.
- In buildings of "skeleton-steel" construction, the steel framing may be left in place during the demolition of masonry. Where this is done, all steel beams, girders, and similar structural supports shall be cleared of all loose material as the masonry demolition progresses downward.
- Walkways or ladders shall be provided to enable employees to safely reach or leave any scaffold or wall.
- Walls, which serve as retaining walls to support earth or adjoining structures, shall not be demolished until such earth has been properly braced or adjoining structures have been properly underpinned.
- Walls, which are to serve as retaining walls against which debris will be piled, shall not be so used unless capable of safely supporting the imposed load.

27.11   Manual Removal of Floors

Part 1926, Subpart T, Standard Number 1926.855 states that:

- Openings cut in a floor shall extend the full span of the arch between supports.
- Before demolishing any floor arch, debris and other material shall be removed from such arch and other adjacent floor area. Planks not less than 2 inches by 10 inches in cross-section, full size undressed, shall be provided for, and shall be used by employees to stand on while breaking down floor arches between beams. Such planks shall be so located as to provide a safe support for the workmen should the arch between the beams collapse. The open space between planks shall not exceed 16 inches.
- Safe walkways, not less than 18 inches wide, formed of planks not less than 2 inches thick if wood, or of equivalent strength if metal, shall be provided and

used by workmen when necessary to enable them to reach any point without walking upon exposed beams.

- Stringers of ample strength shall be installed to support the flooring planks, and the ends of such stringers shall be supported by floor beams or girders, and not by floor arches alone.

- Planks shall be laid together over solid bearings with the ends overlapping at least 1 foot.

- When floor arches are being removed, employees shall not be allowed in the area directly underneath, and such an area shall be barricaded to prevent access to it.

- Demolition of floor arches shall not be started until these, and the surrounding floor area for a distance of 20 feet, have been cleared of debris and any other unnecessary materials.

27.12 Removal of Walls, Floors, and Material with Equipment

Part 1926, Subpart T, Standard Number 1926.856 states that mechanical equipment shall not be used on floors or working surfaces unless such floors or surfaces are of sufficient strength to support the imposed load. Floor openings shall have curbs or stop-logs to prevent equipment from running over the edge.

27.13 Storage of Waste Materials

Part 1926, Subpart T, Standard Number 1926.857 states that:

- The storage of waste material and debris on any floor shall not exceed the allowable floor loads.

- In buildings having wooden floor construction, the flooring boards may be removed from not more than one floor above grade to provide storage space for debris, provided falling material is not permitted to endanger the stability of the structure.

- When wood floor beams serve to brace interior walls or free-standing exterior walls, such beams shall be left in place until other equivalent support can be installed to replace them.

- Floor arches, to an elevation of not more than 25 feet above grade, may be removed to provide storage area for debris: Provided that such removal does not endanger the stability of the structure.

- Storage space into which material is dumped shall be blocked off, except for openings necessary for the removal of material. Such openings shall be kept closed at all times when material is not being removed.

27.14 Removal of Steel Construction

Part 1926, Subpart T, Standard Number 1926.858 states that:

- When floor arches have been removed, planking shall be provided for the workers engaged in razing the steel framing.

- Steel construction shall be dismantled column length by column length, and tier by tier (columns may be in two-story lengths).

- Any structural member being dismembered shall not be overstressed.

27.15 Mechanical Demolition

Part 1926, Subpart T, Standard Number 1926.859 states that:

- No workers shall be permitted in any area, which can be adversely affected by demolition operations, when balling or clamming is being performed. Only those workers necessary for the performance of the operations shall be permitted in this area at any other time.

- The weight of the demolition ball shall not exceed 50% of the crane's rated load, based on the length of the boom and the maximum angle of operation at which the demolition ball will be used, or it shall not exceed 25% of the nominal breaking strength of the line by which it is suspended, whichever results in a lesser value.

- The crane boom and loadline shall be as short as possible.

- The ball shall be attached to the loadline with a swivel-type connection to prevent twisting of the loadline, and shall be attached by positive means in such manner that the weight cannot become accidentally disconnected.

- When pulling over walls or portions thereof, all steel members affected shall have been previously cut free.

- All roof cornices or other such ornamental stonework shall be removed prior to pulling walls over.

- During demolition, continuing inspections by a competent person shall be made as the work progresses to detect hazards resulting from weakened or deteriorated floors, or walls, or loosened material. No employee shall be permitted to work where such hazards exist until they are corrected by shoring, bracing, or other effective means.

Example 27.4

The weight of the demolition ball shall not exceed _____ of the crane's rated load.

    A. 25%

    B. 50%

    C. 75%

    D. 100%

Answer: *B*

27.16 Summary

Construction involves putting up a structure; however, demolition of existing old structure is very often required before putting a new structure. Demolition may be of only interior parts, or of only selective parts, or maybe total demolition or careful demolition to reuse the components. Demolition can be manual, or in mechanical ways such as blasting, crane, etc. The selection of demolition method depends on many factors such as type of structure, size of structure, materials of structure, presence of any hazardous materials, available time, location and surrounding of structure, and so on. Demolition sites pose physical and health hazards to workers such as collapse, falls, hazardous materials, noise, dust, confined space, skin irritation, etc. Steps of demolition includes surveying, removal of hazardous materials, preparation of plan, hazard/risk identifications, safety measures for the potential hazard/risk, execution of the plan, and removal and reuse of the materials. Prior to permitting employees to start demolition operations, an engineering survey shall be made by a competent person. Then the safety instructions by OSHA for stairs, passageways, ladders, chutes, walls, floors, masonry sections, chimneys, etc., must be followed.

27.17   Multiple-Choice Questions

27.1 Which of the following types of demolitions is known as green demolition?

    A.   Hand demolition
    B.   Mechanical demolition
    C.   Implosion demolition
    D.   Wrecking ball demolition

27.2 Where a hazard exists to employees falling through wall openings, the opening shall be protected to a height of approximately:

    A.   24 inches
    B.   32 inches
    C.   42 inches
    D.   48 inches

27.3 Employee entrances to multistory structures being demolished shall be completely protected by sidewalk sheds or canopies, or both, providing protection from the face of the building for a minimum of:

    A.   4 feet
    B.   8 feet
    C.   12 feet
    D.   18 feet

27.4 When debris is dropped through holes in the floor without the use of chutes, the area onto which the material is dropped shall be completely enclosed with barricades not less than:

    A.   24 inches
    B.   32 inches
    C.   42 inches
    D.   48 inches

27.5 When canopies are provided at the employee entrances to multistory structures being demolished, the canopies should be at least _____ feet wider than the building entrances or openings.

    A.   1.0
    B.   1.5
    C.   2.0
    D.   2.5

27.6 When canopies are provided at the employee entrances to multistory structures being demolished, the load capacity of the canopies should be at least:

A. 150 pounds
B. 200 pounds
C. 250 pounds
D. 300 pounds

27.7 The size of the opening cut in a floor for the disposal of materials shall be no larger than:

A. 5% of the aggregate of the total floor area
B. 15% of the aggregate of the total floor area
C. 25% of the aggregate of the total floor area
D. 33% of the aggregate of the total floor area

27.8 Where the material is dumped from mechanical equipment or wheelbarrows, a securely attached toe board or bumper shall be provided at each chute opening not less than:

A. 4 inches thick and 6 inches high
B. 6 inches thick and 8 inches high
C. 4 inches thick and 8 inches high
D. 2 inches thick and 6 inches high

27.9 Floor openings within 10 feet of any wall being demolished shall be:

A. Planked solid
B. Kept open
C. Remain attended
D. Signed properly

27.10 What is the minimum thickness of the safe wood walkways, not less than 18 inches wide, formed of wood planks to be provided for workmen when necessary to enable them to reach any point without walking upon exposed beams?

A. 2 inches
B. 4 inches
C. 5 inches
D. 6 inches

27.18 Practice Problems

27.1 Define demolition and its type.

27.2 Discuss the methods of demolition.

27.3 Discuss the hazards associated with demolition.

27.4 Discuss the steps of demolition.

27.5 List the safety regulations of the preparatory steps of demolition.

27.6 List the safety regulations of the stairs, passageways, and ladders during demolition.

27.7 List the safety regulations of the chutes during demolition.

27.8 List the safety regulations for the removal of materials through floor openings during demolition.

27.9 List the safety regulations for the removal of walls, masonry sections, and chimneys during demolition.

27.10   List the safety regulations for the manual removal of floors during demolition.

27.11   List the safety regulations for the storage of waste materials during demolition.

27.12   List the safety regulations for the manual removal of steel construction during demolition.

27.13   List the safety regulations for the mechanical demolition.

27.19   Critical Thinking and Discussion Topic

You are in charge of a demolition work. There is a powerline nearby. The survey shows that the demolition work is not expected to disturb the powerline by any means. The powerline carries very high-voltage power and a small accident may cost a lot in terms of hazards. As the manager, what will you do in this situation?

(Photo courtesy of darcy-lawrey from pexels.)

CHAPTER **28**

Stairways and Ladders

(Photo courtesy of 123RF.)

28.1 General

A stairway (Fig. 28.1) is a construction designed to bridge a large vertical distance by dividing it into smaller vertical distances, called steps. The stairs may be straight, round, or may consist of two or more straight sections connected at angles. The part of the stairway that has to be stepped on is built to the same thickness as any other flooring. The tread depth is measured from the back of one tread to the back of the next. The width is measured from one side to the other. Handrail is the angled handholding member, as it is distinguished from the vertical balusters that hold it onto stairs that are open on one side; there is often a railing on both sides, sometimes on one side or not at all, on wide staircases, sometimes in the middle or even more. Riser is the vertical

FIGURE 28.1   Stairways in an industrial tower.
(Photo courtesy of Fillipe Gomes.)

portion of the stairway between every tread. A ladder is a structure for climbing up or down, consisting essentially of two long side parts joined at intervals by cross-pieces on which one can step. According to Bureau of Labor statistics, one-third of the 645 construction fatalities in 2009 resulted from falls from ladders and on stairs.

Example 28.1
According to Bureau of Labor statistics, _____% of the 645 construction fatalities in 2009 resulted from falls from ladders and on stairs.

   A. 10

   B. 20

   C. 24

   D. 33

Answer: *D*

28.2   General Safety Regulations
This subpart applies to all stairways and ladders used in construction, alteration, repair (including painting and decorating), and demolition workplaces covered under 29 CFR

part 1926, and also sets forth, in specified circumstances, when ladders and stairways are required to be provided.

Part 1926, Subpart X, Standard Number 1926.1051 states that:

- A stairway or ladder shall be provided at all personnel points of access where there is a break in elevation of 19 inches (48 centimeters) or more, and no ramp, runway, sloped embankment, or personnel hoist is provided.

- Employees shall not use any spiral stairways that will not be a permanent part of the structure on which construction work is being performed.

- A double-cleated ladder or two or more separate ladders shall be provided when ladders are the only means of access or exit from a working area for 25 or more employees, or when a ladder is to serve simultaneous two-way traffic.

- When a building or structure has only one point of access between levels, that point of access shall be kept clear to permit free passage of employees. When work must be performed or equipment must be used such that free passage at that point of access is restricted, a second point of access shall be provided and used.

- When a building or structure has two or more points of access between levels, at least one point of access shall be kept clear to permit free passage of employees.

- Employers shall provide and install all stairway and ladder fall protection systems required by this subpart and shall comply with all other pertinent requirements of this subpart before employees begin the work that necessitates the installation and use of stairways, ladders, and their respective fall protection systems.

Example 28.2

A _____ ladder should be provided when two single ladders aren't available for a working area with 25 or more workers.

    A. Double-cleated

    B. Triple-cleated

    C. Single

    D. Multicleated

Answer: *A*

28.3 Safety Regulations for Stairways

Part 1926, Subpart X, Standard Number 1926.1052 states that:

- Stairways that will not be a permanent part of the structure on which construction work is being performed shall have landings of not less than 30 inches (76 centimeters) in the direction of travel and extend at least 22 inches (56 centimeters) in width at every 12 feet (3.7 meter) or less of vertical rise.

- Stairs shall be installed between 30 and 50 degrees from horizontal.

- Riser height and tread depth shall be uniform within each flight of stairs, including any foundation structure used as one or more treads of the stairs. Variations in riser height or tread depth shall not be over 1/4 inch (0.6 centimeter) in any stairway system.

- Where doors or gates open directly on a stairway, a platform shall be provided, and the swing of the door shall not reduce the effective width of the platform to less than 20 inches (51 centimeters).

- Metal pan landings and metal pan treads, when used, shall be secured in place before filling with concrete or other material.

- All parts of stairways shall be free of hazardous projections, such as protruding nails.

- Slippery conditions on stairways shall be eliminated before the stairways are used to reach other levels.

Example 28.3

Stairway must have uniform rise height and tread depth; variations in rise height or tread depth shall not be over _____ inch in any stairway system.

    A. 1/4

    B. 1/2

    C. 3/4

    D. 5/8

Answer: A

28.4 Safety Regulations for Stairrails and Handrails

Handrails and stairrails are used to protect workers from falling when using stairways. The clearance of temporary handrails must be at least 3 inches between handrail and walls, stairrail systems, and other objects.

According to the Part 1926, Subpart X, Standard Number 1926.1052, the following requirements apply to all stairrails and handrails as indicated:

- Stairways having four or more risers or rising more than 30 inches (76 centimeters), whichever is less, shall be equipped with:
  - At least one handrail
  - One stairrail system along each unprotected side or edge

- Winding and spiral stairways shall be equipped with a handrail offset sufficiently to prevent walking on those portions of the stairways where the tread width is less than 6 inches (15 centimeters).

- The height of stairrails shall be as follows:
  - Stairrails installed after March 15, 1991, shall be not less than 36 inches (91.5 centimeters) from the upper surface of the stairrail system to the surface of the tread, in line with the face of the riser at the forward edge of the tread.

- Stairrails installed before March 15, 1991, shall be not less than 30 inches (76 centimeters) nor more than 34 inches (86 centimeters) from the upper surface of the stairrail system to the surface of the tread, in line with the face of the riser at the forward edge of the tread.

- Midrails, screens, mesh, intermediate vertical members, or equivalent intermediate structural members shall be provided between the top rail of the stairrail system and the stairway steps.

- Midrails, when used, shall be located at a height midway between the top edge of the stairrail system and the stairway steps.

- Screens or mesh, when used, shall extend from the top rail to the stairway step, and along the entire opening between top rail supports.

- When intermediate vertical members, such as balusters, are used between posts, they shall be not more than 19 inches (48 centimeters) apart.

- Other structural members, when used, shall be installed such that there are no openings in the stairrail system that are more than 19 inches (48 centimeters) wide.

- Handrails and the top rails of stairrail systems shall be capable of withstanding, without failure, a force of at least 200 pounds (890 newton) applied within 2 inches (5 centimeters) of the top edge, in any downward or outward direction, at any point along the top edge.

- The height of handrails shall be not more than 37 inches (94 centimeters) or less than 30 inches (76 centimeters) from the upper surface of the handrail to the surface of the tread, in line with the face of the riser at the forward edge of the tread.

- When the top edge of a stairrail system also serves as a handrail, the height of the top edge shall not be more than 37 inches (94 centimeters) or less than 36 inches (91.5 centimeters) from the upper surface of the stairrail system to the surface of the tread, in line with the face of the riser at the forward edge of the tread.

- Stairrail systems and handrails shall be so surfaced as to prevent injury to employees from punctures or lacerations, and to prevent snagging of clothing.

- Handrails shall provide an adequate handhold for employees grasping them to avoid falling.

- The ends of stairrail systems and handrails shall be constructed so as not to constitute a projection hazard.

- Handrails that will not be a permanent part of the structure being built shall have a minimum clearance of 3 inches (8 centimeter) between the handrail and walls, stairrail systems, and other objects.

28.5   Safety Regulations for Ladders

Ladders (Fig. 28.2) must be kept in a safe and good working condition. The following points are important to consider while using or working with ladders:

- The area around the top and bottom of the ladder must be kept clean.

- Always keep ladders away from slipping hazards.

- Ensure that rungs are spaced 10–14 inches from each other. Also, ensure that cleats and steps are uniformly spaced.

FIGURE 28.2 Two ladders in a construction site.
(Photo courtesy of Rene Asmussen.)

Always use ladders only for their designed purposes. Do not lash ladders together to make a long ladder, unless they are designed for that purpose. Never overload ladders beyond their capacities. The manufacturer's rated capacity must be taken into consideration when using ladders. Portable ladders are those ladders that can be readily moved or carried. Before using portable ladders, always inspect for cracks, dents, and missing rungs; rungs must be designed to minimize slipping risk. Always use ladders on stable and level surfaces, unless they are precisely designed for other surfaces. Ladders placed in areas such as passageways, doorways, or where they can be displaced by workplace activities or traffic must be secured to prevent accidental movement, or a barricade must be used to keep traffic or activities away from the ladder. Do not use ladders on slippery surfaces, unless they are adequately protected with slip-resistant feet/material.

Never use the top or top step of a stepladder as a step; otherwise, it could lead to a severe accident. Do not use cross-bracing given on the rear of a stepladder for climbing, unless the ladder is designed for that purpose. A metal spreader or locking device must be provided on each stepladder to hold the front and back sections in an open position when the ladder is being used. It is necessary that a competent person inspect ladders for visible defects, like broken or missing rungs; if a defective ladder is found, immediately mark it defective, discard the ladder in a manner that it will not be recovered and reused, or tag it "Do Not Use."

Defective ladders need to be immediately removed from the service until repaired. Furthermore, ladders must be inspected on a periodic basis and after any incident that could affect their safe use. Ladders must be constructed with nonconductive side rails if they are used in places where the employee or the ladder could contact exposed energized electrical equipment.

Employees should always face the ladder when going up or down. They should grab the ladder with at least one hand while mounting or dismounting, and each employee must never carry any load or object that could cause the employee to lose balance and fall.

Part 1926, Subpart X, Standard Number 1926.1053 states that ladders shall be capable of supporting the following loads without failure:

Each self-supporting portable ladder: At least four times the maximum intended load, except that each extra-heavy-duty type 1A metal or plastic ladder shall sustain at least 3.3 times the maximum intended load. The ability of a ladder to sustain the loads indicated in this paragraph shall be determined by applying or transmitting the requisite load to the ladder in a downward vertical direction.

Each portable ladder that is not self-supporting: At least four times the maximum intended load, except that each extra-heavy-duty type 1A metal or plastic ladders shall sustain at least 3.3 times the maximum intended load. The ability of a ladder to sustain the loads indicated in this paragraph shall be determined by applying or transmitting the requisite load to the ladder in a downward vertical direction when the ladder is placed at an angle of 75-1/2 degrees from the horizontal. Non-self-supporting ladders must be placed or positioned at an angle where the horizontal distance from the top support to the foot of the ladder is one-fourth of the working length of the ladder.

Each fixed ladder: At least two loads of 250 pounds (114 kilograms) each, concentrated between any two consecutive attachments (the number and position of additional concentrated loads of 250 pounds (114 kilograms) each, determined from anticipated usage of the ladder, shall also be included), plus anticipated loads caused by ice buildup, winds, rigging, and impact loads resulting from the use of ladder safety devices. Each step or rung shall be capable of supporting a single concentrated load of at least 250 pounds (114 kilograms) applied in the middle of the step or rung.

The following requirements apply to the use of all ladders, including job-made ladders, except as otherwise indicated:

- When portable ladders are used for access to an upper landing surface, the ladder side rails shall extend at least 3 feet (0.9 meter) above the upper landing surface to which the ladder is used to gain access; or, when such an extension is not possible because of the ladder's length, then the ladder shall be secured at its top to a rigid support that will not deflect, and a grasping device, such as a grabrail, shall be provided to assist employees in mounting and dismounting the ladder. In no case shall the extension be such that ladder deflection under a load would, by itself, cause the ladder to slip off its support.
- Ladders shall be maintained free of oil, grease, and other slipping hazards.
- Ladders shall not be loaded beyond the maximum intended load for which they were built, or beyond their manufacturer's rated capacity.
- Ladders shall be used only for the purpose for which they were designed.

Example 28.4

A _____ spreader or locking device must be provided on each stepladder to hold the front and back sections in an open position when the ladder is being used.

    A. Plastic

    B. Paper

    C. Metal

    D. Wooden

Answer: C

Example 28.5

Ladders placed in areas such as passageways, doorways, or where they can be _____ by workplace activities or traffic must be secured to prevent accidental movement.

    A. Viewed

    B. Seen

    C. Displaced

    D. Hidden

Answer: C

Example 28.6

The height of handrails shall not be more than _____ inches or less than 30 inches from the upper surface of the handrail to the surface of the tread, in line with the face of the rise at the forward edge of the tread.

    A. 50

    B. 37

    C. 44

    D. 20

Answer: B

Example 28.7

Portable ladders are those that are fixed at a specific location.

    • True

    • False

Answer: *False*
Portable ladders are those ladders that can be readily moved or carried. Before using portable ladders always inspect for cracks, dents, and missing rungs; rungs must be designed to minimize slipping risk.

Example 28.8
Non-self-supporting ladders must be placed or positioned at an angle where the horizontal distance from the top support to the foot of the ladder is _____ the working length of the ladder.

    A. 1/2

    B. 1/3

    C. 1/4

    D. 2/3

Answer: C

28.6 Summary

Ladders having a pitch in excess of 90 degrees from horizontal are not permitted. Ladders must be used only for the purposes and in the manner for which they were designed. For instance, non-self-supporting ladders are to be used at an angle such that the horizontal distance from the top support to the foot of the ladder is approximately one-fourth of the working length of the ladder (the distance along the ladder between the foot and top support).

The combined weight of the employee using a portable ladder and any tools and supplies carried by the employee is not to exceed the maximum intended load of the ladder. Ladders with structural or other defects must be immediately tagged with a danger tag reading "Out of Service," "Do Not Use," etc., and be withdrawn from service until repaired. Single-rail ladders must not be used.

28.7 Multiple-Choice Questions

28.1 Handrails must be provided on all stairways that have:

    A.   Three or more rises
    B.   Higher than 2.5 feet
    C.   Higher than 30 feet
    D.   Any number of rise

28.2 Handrails and top rails must be capable of withstanding a load/force of:

    A.   150 pounds
    B.   200 pounds
    C.   250 pounds
    D.   300 pounds

28.3 The normal inclination angle of stairs with respect to the horizontal shall be between:

    A.   20 and 40 degrees
    B.   20 and 50 degrees
    C.   30 and 60 degrees
    D.   30 and 50 degrees

28.4 Stairways that will not be a permanent part of the structure on which construction work is being performed shall have landings of not less than:

A. 20 inches
B. 25 inches
C. 30 inches
D. 36 inches

28.5 Where doors or gates open directly on a stairway, a platform shall be provided, and the swing of the door shall not reduce the effective width of the platform to less than:

A. 12 inches
B. 20 inches
C. 24 inches
D. 30 inches

28.6 When intermediate vertical members, such as balusters, are used between posts, they shall be not more than:

A. 12 inches apart
B. 19 inches apart
C. 24 inches apart
D. 30 inches apart

28.7 While using or working with ladders, ensure that rungs are spaced:

A. 8–12 inches from each other
B. 10–19 inches from each other
C. 12–24 inches from each other
D. 10–14 inches from each other

28.8 Each self-supporting portable ladder shall be capable of supporting:

A. At least 2.5 times the maximum intended load
B. At least 3.0 times the maximum intended load
C. At least 4.0 times the maximum intended load
D. At least 5.0 times the maximum intended load

28.9 When portable ladders are used for access to an upper landing surface, the ladder side rails shall extend at least:

A. 3 feet above the upper landing surface
B. 4 feet above the upper landing surface
C. 5 feet above the upper landing surface
D. 6 feet above the upper landing surface

28.10 Each step or rung shall be capable of supporting a single concentrated load of at least:

A. 114 kg applied in the middle of the step
B. 134 kg applied in the middle of the step
C. 94 kg applied in the middle of the step
D. 154 kg applied in the middle of the step

28.11 Ladders are not required to be inspected for visible defects prior to the first use each work shift, and after any occurrence that could affect their safe use.

- True
- False

28.8   Practice Problems

28.1   List the general safety regulations for stairways and ladders.

28.2   List the safety regulations for stairways.

28.3   List the general safety regulations for stairrails and handrails

28.4   List the general safety regulations for ladders.

28.9   Critical Thinking and Discussion Topic

You installed the safety features required for a construction site with a lot of stairways. However, due to a small rain the stairways looked slippery to you. Workers said that the stairways were perfect and you let them work over there. Suddenly, an accident happened where a worker slipped over and broke his leg. Now, you are required to write a report. What will be your explanation?

(Photo courtesy of ksenia-chernaya from pexels.)

CHAPTER **29**

Toxic and Hazardous
Substances

(Photo courtesy of Pexels.)

29.1   General

Toxic substance means the substance that is harmful or fatal to living organisms when absorbed or ingested. Hazardous substance is the lower level of potentially harmful substances; toxic is higher. Hazardous substance can be but isn't necessarily toxic. All toxic substances are hazardous. For example, a nail on the road is a hazard but not toxic. This chapter discusses the safety regulations for toxic and hazardous substances.

Example 29.1
The term _____ is used to describe the ability of a substance to cause a harmful effect.

 A. Toxicity

 B. Chemical

 C. Hazard

 D. All the above

Answer: *A*

29.2   Asbestos

29.2.1   Generals

Asbestos includes chrysotile, amosite, crocidolite, tremolite asbestos, anthophyllite asbestos, actinolite asbestos, and any of these minerals that has been chemically treated and/or altered. For purposes of this standard, "asbestos" includes presumed asbestos containing material (PACM). It is a hazardous substance as its fiber masses break easily into tiny dust particles and get airborne.

PACM means thermal system insulation and surfacing material found in buildings constructed no later than 1980. Asbestos-containing material (ACM) means any material containing more than 1% asbestos. Surfacing ACM means surfacing material which contains more than 1% asbestos. ACM is more commonly used in insulation material, fireproofing material, and acoustical material. Thermal system insulation (TSI) means ACM applied to pipes, fittings, boilers, breeching, tanks, ducts, or other structural components to prevent heat loss or gain. The likelihood of a person developing an asbestos-related disease is based on all these factors such as smoking, amount, and duration of exposure, age, health conditions, etc.

 a) Class I asbestos work means activities involving the removal of TSI and surfacing ACM and PACM.

 b) Class II asbestos work means activities involving the removal of ACM which is not thermal system insulation or surfacing material. This includes, but is not limited to, the removal of asbestos-containing wallboard, floor tile and sheeting, roofing and siding shingles, and construction mastics.

 c) Class III asbestos work means repair and maintenance operations, where "ACM," including TSI and surfacing ACM and PACM, is likely to be disturbed.

d) Class IV asbestos work means maintenance and custodial activities during which employees contact but do not disturb ACM or PACM and activities to clean up dust, waste, and debris resulting from Class I, II, and III activities.

Part 1926, Subpart Z, Standard Number 1926.1101 provides the specification for the following cases:

- Demolition or salvage of structures where asbestos is present
- Removal or encapsulation of materials containing asbestos
- Construction, alteration, repair, maintenance, or renovation of structures, substrates, or portions thereof that contain asbestos
- Installation of products containing asbestos
- Asbestos spill/emergency cleanup
- Transportation, disposal, storage, containment of and housekeeping activities involving asbestos or products containing asbestos, on the site or location at which construction activities are performed

Coverage under this standard shall be based on the nature of the work operation involving asbestos exposure. This section does not apply to asbestos-containing asphalt roof coatings, cements, and mastics.

Example 29.2
Asbestos-containing materials (ACM) are more commonly used in:

A. Insulation material

B. Fireproofing material

C. Acoustical material

D. All of the above

Answer: D

29.2.2   Permissible Exposure Limits (PELS)

The permissible exposure limits (PELs) of asbestos are as follows:

- **Time-weighted average (TWA) limit.** The employer shall ensure that no employee is exposed to an airborne concentration of asbestos in excess of 0.1 fiber per cubic centimeter of air as an 8-hour TWA.
- **Excursion limit.** The employer shall ensure that no employee is exposed to an airborne concentration of asbestos in excess of 1.0 fiber per cubic centimeter of air (1 f/cc) as averaged over a sampling period of thirty (30) minutes.

29.2.3   Safety Regulations for Asbestos

Asbestos hazards at a multiemployer worksite shall be abated by the contractor who created or controls the source of asbestos contamination. For example, if there is a significant breach of an enclosure containing Class I work, the employer responsible for erecting the enclosure shall repair the breach immediately. In addition, all employers of employees exposed to asbestos hazards shall comply with applicable protective provisions to protect their employees. For example, if employees working immediately adjacent to a Class I asbestos job are exposed to asbestos due to the inadequate containment of such job, their employer shall either remove the employees from the area until the enclosure breach is repaired, or perform an initial exposure assessment. All employers of employees working adjacent to regulated areas established by another employer on a multiemployer worksite shall take steps on a daily basis to ascertain the integrity of the enclosure and/or the effectiveness of the control method relied on by the primary asbestos contractor to ensure that asbestos fibers do not migrate to such adjacent areas. In a high-risk asbestos-exposed area, the worker with the lower risk of exposure is likely the worker who disposes of asbestos-contaminated waste materials. The measures required to control asbestos fibers in the air during high-risk activities include drench all ACM with water, apply a constant, light spray of water near workers, clean exposed surfaces with damp wipes and cover with sealant, and so on.

Example 29.3

The measures required to control asbestos fibers in the air during high-risk activities include all these procedures, except:

    A. Drench all ACM with water.

    B. Apply a constant, light spray of water near workers.

    C. Clean exposed surfaces with damp wipes and cover with sealant.

    D. Keep work area air pressure higher than surrounding pressure.

Answer: D

Example 29.4

In a high-risk asbestos-exposed area, the worker with the lower risk of exposure is likely the worker who:

    A. Disposes of asbestos-contaminated waste materials

    B. Encapsulates asbestos-contaminated materials

    C. Tests for asbestos contamination

    D. Removes asbestos-contaminated construction materials

Answer: A

29.2.4   Respirator Program

No employee shall be assigned to asbestos work that requires respirator (either air-supplying or atmosphere-supplying) use if, based on their most recent medical examination, the examining physician determines that the employee will be unable to function normally while using a respirator, or that the safety or health of the employee or other employees will be impaired by the employee's respirator use. Such employees must be assigned to another job or given the opportunity to transfer to a different position that they can perform. If such a transfer position is available, it must be with the same employer, in the same geographical area, and with the same seniority, status, rate of pay, and other job benefits the employee had just prior to such transfer.

Example 29.5

For their workers involved in high-risk asbestos-exposed jobs, employers must provide all these preventions, except:

    A. Cancer insurance

    B. Clean rooms

    C. Showers

    D. Respirators

Answer: A

29.3 Thirteen Carcinogens

29.3.1 Definitions

Part 1926, Subpart Z, Standard Number 1926.1103 is identical to Part 1910, Subpart Z, Standard Number 1910.1003. This section applies to any area in which the 13 carcinogens addressed by this section are manufactured, processed, repackaged, released, handled, or stored, but shall not apply to transshipment in sealed containers. The 13 carcinogens are the following:

1. 4-Nitrobiphenyl
2. Alpha-naphthylamine
3. Methyl chloromethyl ether
4. 3,3'-Dichlorobenzidine (and its salts)
5. Bis-chloromethyl ether
6. Beta-naphthylamine
7. Benzidine
8. 4-Aminodiphenyl
9. Ethyleneimine
10. Beta-Propiolactone
11. 2-Acetylaminofluorene
12. 4-Dimethylaminoazo-Benzene
13. N-Nitrosodimethylamine

29.3.2 Requirements for Areas Containing a Carcinogen

A regulated area shall be established by an employer where a carcinogen addressed by this section is manufactured, processed, used, repackaged, released, handled, or stored.

All such areas shall be controlled in accordance with the requirements for the following category or categories describing the operation involved:

Isolated systems. Employees working with a carcinogen addressed by this section within an isolated system such as a "glove box" shall wash their hands and arms upon completion of the assigned task and before engaging in other activities not associated with the isolated system.

Closed system operation. Within regulated areas where the carcinogens addressed by this section are stored in sealed containers, or contained in a closed system, including piping systems, with any sample ports or openings closed while the carcinogens addressed by this section are contained within, access shall be restricted to authorized employees only.

Open-vessel system operations. Open-vessel system operations are prohibited. Each operation shall be provided with continuous local exhaust ventilation so that air movement is always from ordinary work areas to the operation. Exhaust air shall not be discharged to regulated areas, nonregulated areas, or the external environment unless decontaminated. Clean makeup air shall be introduced in sufficient volume to maintain the correct operation of the local exhaust system.

Employees shall be provided with, and required to wear, clean, full-body protective clothing (smocks, coveralls, or long-sleeved shirt and pants), shoe covers, and gloves prior to entering the regulated area. Prior to each exit from a regulated area, employees shall be required to remove and leave protective clothing and equipment at the point of exit and at the last exit of the day, to place used clothing and equipment in impervious containers at the point of exit for purposes of decontamination or disposal.

29.4   Vinyl Chloride

Part 1926, Subpart Z, Standard Number 1926.1117 follows 1910.1017. This section applies to the manufacture, reaction, packaging, repackaging, storage, handling, or use of vinyl chloride or polyvinyl chloride, but does not apply to the handling or use of fabricated products made of polyvinyl chloride.

29.4.1   Methods of Compliance

Employee exposures to vinyl chloride shall be controlled to at or below the PEL provided by engineering, work-practice, and personal protective controls as follows:

- Feasible engineering and work-practice controls shall immediately be used to reduce exposures to or below the PEL.
- Wherever feasible engineering and work-practice controls which can be instituted immediately are not sufficient to reduce exposures to or below the PEL, they shall nonetheless be used to reduce exposures to the lowest practicable level, and shall be supplemented by respiratory protection. A program shall be established and implemented to reduce exposures to or below the PEL, or to the greatest extent feasible, solely by means of engineering and work-practice controls, as soon as feasible.
- Written plans for such a program shall be developed and furnished upon request for examination and copying to authorized representatives of the Assistant Secretary and the Director. Such plans must be updated at least annually.

29.4.2 Permissible Exposure Limit

- No employee may be exposed to vinyl chloride at concentrations greater than 1 ppm averaged over any 8-hour period.

- No employee may be exposed to vinyl chloride at concentrations greater than 5 ppm averaged over any period not exceeding 15 minutes.

- No employee may be exposed to vinyl chloride by direct contact with liquid vinyl chloride.

29.4.3 Respiratory Protection

For employees who use respirators required by this section, the employer must provide each employee an appropriate respirator that complies with the requirements of this paragraph.

29.4.4 Hazardous Operations

Employees engaged in hazardous operations, including entry of vessels to clean polyvinyl chloride residue from vessel walls, shall be provided and required to wear and use protective garments to prevent skin contact with liquid vinyl chloride or with polyvinyl chloride residue from vessel walls. The protective garments shall be selected for the operation and its possible exposure conditions.

29.4.5 Emergency Situations

A written operational plan for emergency situations shall be developed for each facility storing, handling, or otherwise using vinyl chloride as a liquid or compressed gas. Appropriate portions of the plan shall be implemented in the event of an emergency. The plan shall specifically provide that:

- Employees engaged in hazardous operations or correcting situations of existing hazardous releases shall be equipped as required

- Other employees not so equipped shall evacuate the area and not return until conditions are controlled by the methods required and the emergency is abated.

29.5 Inorganic Arsenic

Inorganic arsenic means copper aceto-arsenite and all inorganic compounds containing arsenic except arsine, measured as arsenic (As).

Part 1926, Subpart Z, Standard Number 1926.1118 follows 1910.1018.

Permissible exposure limit. The employer shall ensure that no employee is exposed to inorganic arsenic at concentrations greater than 10 micrograms per cubic meter of air (10 µg/m^3), averaged over any 8-hour period.

Respiratory protection. For employees who use respirators required by this section, the employer must provide each employee an appropriate respirator that complies with the requirements of this paragraph. Respirators must be used during:

- Periods necessary to install or implement feasible engineering or work-practice controls

- Work operations, such as maintenance and repair activities, for which the employer establishes that engineering and work-practice controls are not feasible .

- Work operations for which engineering and work-practice controls are not yet sufficient to reduce employee exposures to or below the PEL

Protective work clothing and equipment.   Where the possibility of skin or eye irritation from inorganic arsenic exists, and for all workers working in regulated areas, the employer shall provide at no cost to the employee and ensure that employees use appropriate and clean protective work clothing and equipment such as, but not limited to:

- Coveralls or similar full-body work clothing

- Gloves, and shoes or coverlets

- Face shields or vented goggles when necessary to prevent eye irritation

- Impervious clothing for employees subject to exposure to arsenic trichloride

29.6   Beryllium

Part 1926, Subpart Z, Standard Number 1926.1124.
   Permissible exposure limits (PELs):

- TWA PEL: The employer must ensure that no employee is exposed to an airborne concentration of beryllium in excess of 0.2 $\mu g/m^3$ calculated as an 8-hour TWA.

- Short-term exposure limit (STEL): The employer must ensure that no employee is exposed to an airborne concentration of beryllium in excess of 2.0 $\mu g/m^3$ as determined over a sampling period of 15 minutes.

Engineering and work-practice controls.   Where exposures are, or can reasonably be expected to be, at or above the action level, the employer must ensure that at least one of the following is in place to reduce airborne exposure:

- Material and/or process substitution

- Isolation, such as ventilated partial or full enclosures

- Local exhaust ventilation, such as at the points of operation, material handling, and transfer

- Process control, such as wet methods and automation

Respiratory protection.   The employer must provide respiratory protection at no cost to the employee and ensure that each employee uses respiratory protection:

- During periods necessary to install or implement feasible engineering and work-practice controls where airborne exposure exceeds, or can reasonably be expected to exceed, the TWA PEL or STEL

- During operations, including maintenance and repair activities and nonroutine tasks, when engineering and work-practice controls are not feasible and airborne exposure exceeds, or can reasonably be expected to exceed, the TWA PEL or STEL

- During operations for which an employer has implemented all feasible engineering and work-practice controls when such controls are not sufficient to reduce airborne exposure to or below the TWA PEL or STEL

- During emergencies

Personal protective clothing and equipment.   The employer must provide at no cost, and ensure that each employee uses, appropriate personal protective clothing and equipment in accordance with the written exposure control plan required and OSHA's Personal Protective and Life Saving Equipment standards for construction:

- Where airborne exposure exceeds, or can reasonably be expected to exceed, the TWA PEL or STEL

- Where there is a reasonable expectation of dermal contact with beryllium

29.7   Chromium (VI)

Part 1926, Subpart Z, Standard Number 1926.1126 provides:

Chromium (VI) [hexavalent chromium or Cr(VI)] means chromium with a valence of positive six, in any form and in any compound.

Permissible exposure limit (PEL).   The employer shall ensure that no employee is exposed to an airborne concentration of chromium (VI) in excess of 5 micrograms per cubic meter of air (5 $\mu g/m^3$), calculated as an 8-hour TWA. Each employer who has a workplace or work operation covered by this section shall determine the 8-hour TWA exposure for each employee exposed to chromium (VI).

Engineering and work-practice controls.   The employer shall use engineering and work-practice controls to reduce and maintain employee exposure to chromium (VI) to or below the PEL unless the employer can demonstrate that such controls are not feasible. Wherever feasible engineering and work-practice controls are not sufficient to reduce employee exposure to or below the PEL, the employer shall use them to reduce employee exposure to the lowest levels achievable, and shall supplement them by the use of respiratory protection.

Where the employer can demonstrate that a process or task does not result in any employee exposure to chromium (VI) above the PEL for 30 or more days per year (12 consecutive months), the requirement to implement engineering and work-practice controls to achieve the PEL does not apply to that process or task.

Respiratory protection.   Where respiratory protection is required by this section, the employer must provide each employee an appropriate respirator that complies with the requirements of this paragraph. Respiratory protection is required during:

- Periods necessary to install or implement feasible engineering and work-practice controls

- Work operations, such as maintenance and repair activities, for which engineering and work-practice controls are not feasible

- Work operations for which an employer has implemented all feasible engineering and work-practice controls and such controls are not sufficient to reduce exposures to or below the PEL

- Work operations where employees are exposed above the PEL for fewer than 30 days per year, and the employer has elected not to implement engineering and work-practice controls to achieve the PEL

- Emergencies

Protective work clothing and equipment. Where a hazard is present or is likely to be present from skin or eye contact with chromium (VI), the employer shall provide appropriate personal protective clothing and equipment at no cost to employees, and shall ensure that employees use such clothing and equipment.

29.8 Cadmium

Part 1926, Subpart Z, Standard Number 1926.1127 provides:

Permissible exposure limit (PEL). The employer shall ensure that no employee is exposed to an airborne concentration of cadmium in excess of five micrograms per cubic meter of air (5 μg/m^3), calculated as an 8-hour TWA exposure.

Respirator protection. For employees who use respirators required by this section, the employer must provide each employee an appropriate respirator that complies with the requirements of this paragraph. Respirators must be used during:

- Periods necessary to install or implement feasible engineering and work-practice controls when employee exposures exceed the PEL

- Maintenance and repair activities, and brief or intermittent work operations, for which employee exposures exceed the PEL and engineering and work-practice controls are not feasible or are not required

- Work operations for which the employer has implemented all feasible engineering and work-practice controls, and such controls are not sufficient to reduce employee exposures to or below the PEL

- Work operations for which an employee, who is exposed to cadmium at or above the action level, requests a respirator

- Work operations for which engineering controls are not required by paragraph (f)(1)(ii) of this section to reduce employee exposures that exceed the PEL

Protective work clothing and equipment. If an employee is exposed to airborne cadmium above the PEL or where skin or eye irritation is associated with cadmium exposure at any level, the employer shall provide at no cost to the employee, and ensure that the employee uses, appropriate protective work clothing and equipment that prevents contamination of the employee and the employee's garments. Protective work clothing and equipment includes, but is not limited to:

- Coveralls or similar full-body work clothing

- Gloves, head coverings, and boots or foot coverings

- Face shields, vented goggles, or other appropriate protective equipment

29.9 Benzene

Part 1926, Subpart Z, Standard Number 1926.1128 follows 1910.1028.

Benzene (C_6H_6) (CAS Registry No. 71-43-2) means liquefied or gaseous benzene. It includes benzene contained in liquid mixtures and the benzene vapors released by these liquids. It does not include trace amounts of unreacted benzene contained in solid materials.

Permissible exposure limits (PELs):

- TWA limit: The employer shall ensure that no employee is exposed to an airborne concentration of benzene in excess of one part of benzene per million parts of air (1 ppm) as an 8-hour TWA.

- STEL. The employer shall ensure that no employee is exposed to an airborne concentration of benzene in excess of five (5) ppm as averaged over any 15-minute period.

The employer shall establish a regulated area wherever the airborne concentration of benzene exceeds or can reasonably be expected to exceed the PELs, either the 8-hour TWA exposure of 1 ppm or the STEL of 5 ppm for 15 minutes.

Engineering controls and work practices:

- The employer shall institute engineering controls and work practices to reduce and maintain employee exposure to benzene at or below the PELs, except to the extent that the employer can establish that these controls are not feasible.

- Wherever the feasible engineering controls and work practices which can be instituted are not sufficient to reduce employee exposure to or below the PELs, the employer shall use them to reduce employee exposure to the lowest levels achievable by these controls and shall supplement them by the use of respiratory protection which complies with the requirements of paragraph (g) of this section.

- Where the employer can document that benzene is used in a workplace less than a total of 30 days per year, the employer shall use engineering controls, work-practice controls, or respiratory protection or any combination of these controls to reduce employee exposure to benzene to or below the PELs, except that employers shall use engineering and work-practice controls, if feasible, to reduce exposure to or below 10 ppm as an 8-hour TWA.

Respiratory protection.   For employees who use respirators required by this section, the employer must provide each employee an appropriate respirator that complies with the requirements of this paragraph. Respirators must be used during:

- Periods necessary to install or implement feasible engineering and work-practice controls

- Work operations for which the employer establishes that compliance with either the TWA or STEL through the use of engineering and work-practice controls is not feasible; for example, some maintenance and repair activities, vessel cleaning, or other operations for which engineering and work-practice controls are infeasible because exposures are intermittent and limited in duration

- Work operations for which feasible engineering and work-practice controls are not yet sufficient, or are not required under paragraph (f)(1)(iii) of this section, to reduce employee exposure to or below the PELs

- Emergencies

29.10 1,2-Dibromo-3-Chloropropane

Part 1926, Subpart Z, Standard Number 1926.1144 follows 1910.1144.

Permissible exposure limit:

- Inhalation: The employer shall ensure that no employee is exposed to an airborne concentration of DBCP in excess of 1 part DBCP per billion parts of air (ppb) as an 8-hour TWA.

- Dermal and eye exposure: The employer shall ensure that no employee is exposed to eye or skin contact with DBCP.

Respiratory protection.   For employees who are required to use respirators by this section, the employer must provide each employee an appropriate respirator that complies with the requirements of this paragraph. Respirators must be used during:

- Periods necessary to install or implement feasible engineering and work-practice controls

- Maintenance and repair activities for which engineering and work-practice controls are not feasible

- Work operations for which feasible engineering and work-practice controls are not yet sufficient to reduce employee exposure to or below the PEL

- Emergencies

Protective clothing and equipment.   Where there is any possibility of eye or dermal contact with liquid or solid DBCP, the employer shall provide, at no cost to the employee, and ensure that the employee wears impermeable protective clothing and equipment to protect the area of the body which may come in contact with DBCP. Eye and face protection shall meet the requirements of §1910.133 of this part.

29.11 Acrylonitrile

Part 1926, Subpart Z, Standard Number 1926.1145 follows 1910.1045.

Acrylonitrile or AN means acrylonitrile monomer with chemical formula $CH_2 = CHCN$.

Permissible exposure limits:

- TWA limit: The employer shall ensure that no employee is exposed to an airborne concentration of acrylonitrile in excess of two (2) parts acrylonitrile per million parts of air (2 ppm) as an 8-hour TWA.

- Ceiling limit: The employer shall ensure that no employee is exposed to an airborne concentration of acrylonitrile in excess of ten (10) ppm as averaged over any fifteen (15)-minute period during the work day.

- Dermal and eye exposure: The employer shall ensure that no employee is exposed to skin contact or eye contact with liquid AN.

Determinations of airborne exposure levels shall be made from air samples that are representative of each employee's exposure to AN over an 8-hour period.

Engineering and work-practice controls:

- By November 2, 1980, the employer shall institute engineering and work-practice controls to reduce and maintain employee exposures to AN, to or below the PELs, except to the extent that the employer establishes that such controls are not feasible.

- Wherever the engineering and work-practice controls which can be instituted are not sufficient to reduce employee exposures to or below the PELs, the employer shall nonetheless use them to reduce exposures to the lowest levels achievable by these controls, and shall supplement them by the use of respiratory protection which complies with the requirements of paragraph (h) of this section.

Respiratory protection.   For employees who use respirators required by this section, the employer must provide each employee an appropriate respirator that complies with the requirements of this paragraph. Respirators must be used during:

- Periods necessary to install or implement feasible engineering and work-practice controls

- Work operations, such as maintenance and repair activities or reactor cleaning, for which the employer establishes that engineering and work-practice controls are not feasible

- Work operations for which feasible engineering and work-practice controls are not yet sufficient to reduce employee exposure to or below the PELs

- Emergencies

Protective clothing and equipment.   Where eye or skin contact with liquid AN may occur, the employer shall provide at no cost to the employee, and ensure that employees wear, impermeable protective clothing or other equipment to protect any area of the body which may come in contact with liquid AN.

29.12   Ethylene Oxide

Part 1926, Subpart Z, Standard Number 1926.1147 follows 1910.1047.

Ethylene oxide or EtO means the three-membered ring organic compound with chemical formula C_2H_4O.

Permissible exposure limits:

- Eight-hour TWA: The employer shall ensure that no employee is exposed to an airborne concentration of EtO in excess of one (1) part EtO per million parts of air (1 ppm) as an 8-hour TWA.

- Excursion limit: The employer shall ensure that no employee is exposed to an airborne concentration of EtO in excess of 5 parts of EtO per million parts of air (5 ppm) as averaged over a sampling period of 15 minutes.

Engineering controls and work practices:

- The employer shall institute engineering controls and work practices to reduce and maintain employee exposure to or below the TWA and to or below the excursion limit, except to the extent that such controls are not feasible.

- Wherever the feasible engineering controls and work practices that can be instituted are not sufficient to reduce employee exposure to or below the TWA and to or below the excursion limit, the employer shall use them to reduce employee exposure to the lowest levels achievable by these controls and shall supplement them by the use of respiratory protection.

- Engineering controls are generally infeasible for the following operations: collection of quality assurance sampling from sterilized materials removal of biological indicators from sterilized materials: loading and unloading of tank cars; changing of ethylene oxide tanks on sterilizers; and vessel cleaning. For these operations, engineering controls are required only where the Assistant Secretary demonstrates that such controls are feasible.

Respiratory protection and personal protective equipment.   For employees who use respirators required by this section, the employer must provide each employee an appropriate respirator that complies with the requirements of this paragraph. Respirators must be used during:

- Periods necessary to install or implement feasible engineering and work-practice controls

- Work operations, such as maintenance and repair activities and vessel cleaning, for which engineering and work-practice controls are not feasible

- Work operations for which feasible engineering and work-practice controls are not yet sufficient to reduce employee exposure to or below the TWA

- Emergencies

29.13   Formaldehyde

Part 1926, Subpart Z, Standard Number 1926.1148 follows 1910.1048.

Formaldehyde means the chemical substance, HCHO.

Permissible exposure limit (PEL):

- TWA: The employer shall ensure that no employee is exposed to an airborne concentration of formaldehyde which exceeds 0.75 parts formaldehyde per million parts of air (0.75 ppm) as an 8-hour TWA.

- STEL: The employer shall ensure that no employee is exposed to an airborne concentration of formaldehyde which exceeds two parts formaldehyde per million parts of air (2 ppm) as a 15-minute STEL.

Engineering controls and work practices.   The employer shall institute engineering and work-practice controls to reduce and maintain employee exposures to formaldehyde at or below the TWA and the STEL.

Respiratory protection. For employees who use respirators required by this section, the employer must provide each employee an appropriate respirator that complies with the requirements of this paragraph. Respirators must be used during:

- Periods necessary to install or implement feasible engineering and work-practice controls.

- Work operations, such as maintenance and repair activities or vessel cleaning, for which the employer establishes that engineering and work-practice controls are not feasible.

- Work operations for which feasible engineering and work-practice controls are not yet sufficient to reduce employee exposure to or below the PELs.

- Emergencies

Protective equipment and clothing. Employers shall comply with the provisions of 29 CFR 1910.132 and 29 CFR 1910.133. When protective equipment or clothing is provided under these provisions, the employer shall provide these protective devices at no cost to the employee and ensure that the employee wears them. The employer shall select protective clothing and equipment based upon the form of formaldehyde to be encountered, the conditions of use, and the hazard to be prevented.

- All contact of the eyes and skin with liquids containing 1% or more formaldehyde shall be prevented by the use of chemical protective clothing made of material impervious to formaldehyde and the use of other personal protective equipment, such as goggles and face shields, as appropriate to the operation.

- Contact with irritating or sensitizing materials shall be prevented to the extent necessary to eliminate the hazard.

- Where a face shield is worn, chemical safety goggles are also required if there is a danger of formaldehyde reaching the area of the eye.

- Full-body protection shall be worn for entry into areas where concentrations exceed 100 ppm and for emergency reentry into areas of unknown concentration.

29.14   Methylene Chloride

Part 1926, Subpart Z, Standard Number 1926.1152 follows 1910.1052.
   Permissible exposure limits (PELs):

- Eight-hour TWA PEL: The employer shall ensure that no employee is exposed to an airborne concentration of MC in excess of twenty-five parts of MC per million parts of air (25 ppm) as an 8-hour TWA.

- STEL: The employer shall ensure that no employee is exposed to an airborne concentration of MC in excess of one hundred and twenty-five parts of MC per million parts of air (125 ppm) as determined over a sampling period of 15 minutes.

Engineering and work-practice controls. The employer shall institute and maintain the effectiveness of engineering and work-practice controls to reduce employee exposure to or below the PELs except to the extent that the employer can demonstrate that such controls are not feasible. Wherever the feasible engineering and work-practice controls which can be instituted are not sufficient to reduce employee exposure to or below the 8-TWA PEL or STEL, the employer shall use them to reduce employee exposure to the lowest levels achievable by these controls and shall supplement them by the use of respiratory protection that complies with the requirements of paragraph (g) of this section.

Respiratory protection.   For employees who use respirators required by this section, the employer must provide each employee an appropriate respirator that complies with the requirements of this paragraph. Respirators must be used during:

- Periods when an employee's exposure to MC exceeds the 8-hour TWA PEL, or STEL (e.g., when an employee is using MC in a regulated area)
- Periods necessary to install or implement feasible engineering and work-practice controls
- A few work operations, such as some maintenance operations and repair activities, for which the employer demonstrates that engineering and work-practice controls are infeasible
- Work operations for which feasible engineering and work-practice controls are not sufficient to reduce employee exposures to or below the PELs
- Emergencies

Protective work clothing and equipment:

- Where needed to prevent MC-induced skin or eye irritation, the employer shall provide clean protective clothing and equipment which is resistant to MC, at no cost to the employee, and shall ensure that each affected employee uses it. Eye and face protection shall meet the requirements of 29 CFR 1910.133 or 29 CFR 1915.153, as applicable.
- The employer shall clean, launder, repair, and replace all protective clothing and equipment mentioned in 29 CFR 1910.133 or 29 CFR 1915.153 as needed to maintain their effectiveness.
- The employer shall be responsible for the safe disposal of such clothing and equipment.

29.15   Respirable Crystalline Silica

Part 1926, Subpart Z, Standard Number 1926.1153 provides:

Respirable crystalline silica means quartz, cristobalite, and/or tridymite contained in airborne particles that are determined to be respirable by a sampling device designed to meet the characteristics for respirable-particle-size-selective samplers specified in the International Organization for Standardization (ISO) 7708:1995: Air Quality—Particle Size Fraction Definitions for Health-Related Sampling. Health hazards related to overexposure to respirable crystalline silica are chronic obstructive pulmonary disease (COPD), chronic kidney disease (CKD), silicosis, etc. OSHA estimates that about 2,300,000 workers are exposed to respirable crystalline silica in the United States.

Permissible exposure limit (PEL).   The employer shall ensure that no employee is exposed to an airborne concentration of respirable crystalline silica in excess of 50 µg/m³, calculated as an 8-hour TWA. At the 25 µg/m³ averaged over one work day, the employer must act to reduce the exposure to the worker. The employer must prove that a workplace with over 25 µg/m³ (action level) has been assessed.

Engineering and work-practice controls.   The employer shall use engineering and work-practice controls to reduce and maintain employee exposure to respirable crystalline silica to or below the PEL, unless the employer can demonstrate that such controls are not feasible. Wherever such feasible engineering and work-practice controls are not sufficient to reduce employee exposure to or below the PEL, the employer shall nonetheless use them to reduce employee exposure to the lowest feasible level and shall supplement them with the use of respiratory protection. Wet-cutting methods, vacuums equipped with a 0.3-micron Hepa filter, substrate substitutes for silica sand, etc., are the examples of engineering controls.

Respiratory protection.   Where respiratory protection is required by this section, the employer must provide each employee an appropriate respirator that complies with the requirements. Respiratory protection is required:

- Where exposures exceed the PEL during periods necessary to install or implement feasible engineering and work-practice controls
- Where exposures exceed the PEL during tasks, such as certain maintenance and repair tasks, for which engineering and work-practice controls are not feasible
- During tasks for which an employer has implemented all feasible engineering and work-practice controls and such controls are not sufficient to reduce exposures to or below the PEL

Example 29.6
Health hazards related to overexposure to respirable crystalline silica are:

A. Chronic obstructive pulmonary disease (COPD)

B. Chronic kidney disease (CKD)

C. Silicosis

D. All of the above

Answer: D

Example 29.7
OSHA has revised its silica standard to reflect current research which showed the previous standard wasn't protective enough for workers. Type the correct values in the blanks.

A. The permissible exposure limit (PEL) for silica has been lowered to _____micrograms of respirable crystalline silica per cubic meter of air (µg/m³).

B. OSHA's permissible exposure limit (PEL) for silica exposure is over a TWA of an _____ hour period.

C. At the _____ µg/m³ averaged over one work day the employer must act to reduce the exposure to the worker.

D. The employer must prove that a workplace with over _____ µg/m³ (action level) has been assessed.

E. When workers use handheld power saws outdoors for more than _____ hours/shift, a respirator with a minimum assigned protection factor (APF) of 10 is required.

F. For handheld and stand-mounted drills, a dust collector must be provided with a filter with _____ % or greater efficiency.

G. OSHA estimates that about _____ workers are exposed to respirable crystalline silica in the United States.

Answers:

A. 50

B. 8

C. 25

D. 25

E. 4

F. 99

G. 2,300,000

29.16 Summary

All workers who are required to perform asbestos-related tasks must wear the appropriate protective equipment. Protective clothing should be made with a material that does not allow asbestos fibers to penetrate. The protective clothing should cover the whole body, and should fit comfortably at the neck, wrists, and ankles. Headgear and boots that resist the penetration of asbestos fibers must also be worn, and if protective clothing gets damaged or torn, it must be replaced immediately. Similar types of safety precautions must be followed for the 13 carcinogens. A regulated area shall be established by an employer where a carcinogen is addressed and controlled in accordance with the requirements for the category or categories describing the operation involved.

29.17 Multiple-Choice Questions

29.1 Asbestos-containing material (ACM) means any material containing more than:

A. 0.01% asbestos
B. 0.1% asbestos
C. 1% asbestos
D. 1.5% asbestos

29.2 The employer shall ensure that no employee is exposed to an airborne concentration of asbestos as an 8-hour TWA in excess of:

A. 0.001 fiber per cubic centimeter of air
B. 1.0 fiber per cubic centimeter of air
C. 0.01 fiber per cubic centimeter of air
D. 0.1 fiber per cubic centimeter of air

29.3 The employer shall ensure that no employee is exposed to an airborne concentration of asbestos as averaged over a sampling period of thirty (30) minutes in excess of:

A. 0.001 fiber per cubic centimeter of air
B. 1.0 fiber per cubic centimeter of air
C. 0.01 fiber per cubic centimeter of air
D. 0.1 fiber per cubic centimeter of air

29.4 The 13 carcinogens does not include:

A. Hydrogen sulfide
B. Benzidine
C. 4-Aminodiphenyl
D. Ethyleneimine

29.5 No employee may be exposed to vinyl chloride at concentrations greater than 1 ppm averaged:

A. Over any 0.25-hour period
B. Over any 4-hour period
C. Over any 8-hour period
D. Over any 24-hour period

29.6 No employee may be exposed to vinyl chloride at concentrations greater than 5 ppm averaged:

A. Over any 0.25-hour period
B. Over any 4-hour period
C. Over any 8-hour period
D. Over any 24-hour period

29.7 The employer shall ensure that no employee is exposed to inorganic arsenic at concentrations based on averaged over any 8-hour period greater than:

A. 0.1 micrograms per cubic meter of air
B. 1.0 micrograms per cubic meter of air
C. 10 micrograms per cubic meter of air
D. 15 micrograms per cubic meter of air

29.8 The employer must ensure that no employee is exposed to an airborne concentration of beryllium calculated as an 8-hour TWA in excess of:

A. $0.2\,\mu g/m^3$
B. $2.0\,\mu g/m^3$
C. $1.0\,\mu g/m^3$
D. $10.0\,\mu g/m^3$

29.9 The employer must ensure that no employee is exposed to an airborne concentration of beryllium calculated over a sampling period of 15 minutes in excess of:

A. $0.2\,\mu g/m^3$
B. $2.0\,\mu g/m^3$
C. $1.0\,\mu g/m^3$
D. $10.0\,\mu g/m^3$

29.10 The employer shall ensure that no employee is exposed to an airborne concentration of benzene as averaged over any 15-minute period in excess of:

    A.   1 ppm

    B.   5 ppm

    C.   2 ppm

    D.   10 ppm

29.18   Practice Problems

29.1   Define PACM, ACM, TSI, and surfacing ACM.

29.2   List the permissible exposure limits of asbestos.

29.3   List the safety regulations for asbestos.

29.4   List the 13 carcinogens.

29.5   List the permissible exposure limits of vinyl chloride.

29.6   List the permissible exposure limits of inorganic arsenic.

29.7   List the permissible exposure limits of airborne concentration of beryllium.

29.8   List the permissible exposure limits of chromium.

29.9   List the permissible exposure limits of cadmium.

29.10 List the permissible exposure limits of benzene.

29.11 List the permissible exposure limits of 1,2-dibromo-3-chloropropane.

29.12 List the permissible exposure limits of acrylonitrile.

29.13 List the permissible exposure limits of ethylene oxide.

29.14 List the permissible exposure limits of formaldehyde.

29.15 List the permissible exposure limits of methylene chloride.

29.19   Critical Thinking and Discussion Topic

Some workers are brave and do not think of so much of the consequences of any activities. Assume you have a few construction workers who deal with toxic and hazardous substances without proper protection. They are practicing it over years and no accident or health issue was recognized. OSHA team visited the company several times but did not notice it or it was skipped somehow. Now, this news came to your attention when you are in charge of the safety issues. How will you handle it?

CHAPTER 30

Confined Spaces in Construction

(Photo courtesy of swww.lhsfna.org.)

30.1　General

A confined space (Fig. 30.1) is defined as:

- A space that is large enough for an employee to enter
- Has restricted means of entry or exit
- Is not designed for continuous employee occupancy by one or more employees

FIGURE 30.1 A worker inspecting in a confined space.
(Photo courtesy of Norbord.)

A permit-required confined space (permit space) is confined space that presents or has the potential for hazards related to:

- Atmospheric conditions (toxic, flammable, asphyxiating)
- Engulfment
- Configuration
- Any other recognized serious hazard

Examples of confined spaces may include, but are not limited to, the following:

- Storage tanks
- Compartments of ships
- Process vessels
- Pits
- Silos
- Vats
- Wells
- Sewers
- Digesters
- Degreasers

- Reaction vessels
- Boilers
- Ventilation and exhaust ducts
- Tunnels
- Underground utility vaults
- Pipelines

Example 30.1

_____ in liquids or loose materials is one of the leading causes of death from physical hazards in confined space.

    A. Elevation

    B. Engulfment

    C. Fire

    D. Electrocutions

Answer: *B*

30.2   Confined Space Hazards

Some potential hazards you might encounter in confined space are as follows:

Physical hazards.   Physical hazards may result from mechanical equipment or moving parts like agitators, blenders, and stirrers. Dangers may also be present from gases, liquids, or fluids entering the space from connecting pipes. Before entering a permit space, all mechanical equipment must be locked out/tagged out. All lines containing hazardous materials such as steam, gases, or coolants must also be shut off. Other physical hazards include heat and excessive noise. Temperature can build up quickly in a permit space and cause exhaustion or dizziness. Sounds may reverberate and make it hard to hear important directions or warnings.

Oxygen deficiency.   In general, the primary atmospheric hazard associated with permit spaces is oxygen deficiency. Normal air contains 20.8% oxygen by volume. The minimum safe level as indicted by OSHA is 19.5%; OSHA defines the maximum safe level as 23.5%. A low oxygen level can quickly cause death. Oxygen can be displaced by other gases such as argon, nitrogen, or methane. Oxygen can be consumed by chemical reactions such as rusting, rotting, fermentation, or burning of flammable substances.

Combustibility.   Flammable and combustible gases or vapors may be present from previous cargoes, tank coatings, preservatives, and welding gases. These built-up vapors and gases can be ignited by faulty electrical equipment, static electricity, sparks from welding, or cigarettes. The atmosphere is hazardous if tests show the presence of flammable gas, vapor, or mist that is more than 10% of its lower flammable limit.

Toxic air contaminants.   These contaminants occur from material previously stored in the tank or as a result of the use of coatings, cleaning solvents, or preservatives.

Example 30.2

_____ deficiency occurs from chemical or biological reactions which displace or consume oxygen from within a confined space.

A. Nitrogen

B. Oxygen

C. Hydrogen

D. Silica

Answer: B

30.3 Before Entering a Confined Space

The hazards must be identified before anyone enters a permit-required confined space. A written entry permit must be used for entry into a permit-required confined space. The permit outlines the conditions that make the entry safe. The permit must be posted at the space during an entry, and it must be signed by the entry supervisor. Before entering a confined space, the following checks must be performed to ensure safety:

- Control any hazardous energy: Use looks and tags to prevent accidental start-up of equipment while you are working in the permit space. Cut off steam, water, gas, or power lines that enter the space.

- Test the air: Use special instruments for testing the levels of oxygen, combustibility, and toxicity in confined spaces.

- Test before you open the space by probing with test instruments near the entry.

- Once the space is opened, test the air from top to bottom. Some gases like propane and butane are heavy, and they will sink to the bottom of the space. Light gases like methane will rise to the top. So you need to be sure to check all levels.

- After you are sure that the oxygen level is adequate and there is nothing combustible in the space, test for toxicity. Periodic or continuous follow-up testing may be needed during the entry.

- Purge and ventilate the space: Some confined spaces may contain water, sediment, hazardous atmospheres, or other unwanted substances. These substances generally must be purged, that is, pumped out or otherwise removed, before entry.

- Use ventilating equipment to maintain an oxygen level between 19.5 and 23.5%. It also should keep toxic gases and vapors to within accepted levels prescribed by OSHA.

- Use only safe, grounded, explosion-proof equipment, and fans if a combustible atmosphere could develop in the space.

- If ventilation does not eliminate the atmospheric hazards, you will need to wear appropriate respiratory protection. You might also need eye and hearing protection and protective clothing.

Example 30.3

A confined space entry permit is an authorized approval in _____ form.

 A. Oral

 B. Spoken

 C. Written

 D. Ritual

Answer: C

Example 30.4

A flammable atmosphere generally results from vaporization of flammable liquids, by-products of chemical reaction, enriched oxygen atmospheres, or concentration of flammable gases or combustible dusts.

 • True

 • False

Answer:
True. A flammable atmosphere generally results from vaporization of flammable liquids, by-products of chemical reaction, enriched oxygen atmospheres, or concentration of flammable gases or combustible dusts. Three components are necessary for an atmosphere to become flammable: fuel and oxygen, the proper mixture of fuel and oxygen, and a source of ignition.

30.4   Rescue Procedures

Almost two-thirds of the permits-required confined space deaths are the result of people trying to rescue them (Fig. 30.2). When workers enter a permit space, at least one person must stay outside to call for assistance or to offer assistance. In addition to monitoring the safety of the confined space entrant, when the employer designates this attendant to perform rescue procedures, he or she should be equipped with the necessary PPE and rescue equipment and trained in first aid and cardiopulmonary resuscitation (CPR). He or she should keep constant communication with those within space. If a situation arises that requires emergency rescue, the attendant must not enter until additional help arrives. The entrant must wear a full-body harness and lifeline so he or she can be pulled easily out of space. It can be attached to a block and tackle or a winch system that can be operated from outside the space by a single rescuer/attendant. The entrant must be trained to recognize hazardous situations, so that he or she knows when to exit a confined space before rescue is needed.

Figure 30.2   Rescuing a construction worker from a confined space.
(Photo courtesy of Hammer Hanford.)

Example 30.5
The employer of the members of a designated rescue team is not required to ensure that the team members have received all training required for authorized entrants.

- True
- False

Answer:
False. The employer of the members of a designated rescue team is required to ensure that the team members have received all training required for authorized entrants and have also been trained to perform their assigned rescue duties.

30.5   General Safety Regulations

Part 1926, Subpart AA, Standard Number 1926.1203 provides the following specifications:

- Before it begins work at a worksite, each employer must ensure that a competent person identifies all confined spaces in which one or more of the employees it directs may work, and identifies each space that is a permit space, through consideration and evaluation of the elements of that space, including testing as necessary.

- If the workplace contains one or more permit spaces, the employer who identifies, or who receives notice of, a permit space must:
  - Inform exposed employees by posting danger signs or by any other equally effective means, of the existence and location of, and the danger posed by, each permit space.
  - Inform, in a timely manner and in a manner other than posting, its employees' authorized representatives and the controlling contractor of the existence and location of, and the danger posed by, each permit space.
- Each employer who identifies, or receives notice of, a permit space and has not authorized employees it directs to work in that space must take effective measures to prevent those employees from entering that permit space, in addition to complying with all other applicable requirements of this standard.
- If any employer decides that employees it directs will enter a permit space, that employer must have a written permit space program implemented at the construction site. The written program must be made available prior to and during entry operations for inspection by employees and their authorized representatives.

Example 30.6
A controlling contractor is that employer with overall responsibility for construction at the worksite.

- True
- False

Answer:
True. A controlling contractor is that employer with overall responsibility for construction at the worksite. This constructor is responsible for coordinating the entry operations when more than one employer will have employees in the permit space and when other activities on the site could result in a hazard in the space.

30.6   Permit-Required Confined Space Program
Part 1926, Subpart AA, Standard Number 1926.1204 states:
Each entry employer must:

- Implement the measures necessary to prevent unauthorized entry.
- Identify and evaluate the hazards of permit spaces before employees enter them.
- Develop and implement the means, procedures, and practices necessary for safe permit space entry operations, including, but not limited to, the following:
  - Specifying acceptable entry conditions
  - Providing each authorized entrant or that employee's authorized representative with the opportunity to observe any monitoring or testing of permit spaces

- Isolating the permit space and physical hazard(s) within the space
- Purging, inverting, flushing, or ventilating the permit space as necessary to eliminate or control atmospheric hazards. Determining that, in the event the ventilation system stops working, the monitoring procedures will detect an increase in atmospheric hazard levels in sufficient time for the entrants to safely exit the permit space
- Providing pedestrian, vehicle, or other barriers as necessary to protect entrants from external hazards
- Verifying that conditions in the permit space are acceptable for entry throughout the duration of an authorized entry, and ensuring that employees are not allowed to enter into, or remain in, a permit space with a hazardous atmosphere unless the employer can demonstrate that personal protective equipment (PPE) will provide effective protection for each employee in the permit space and provides the appropriate PPE to each employee
- Eliminating any conditions (e.g., high pressure) that could make it unsafe to remove an entrance cover

- Provide the following equipment at no cost to each employee, maintain that equipment properly, and ensure that each employee uses that equipment properly:

- Testing and monitoring equipment needed to comply with this section
- Ventilating equipment needed to obtain acceptable entry conditions
- PPE insofar as feasible engineering and work-practice controls do not adequately protect employees
- Lighting equipment that meets the minimum illumination requirements that is approved for the ignitable or combustible properties of the specific gas, vapor, dust, or fiber that will be present, and that is sufficient to enable employees to see well enough to work safely and to exit the space quickly in an emergency
- Equipment, such as ladders, needed for safe ingress and egress by authorized entrants
- Rescue and emergency equipment needed to comply with the requirements, except to the extent that the equipment is provided by rescue services
- Any other equipment necessary for safe entry into, safe exit from, and rescue from permit spaces

---

Example 30.7

An _____ employer is an employer whose employees actually enter a permit space.

   A. Entry

   B. Monitoring

   C. Host

   D. Subcontractor

Answer: *A*

30.7 Permitting Process

Part 1926, Subpart AA, Standard Number 1926.1205 states:

- Before entry is authorized, each entry employer must document the completion of measures by preparing an entry permit.

- Before entry begins, the entry supervisor identified on the permit must sign the entry permit to authorize entry.

- The completed permit must be made available at the time of entry to all authorized entrants or their authorized representatives, by posting it at the entry portal or by any other equally effective means, so that the entrants can confirm that pre-entry preparations have been completed.

- The duration of the permit may not exceed the time required to complete the assigned task or job identified on the permit.

- The entry supervisor must terminate entry and take the following action when any of the following apply:

    - Cancel the entry permit when the entry operations covered by the entry permit have been completed.

    - Suspend or cancel the entry permit and fully reassess the space before allowing reentry when a condition that is not allowed under the entry permit arises in or near the permit space and that condition is temporary in nature and does not change the configuration of the space or create any new hazards within it.

    - Cancel the entry permit when a condition that is not allowed under the entry permit arises in or near the permit space

- The entry employer must retain each canceled entry permit for at least 1 year to facilitate the review of the permit-required confined space program. Any problems encountered during an entry operation must be noted on the pertinent permit so that appropriate revisions to the permit space program can be made.

30.8   Entry Permit

Part 1926, Subpart AA, Standard Number 1926.1206 states that the entry permit that documents compliance with this section and authorizes entry to a permit space must identify:

- The permit space to be entered

- The purpose of the entry

- The date and the authorized duration of the entry permit

- The authorized entrants within the permit space, by name or by such other means (e.g., through the use of rosters or tracking systems) as will enable the attendant to determine quickly and accurately, for the duration of the permit, which authorized entrants are inside the permit space

- Means of detecting an increase in atmospheric hazard levels in the event the ventilation system stops working
- Each person, by name, currently serving as an attendant
- The individual, by name, currently serving as entry supervisor, and the signature or initials of each entry supervisor who authorizes entry
- The hazards of the permit space to be entered
- The measures used to isolate the permit space and to eliminate or control permit space hazards before entry

The acceptable entry conditions:

- The results of tests and monitoring performed, accompanied by the names or initials of the testers and by an indication of when the tests were performed
- The rescue and emergency services that can be summoned and the means (such as the equipment to use and the numbers to call) for summoning those services
- The communication procedures used by authorized entrants and attendants to maintain contact during the entry
- Equipment, such as PPE, testing equipment, communications equipment, alarm systems, and rescue equipment, to be provided for compliance with this standard
- Any other information necessary, given the circumstances of the particular confined space, to ensure employee safety
- Any additional permits, such as for hot work, that have been issued to authorize work in the permit space

30.9 Training

Part 1926, Subpart AA, Standard Number 1926.1207 states:

- The employer must provide training to each employee whose work is regulated by this standard, at no cost to the employee, and ensure that the employee possesses the understanding, knowledge, and skills necessary for the safe performance of the duties assigned under this standard. This training must result in an understanding of the hazards in the permit space and the methods used to isolate, control, or in other ways protect employees from these hazards, and for those employees not authorized to perform entry rescues, in the dangers of attempting such rescues.
- Training required by this section must be provided to each affected employee:
  - In both a language and vocabulary that the employee can understand
  - Before the employee is first assigned duties under this standard
  - Before there is a change in assigned duties
  - Whenever there is a change in permit space entry operations that presents a hazard about which an employee has not previously been trained
  - Whenever there is any evidence of a deviation from the permit space entry procedures required or there are inadequacies in the employee's knowledge or use of these procedures

- The training must establish employee proficiency in the duties required by this standard and must introduce new or revised procedures, as necessary, for compliance with this standard.
- The employer must maintain training records to show it to OSHA. The training records must contain each employee's name, the name of the trainers, and the dates of training. The documentation must be available for inspection by employees and their authorized representatives, for the period of time the employee is employed by that employer.

30.10   Duties of Authorized Entrants

Part 1926, Subpart AA, Standard Number 1926.1208 states that the entry employer must ensure that all authorized entrants:

- Are familiar with and understand the hazards that may be faced during entry, including information on the mode, signs or symptoms, and consequences of the exposure
- Properly use equipment
- Communicate with the attendant as necessary to enable the attendant to assess entrant status and to enable the attendant to alert entrants of the need to evacuate the space
- Alert the attendant whenever:
  - There is any warning sign or symptom of exposure to a dangerous situation
  - The entrant detects a prohibited condition
- Exit from the permit space as quickly as possible whenever:
  - An order to evacuate is given by the attendant or the entry supervisor
  - There is any warning sign or symptom of exposure to a dangerous situation
  - The entrant detects a prohibited condition
  - An evacuation alarm is activated

30.11   Duties of Attendants

Part 1926, Subpart AA, Standard Number 1926.1209 states:
The entry employer must ensure that each attendant:

- Is familiar with and understands the hazards that may be faced during entry, including information on the mode, signs or symptoms, and consequences of the exposure
- Is aware of possible behavioral effects of hazard exposure in authorized entrants
- Continuously maintains an accurate count of authorized entrants in the permit space and ensures that the means used to identify authorized entrants accurately identifies who is in the permit space
- Remains outside the permit space during entry operations until relieved by another attendant

- Communicates with authorized entrants as necessary to assess entrant status and to alert entrants of the need to evacuate the space
- Assesses activities and conditions inside and outside the space to determine if it is safe for entrants to remain in the space and orders the authorized entrants to evacuate the permit space immediately under any of the following conditions:
  - If there is a prohibited condition
  - If the behavioral effects of hazard exposure are apparent in an authorized entrant
  - If there is a situation outside the space that could endanger the authorized entrants
  - If the attendant cannot effectively and safely perform all the duties required under this section
- Summons rescue and other emergency services as soon as the attendant determines that authorized entrants may need assistance to escape from permit space hazards
- Takes the following actions when unauthorized persons approach or enter a permit space while entry is underway:
  - Warns the unauthorized persons that they must stay away from the permit space
  - Advises the unauthorized persons that they must exit immediately if they have entered the permit space
  - Informs the authorized entrants and the entry supervisor if unauthorized persons have entered the permit space
- Performs nonentry rescues as specified by the employer's rescue procedure
- Performs no duties that might interfere with the attendant's primary duty to assess and protect the authorized entrants

30.12   Duties of Entry Supervisors

Part 1926, Subpart AA, Standard Number 1926.1210 states:

The entry employer must ensure that each entry supervisor:

- Is familiar with and understands the hazards that may be faced during entry, including information on the mode, signs or symptoms, and consequences of the exposure
- Verifies, by checking that the appropriate entries have been made on the permit, that all tests specified by the permit have been conducted and that all procedures and equipment specified by the permit are in place before endorsing the permit and allowing entry to begin
- Terminates the entry and cancels or suspends the permit
- Verifies that rescue services are available and that the means for summoning them are operable, and that the employer will be notified as soon as the services become unavailable
- Removes unauthorized individuals who enter or who attempt to enter the permit space during entry operations

- Determines, whenever responsibility for a permit space entry operation is transferred, and at intervals dictated by the hazards and operations performed within the space, that entry operations remain consistent with terms of the entry permit and that acceptable entry conditions are maintained

30.13   Rescue and Emergency Services

Part 1926, Subpart AA, Standard Number 1926.1211 states:

An employer who designates rescue and emergency services must:

- Evaluate a prospective rescuer's ability to respond to a rescue summons in a timely manner, considering the hazard(s) identified.

- Evaluate a prospective rescue service's ability, in terms of proficiency with rescue-related tasks and equipment, to function appropriately while rescuing entrants from the particular permit space or types of permit spaces identified.

- Select a rescue team or service from those evaluated that:
  - Has the capability to reach the victim(s) within a time frame that is appropriate for the permit space hazard(s) identified
  - Is equipped for, and proficient in, performing the needed rescue services
  - Agrees to notify the employer immediately in the event that the rescue service becomes unavailable

- Inform each rescue team or service of the hazards they may confront when called on to perform rescue at the site.

- Provide the rescue team or service selected with access to all permit spaces from which rescue may be necessary so that the rescue team or service can develop appropriate rescue plans and practice rescue operations.

- An employer whose employees have been designated to provide permit space rescue and/or emergency services must take the following measures and provide all equipment and training at no cost to those employees:
  - Provide each affected employee with the PPE needed to conduct permit space rescues safely and train each affected employee so the employee is proficient in the use of that PPE.
  - Train each affected employee to perform assigned rescue duties. The employer must ensure that such employees successfully complete the training required and establish proficiency as authorized entrants.
  - Train each affected employee in basic first aid and CPR. The employer must ensure that at least one member of the rescue team or service holding a current certification in basic first aid and CPR is available.
  - Ensure that affected employees practice making permit space rescues before attempting an actual rescue, and at least once every 12 months, by means of simulated rescue operations in which they remove dummies, manikins, or actual persons from the actual permit spaces or from representative permit spaces, except practice rescue is not required where the affected employees properly performed a rescue operation during the last 12 months in the same permit space

the authorized entrant will enter, or in a similar permit space. Representative permit spaces must, with respect to opening size, configuration, and accessibility, simulate the types of permit spaces from which rescue is to be performed.

- Nonentry rescue is required unless the retrieval equipment would increase the overall risk of entry or would not contribute to the rescue of the entrant. The employer must designate an entry rescue service whenever nonentry rescue is not selected. Whenever nonentry rescue is selected, the entry employer must ensure that retrieval systems or methods are used whenever an authorized entrant enters a permit space, and must confirm, prior to entry, that emergency assistance would be available in the event that nonentry rescue fails. Retrieval systems must meet the following requirements:

  - Each authorized entrant must use a chest or full-body harness, with a retrieval line attached at the center of the entrant's back near shoulder level, above the entrant's head, or at another point which the employer can establish presents a profile small enough for the successful removal of the entrant. Wristlets or anklets may be used in lieu of the chest or full-body harness if the employer can demonstrate that the use of a chest or full-body harness is infeasible or creates a greater hazard and that the use of wristlets or anklets is the safest and most effective alternative.

  - The other end of the retrieval line must be attached to a mechanical device or fixed point outside the permit space in such a manner that rescue can begin as soon as the rescuer becomes aware that rescue is necessary. A mechanical device must be available to retrieve personnel from vertical-type permit spaces more than 5 feet (1.52 meters) deep.

  - Equipment that is unsuitable for retrieval must not be used, including, but not limited to, retrieval lines that have a reasonable probability of becoming entangled with the retrieval lines used by other authorized entrants, or retrieval lines that will not work due to the internal configuration of the permit space.

- If an injured entrant is exposed to a substance for which a Safety Data Sheet (SDS) or other similar written information is required to be kept at the worksite, that SDS or written information must be made available to the medical facility treating the exposed entrant.

Example 30.8
Each authorized entrant shall use a chest or full-body harness, with a retrieval line attached at the center of the entrant's back, near shoulder level, above the entrant's head, or at another point from which the employer can establish the ability to successfully remove the entrant.

- True
- False

Answer:
True. Each authorized entrant shall use a chest or full-body harness, with a retrieval line attached at the center of the entrant's back, near shoulder level, above the entrant's head, or at another point from which the employer can establish the ability to successfully remove the entrant.

30.14   Employee Participation

Part 1926, Subpart AA, Standard Number 1926.1212 states:

- Employers must consult with affected employees and their authorized representatives on the development and implementation of all aspects of the permit space program required.

- Employers must make available to each affected employee and his/her authorized representatives all information required to be developed by this standard.

30.15   Summary

Material being drawn from the bottom of storage bins can cause the surface to act like quicksand. When a storage bin is emptied from the bottom, the flow of material may form a funnel-shaped path over the outlet. The rate of material flow increases toward the center of the funnel. During an unloading operation, the flow rate can become so great that once a worker is drawn into the flow path, escape is virtually impossible.

Testing and monitoring the air quality in a confined space to ensure that the oxygen level is between 19.5 and 23.5% by volume and the flammable range is less than 10% of the LFL (lower flammable limit) of any flammable materials is required for a permit to be issued.

Oxygen deficiency occurs from many sources, including chemical or biological reactions that displace or consume oxygen from within a confined space. The consumption of oxygen takes place during combustion of flammable substances, as occurs in welding, cutting, or brazing operations. A more subtle form of consumption of oxygen occurs during bacterial action, such as in the fermentation process.

A rescue retrieval line should be attached at one end to a mechanical device or a fixed point outside of the permit space in such a manner that rescue can begin as soon as the rescuer becomes aware that it is necessary and to a full-body harness worn by the entrant on the other end. A mechanical device shall be available to retrieve personnel from vertical-type permit spaces more than 5 feet deep.

An authorized entrant must alert the attendant if he or she recognizes any warning sign or symptom of exposure to a dangerous situation or if the entrant detects a prohibited condition. An entrant must also exit the space if the attendant detects a prohibited condition, the behavioral effects of hazard exposure, or any situation outside of the space that could endanger the entrant.

30.16   Multiple-Choice Questions

30.1 A confined space is a space that:

  A. Isn't designed for continuous employee occupancy
  B. Is large enough for an employee to enter
  C. Has restricted means to entry or exit
  D. All of the above

30.2 A permit-required confined space:

    A. Must be a confined space

    B. Has hazard within it or has the potential for hazard to develop

    C. Might need to be entered to do certain job

    D. All of the above

30.3 Hazardous moving machine parts in a confident space:

    A. Are a physical hazard

    B. Cause the space to meet the definition of a permit-required confidence space

    C. Must be locked out or tagged out before entry

    D. All of the above

30.4 Overtime, rusting metal inside of a confined space can lead to:

    A. A strong order

    B. A dusty atmosphere

    C. An oxygen-deficient atmosphere

    D. Engulfment

30.5 The use of coatings or cleaning solvents inside of a permit space:

    A. Can cause a toxic atmosphere

    B. Is an example of a physical hazard

    C. Is only safe if the space is locked out

    D. Will not cause any hazards

30.6 When you test the air inside of a permit-required confined space:

    A. Check only the bottom of the space

    B. Check only the opening of the space

    C. Test the air from top to bottom

    D. Check only at the top of the space

30.7 An entry permit is signed by

    A. OSHA

    B. The entry supervisor

    C. The entrants

    D. The attendant

30.8 Test the air first for:

    A. Noise

    B. Oxygen

    C. Combustibility

    D. Toxicity

30.9 If ventilation doesn't eliminate the atmospheric hazards in the space:

    A. The entry must be done very quickly.

    B. There must be two teams of attendants outside of the space.

    C. The entrants will need to wear respirators.

    D. None of the above.

30.10 The entrant wears a full-body harness and lifeline so he or she:

    A. Can be quickly pulled from the space in case of an emergency

    B. Will not get lost

    C. Can't get close to moving matching parts

    D. Will not fall down

30.17   Practice Problems

30.1   Define confined space and permit-required confined space.

30.2   List some confined space hazards.

30.3   What are the tasks commonly to be done before entering a confined space?

30.4   What are the rescue procedures from a confined space?

30.5   List some OSHA general safety regulations for using a confined space.

30.6   List some OSHA permit-required confined space program.

30.7   List some duties of authorized entrants in a confined space.

30.18   Critical Thinking and Discussion Topic

Consider two types of confined spaces: artificial and natural. Natural confined space may be a cave or a form of landscape which exists for a long time. Artificial one may be under a relatively new building where access is limited. Considering these two scenarios, what should be the difference in safety adoption to access these two types of confined spaces?

(Photo courtesy of pixabay from pexels.)

CHAPTER 31

Blasting and the Use of Explosives

(Photo courtesy of 123RF.)

31.1   General

Blasting is the controlled use of explosives and other methods, such as gas pressure blast pyrotechnics, to break rock for excavation. It is most commonly practiced in mining, quarrying, and civil engineering such as dam construction, tunnel construction, or road construction. Often known as a rock cut, it is the result of rock blasting. A blasting agent is any material or mixture consisting of a fuel and an oxidizer used for blasting but not classified as an explosive and in which none of the ingredients are classified as an explosive, provided that the furnished (mixed) product cannot be detonated with a No. 8 test blasting cap when confined. Dynamite, ammonium nitrate, ammonium nitrate in fuel

Figure 31.1   Blasting operation.
(Photo courtesy of STELR Australia.)

oil (ANFO), and slurries are the major explosives used for rock excavation. Dynamite has been largely replaced by ammonium nitrate, ANFO, and slurries for construction use because these explosives are cheaper in cost and easier to handle than dynamite. Ammonium nitrate and ANFO are among the least expensive of the listed explosives. ANFO is especially easy to handle since it is a liquid that can be simply poured into the blasthole. However, ammonium nitrate explosives are not water resistant and require an auxiliary explosive (primer) for detonation. Gels, explosives, and water are mixtures of slurries. They may also contain powdered metals (metalized slurries) to increase blast energy. Slurries are cheaper than dynamite but more expensive than explosives on ammonium nitrate. Their main advantages over explosives with ammonium nitrate are water resistance and greater power. Slurries in plastic bags are available as liquids or packaged in. Slurries also require a primer for detonation (Fig. 31.1).

31.2   General Safety Regulations

Part 1926, Subpart U, Standard Number 1926.900 provides some general safety regulations as follows:

- The employer shall permit only authorized and qualified persons to handle and use explosives.

- Smoking, firearms, matches, open flame lamps, and other fires, flame, or heat-producing devices and sparks shall be prohibited in or near explosive magazines or while explosives are being handled, transported, or used.

- No person shall be allowed to handle or use explosives while under the influence of intoxicating liquors, narcotics, or other dangerous drugs.

- All explosives shall be accounted for at all times. Explosives not being used shall be kept in a locked magazine, unavailable to persons not authorized to handle them. The employer shall maintain an inventory and use record of all explosives. Appropriate authorities shall be notified of any loss, theft, or unauthorized entry into a magazine.

- The prominent display of adequate signs, warning against the use of mobile radio transmitters, on all roads within 1000 feet of blasting operations. Whenever adherence to the 1000-foot distance would create an operational handicap, a competent person shall be consulted to evaluate the particular situation, and alternative provisions may be made which are adequately designed to prevent any premature firing of electric blasting caps.

31.3 Blaster Qualifications

Part 1926, Subpart U, Standard Number 1926.901 provides the blaster qualifications as follows:

- A blaster shall be able to understand and give written and oral orders.

- A blaster shall be in good physical condition and not be addicted to narcotics, intoxicants, or similar types of drugs.

- A blaster shall be qualified, by reason of training, knowledge, or experience, in the field of transporting, storing, handling, and use of explosives, and have a working knowledge of state and local laws and regulations which pertain to explosives.

- Blasters shall be required to furnish satisfactory evidence of competency in handling explosives and performing in a safe manner the type of blasting that will be required.

- The blaster shall be knowledgeable and competent in the use of each type of blasting method used.

31.4 Surface Transportation of Explosives

According to Part 1926, Subpart U, Standard Number 1926.902, some major regulations for surface transportation of explosives are as follows:

- Motor vehicles or conveyances transporting explosives shall only be driven by, and be in the charge of, a licensed driver who is physically fit. He shall be familiar with the local, state, and federal regulation governing the transportation of explosives.

- No person shall smoke, or carry matches or any other flame-producing device, nor shall firearms or loaded cartridges be carried while in or near a motor vehicle or conveyance transporting explosives.

- Explosives, blasting agents, and blasting supplies shall not be transported with other materials or cargoes. Blasting caps (including electric) shall not be transported in the same vehicle with other explosives.

- Vehicles used for transporting explosives shall be strong enough to carry the load without difficulty, and shall be in good mechanical condition.

- When explosives are transported by a vehicle with an open body, a Class II magazine or original manufacturer's container shall be securely mounted on the bed to contain the cargo.

- All vehicles used for the transportation of explosives shall have tight floors, and any exposed spark-producing metal on the inside of the body shall be covered with wood, or other nonsparking material, to prevent contact with containers of explosives.

- Every motor vehicle or conveyance used for transporting explosives shall be marked or placarded on both sides, the front, and the rear with the word "Explosives" in red letters, not less than 4 inches in height, on white background. In addition to such marking or placarding, the motor vehicle or conveyance may display, in such a manner that it will be readily visible from all directions, a red flag 18 inches by 30 inches, with the word "Explosives" painted, stamped, or sewed thereon, in white letters, at least 6 inches in height.

- Each vehicle used for transportation of explosives shall be equipped with a fully charged fire extinguisher, in good condition. An Underwriters Laboratory-approved extinguisher of not less than 10-ABC rating will meet the minimum requirement. The driver shall be trained in the use of the extinguisher on his vehicle.

- Motor vehicles or conveyances carrying explosives, blasting agents, or blasting supplies shall not be taken inside a garage or shop for repairs or servicing.

- No motor vehicle transporting explosives shall be left unattended.

31.5   Underground Transportation of Explosives

According to the Part 1926, Subpart U, Standard Number 1926.903, some major regulations for underground transportation of explosives are as follows:

- All explosives or blasting agents in transit underground shall be taken to the place of use or storage without delay.

- The quantity of explosives or blasting agents taken to an underground loading area shall not exceed the amount estimated to be necessary for the blast.

- Explosives in transit shall not be left unattended.

- The hoist operator shall be notified before explosives or blasting agents are transported in a shaft conveyance.

- Trucks used for the transportation of explosives underground shall have the electrical system checked weekly to detect any failures which may constitute an electrical hazard. A certification record which includes the date of the inspection; the signature of the person who performed the inspection; and a serial number, or other identifier, of the truck inspected shall be prepared and the most recent certification record shall be maintained on file.

- The installation of auxiliary lights on truck beds, which are powered by the truck's electrical system, shall be prohibited.

- Explosives and blasting agents shall be hoisted, lowered, or conveyed in a powder car. No other materials, supplies, or equipment shall be transported in the same conveyance at the same time.

- No one, except the operator, his helper, and the powder man, shall be permitted to ride on a conveyance transporting explosives and blasting agents.

- No person shall ride in any shaft conveyance transporting explosives and blasting agents.

- No explosives or blasting agents shall be transported on any locomotive. At least two car lengths shall separate the locomotive from the powder car.

- No explosives or blasting agents shall be transported on a man haul trip.

- The car or conveyance containing explosives or blasting agents shall be pulled, not pushed, whenever possible.

- The powder car or conveyance especially built for the purpose of transporting explosives or blasting agents shall bear a reflectorized sign on each side with the word "Explosives" in letters, not less than 4 inches in height; upon a background of sharply contrasting color.

- Compartments for transporting detonators and explosives in the same car or conveyance shall be physically separated by a distance of 24 inches or by a solid partition at least 6 inches thick.

- Detonators and other explosives shall not be transported at the same time in any shaft conveyance.

- Explosives, blasting agents, or blasting supplies shall not be transported with other materials.

- Explosives or blasting agents, not in original containers, shall be placed in a suitable container when transported manually.

- Detonators, primers, and other explosives shall be carried in separate containers when transported manually.

31.6   Storage of Explosives and Blasting Agents

According to Part 1926, Subpart U, Standard Number 1926.904, some major regulations for storage of explosives and blasting agents are as follows:

- Blasting caps, electric blasting caps, detonating primers, and primed cartridges shall not be stored in the same magazine with other explosives or blasting agents.

- Smoking and open flames shall not be permitted within 50 feet of explosives and detonator storage magazine.

- No explosives or blasting agents shall be permanently stored in any underground operation until the operation has been developed to the point where at least two modes of exit have been provided.

- Permanent underground storage magazines shall be at least 300 feet from any shaft, audit, or active underground working area.

- Permanent underground magazines containing detonators shall not be located closer than 50 feet to any magazine containing other explosives or blasting agents.

31.7   Loading of Explosives or Blasting Agents

According to Part 1926, Subpart U, Standard Number 1926.905, some major regulations for loading of explosives or blasting agents are as follows:

- Procedures that permit safe and efficient loading shall be established before loading is started.

- All drill holes shall be sufficiently large to admit freely the insertion of the cartridges of explosives.

- Tamping shall be done only with wood rods or plastic tamping poles without exposed metal parts, but nonsparking metal connectors may be used for jointed poles. Violent tamping shall be avoided. The primer shall never be tamped.

- No holes shall be loaded except those to be fired in the next round of blasting. After loading, all remaining explosives and detonators shall be immediately returned to an authorized magazine.

- Drilling shall not be started until all remaining butts of old holes are examined for unexploded charges, and if any are found, they shall be refired before work proceeds.

- No person shall be allowed to deepen drill holes which have contained explosives or blasting agents.

- No explosives or blasting agents shall be left unattended at the blast site.

- Machines and all tools not used for loading explosives into bore holes shall be removed from the immediate location of holes before explosives are delivered. Equipment shall not be operated within 50 feet of loaded holes.

- No activity of any nature other than that which is required for loading holes with explosives shall be permitted in a blast area.

- Powerlines and portable electric cables for equipment being used shall be kept at a safe distance from explosives or blasting agents being loaded into drill holes. Cables in the proximity of the blast area shall be deenergized and locked out by the blaster.

- Holes shall be checked prior to loading to determine depth and conditions. Where a hole has been loaded with explosives but the explosives have failed to detonate, there shall be no drilling within 50 feet of the hole.

- When loading a long line of holes with more than one loading crew, the crews shall be separated by practical distance consistent with efficient operation and supervision of crews.

- No explosive shall be loaded or used underground in the presence of combustible gases or combustible dusts.

- No explosives other than those in Fume Class 1, as set forth by the Institute of Makers of Explosives, shall be used; however, explosives complying with the requirements of Fume Class 2 and Fume Class 3 may be used if adequate ventilation has been provided.

- All blastholes in open work shall be stemmed to the collar or to a point which will confine the charge.

- Warning signs, indicating a blast area, shall be maintained at all approaches to the blast area. The warning sign lettering shall not be less than 4 inches in height on a contrasting background.

- A bore hole shall never be sprung when it is adjacent to or near a hole that is loaded. Flashlight batteries shall not be used for springing holes.

- Drill holes which have been sprung or chambered, and which are not water-filled, shall be allowed to cool before explosives are loaded.

- No loaded holes shall be left unattended or unprotected.

- The blaster shall keep an accurate, up-to-date record of explosives, blasting agents, and blasting supplies used in a blast and shall keep an accurate running inventory of all explosives and blasting agents stored on the operation.

- When loading blasting agents pneumatically over electric blasting caps, semiconductive delivery hose shall be used and the equipment shall be bonded and grounded.

31.8 Initiation of Explosive Charges-Electric Blasting

Based on Part 1926, Subpart U, Standard Number 1926.906, some major regulations for initiation of explosive charges-electric blasting are as follows:

- Electric blasting caps shall not be used where sources of extraneous electricity make the use of electric blasting caps dangerous. Blasting cap leg wires shall be kept short-circuited (shunted) until they are connected into the circuit for firing.

- Before adopting any system of electrical firing, the blaster shall conduct a thorough survey for extraneous currents, and all dangerous currents shall be eliminated before any holes are loaded.

- In any single blast using electric blasting caps, all caps shall be of the same style or function, and of the same manufacture.

- Electric blasting shall be carried out by using blasting circuits or power circuits in accordance with the electric blasting cap manufacturer's recommendations, or an approved contractor or his designated representative.

- When firing a circuit of electric blasting caps, care must be exercised to ensure that an adequate quantity of delivered current is available, in accordance with the manufacturer's recommendations.

- Connecting wires and lead wires shall be insulated single solid wires of sufficient current-carrying capacity.

- Bus wires shall be solid single wires of sufficient current-carrying capacity.

- When firing electrically, the insulation on all firing lines shall be adequate and in good condition.

- A power circuit used for firing electric blasting caps shall not be grounded.

- In underground operations when firing from a power circuit, a safety switch shall be placed in the permanent firing line at intervals. This switch shall be made so it can be locked only in the "Off" position and shall be provided with a short-circuiting arrangement of the firing lines to the cap circuit.

- In underground operations there shall be a "lightning" gap of at least 5 feet in the firing system ahead of the main firing switch; that is, between this switch and the source of power. This gap shall be bridged by a flexible jumper cord just before firing the blast.

- When firing from a power circuit, the firing switch shall be locked in the open or "Off" position at all times, except when firing. It shall be so designed that the firing lines to the cap circuit are automatically short-circuited when the switch is in the "Off" position. Keys to this switch shall be entrusted only to the blaster.

- Blasting machines shall be in good condition and the efficiency of the machine shall be tested periodically to make certain that it can deliver power at its rated capacity.

- When firing with blasting machines, the connections shall be made as recommended by the manufacturer of the electric blasting caps used.

- The number of electric blasting caps connected to a blasting machine shall not be in excess of its rated capacity. Furthermore, in primary blasting, a series circuit shall contain no more caps than the limits recommended by the manufacturer of the electric blasting caps in use.

- The blaster shall be in charge of the blasting machines, and no other person shall connect the leading wires to the machine.

- Blasters, when testing circuits to charged holes, shall use only blasting galvanometers or other instruments that are specifically designed for this purpose.

- Whenever the possibility exists that a leading line or blasting wire might be thrown over a live powerline by the force of an explosion, care shall be taken to see that the total length of wires are kept too short to hit the lines, or that the wires are securely anchored to the ground. If neither of these requirements can be satisfied, a nonelectric system shall be used.

- In electrical firing, only the man making leading wire connections shall fire the shot. All connections shall be made from the bore hole back to the source of firing current, and the leading wires shall remain shorted and not be connected to the blasting machine or other source of current until the charge is to be fired.

- After firing an electric blast from a blasting machine, the leading wires shall be immediately disconnected from the machine and short-circuited.

31.9   Use of Safety Fuse

Based on Part 1926, Subpart U, Standard Number 1926.907, some major regulations for use of safety fuse are as follows:

- Safety fuse shall only be used where sources of extraneous electricity make the use of electric blasting caps dangerous. The use of a fuse that has been hammered or injured in any way shall be forbidden.

- The hanging of a fuse on nails or other projections which will cause a sharp bend to be formed in the fuse is prohibited.

- Before capping safety fuse, a short length shall be cut from the end of the supply reel so as to assure a fresh cut end in each blasting cap.

- Only a cap crimper of approved design shall be used for attaching blasting caps to safety fuse. Crimpers shall be kept in good repair and accessible for use.

- No unused cap or short capped fuse shall be placed in any hole to be blasted; such unused detonators shall be removed from the working place and destroyed.

- No fuse shall be capped, or primers made up, in any magazine or near any possible source of ignition.

- No one shall be permitted to carry detonators or primers of any kind on his person.

- The minimum length of safety fuse to be used in blasting shall be as required by State law, but shall not be less than 30 inches.

- At least two men shall be present when multiple cap and fuse blasting is done by hand-lighting methods.

- Not more than 12 fuses shall be lighted by each blaster when hand-lighting devices are used. However, when two or more safety fuses in a group are lighted as one by means of igniter cord, or other similar fuse-lighting devices, they may be considered as one fuse.

- The so-called "drop fuse" method of dropping or pushing a primer or any explosive with a lighted fuse attached is forbidden.

- Cap and fuse shall not be used for firing mudcap charges unless charges are separated sufficiently to prevent one charge from dislodging other shots in the blast.

- When blasting with safety fuses, consideration shall be given to the length and burning rate of the fuse. Sufficient time, with a margin of safety, shall always be provided for the blaster to reach a place of safety.

31.10   Use of Detonating Cord

Part 1926, Subpart U, Standard Number 1926.908 states that:

- Care shall be taken to select a detonating cord consistent with the type and physical condition of the bore hole and stemming and the type of explosives used.

- Detonating cord shall be handled and used with the same respect and care as given to other explosives.

- The line of detonating cord extending out of a bore hole or from a charge shall be cut from the supply spool before loading the remainder of the bore hole or placing additional charges.

- Detonating cord shall be handled and used with care to avoid damaging or severing the cord during and after loading and hooking-up.

- Detonating cord connections shall be competent and positive in accordance with approved and recommended methods. Knot-type or other cord-to-cord connections shall be made only with detonating cord in which the explosive core is dry.
- All detonating cord trunklines and branchlines shall be free of loops, sharp kinks, or angles that direct the cord back toward the oncoming line of detonation.
- All detonating cord connections shall be inspected before firing the blast.
- When detonating cord millisecond-delay connectors or short-interval-delay electric blasting caps are used with detonating cord, the practice shall conform strictly to the manufacturer's recommendations.
- When connecting a blasting cap or an electric blasting cap to detonating cord, the cap shall be taped or otherwise attached securely along the side or the end of the detonating cord, with the end of the cap containing the explosive charge pointed in the direction in which the detonation is to proceed.
- Detonators for firing the trunkline shall not be brought to the loading area nor attached to the detonating cord until everything else is in readiness for the blast.

31.11 Firing the Blast

Part 1926, Subpart U, Standard Number 1926.909 states that:

- Before a blast is fired, a loud warning signal shall be given by the blaster in charge, who has made certain that all surplus explosives are in a safe place and all employees, vehicles, and equipment are at a safe distance, or under sufficient cover.
- Flagmen shall be safely stationed on highways which pass through the danger zone so as to stop traffic during blasting operations.
- It shall be the duty of the blaster to fix the time of blasting.
- Before firing an underground blast, warning shall be given, and all possible entries into the blasting area, and any entrances to any working place where a drift, raise, or other opening is about to hole through, shall be carefully guarded. The blaster shall make sure that all employees are out of the blast area before firing a blast.

31.12 Inspection after Blasting

According to Part 1926, Subpart U, Standard Number 1926.910, some major regulations for inspection after blasting are as follows:

- Immediately after the blast has been fired, the firing line shall be disconnected from the blasting machine, or where power switches are used, they shall be locked open or in the off position.
- Sufficient time shall be allowed, not less than 15 minutes in tunnels, for the smoke and fumes to leave the blasted area before returning to the shot. An inspection of the area and the surrounding rubble shall be made by the blaster to determine if all charges have been exploded before employees are allowed to return to the operation, and in tunnels, after the muck pile has been wetted down.

31.13   Misfires

According to Part 1926, Subpart U, Standard Number 1926.911, some major regulations for blasting misfires are as follows:

- If a misfire is found, the blaster shall provide proper safeguards for excluding all employees from the danger zone.

- No other work shall be done except that necessary to remove the hazard of the misfire and only those employees necessary to do the work shall remain in the danger zone.

- No attempt shall be made to extract explosives from any charged or misfired hole; a new primer shall be put in and the hole reblasted. If refiring of the misfired hole presents a hazard, the explosives may be removed by washing out with water or, where the misfire is under water, blown out with air.

- If there are any misfires while using cap and fuse, all employees shall remain away from the charge for at least 1 hour. Misfires shall be handled under the direction of the person in charge of the blasting. All wires shall be carefully traced and a search made for unexploded charges.

- No drilling, digging, or picking shall be permitted until all missed holes have been detonated or the authorized representative has approved that work can proceed.

Example 31.1

If there are any misfires while using cap and fuse, all employees shall remain away from the charge for at least:

    A. 15 minutes

    B. 30 minutes

    C. 60 minutes

    D. 90 minutes

Answer: C

31.14   Underwater Blasting

According to Part 1926, Subpart U, Standard Number 1926.912, some major regulations for underwatering blasting are as follows:

- A blaster shall conduct all blasting operations, and no shot shall be fired without his approval.

- Loading tubes and casings of dissimilar metals shall not be used because of possible electric transient currents from galvanic action of the metals and water.

- Only water-resistant blasting caps and detonating cords shall be used for all marine blasting. Loading shall be done through a nonsparking metal-loading tube when tube is necessary.

- No blast shall be fired while any vessel under way is closer than 1500 feet to the blasting area. Those on board vessels or craft moored or anchored within 1500 feet shall be notified before a blast is fired.

- No blast shall be fired while any swimming or diving operations are in progress in the vicinity of the blasting area. If such operations are in progress, signals and arrangements shall be agreed upon to assure that no blast shall be fired while any person is in the water.

- Blasting flags shall be displayed.

- The storage and handling of explosives aboard vessels used in underwater blasting operations shall be according to provisions outlined herein on handling and storing explosives.

- When more than one charge is placed under water, a float device shall be attached to an element of each charge in such manner that it will be released by the firing. Misfires shall be handled in accordance with the requirements of 1926.911.

Example 31.2
No blast shall be fired while any vessel under way is closer than _____ feet to the blasting area.

   A. 500

   B. 1000

   C. 1500

   D. 2000

Answer: C

31.15 Blasting in Excavation Work under Compressed Air

According to Part 1926, Subpart U, Standard Number 1926.913, some major regulations for blasting in excavation work under compressed air are as follows:

- Detonators and explosives shall not be stored or kept in tunnels, shafts, or caissons. Detonators and explosives for each round shall be taken directly from the magazines to the blasting zone and immediately loaded. Detonators and explosives left over after loading a round shall be removed from the working chamber before the connecting wires are connected up.

- When detonators or explosives are brought into an air lock, no employee except the powderman, blaster, lock tender, and the employees necessary for carrying shall be permitted to enter the air lock. No other material, supplies, or equipment shall be locked through with the explosives.

- Detonators and explosives shall be taken separately into pressure working chambers.

- The blaster or powderman shall be responsible for the receipt, unloading, storage, and on-site transportation of explosives and detonators.

- All metal pipes, rails, air locks, and steel tunnel lining shall be electrically bonded together and grounded at or near the portal or shaft, and such pipes and rails shall be cross-bonded together at not less than 1000-foot intervals throughout the length of the tunnel. In addition, each low air supply pipe shall be grounded at its delivery end.

- The explosives suitable for use in wet holes shall be water-resistant and shall be Fume Class 1.

- When tunnel excavation in rock face is approaching mixed face, and when tunnel excavation is in mixed face, blasting shall be performed with light charges and with light burden on each hole. Advance drilling shall be performed as tunnel excavation in rock face approaches mixed face, to determine the general nature and extent of rock cover and the remaining distance ahead to soft ground as excavation advances.

Example 31.3

Detonators and explosives shall be taken _____ into pressure working chambers.

    A. Separately

    B. Together

    C. Combined

    D. Rapidly

Answer: *A*

31.16   Summary

During the drilling and blasting in construction, a number of holes are drilled into the rock, which are then filled with explosives. Then, detonating the explosive causes the rock to collapse. Rubble is removed and the new tunnel surface is reinforced. This process is repeated until the desired excavation is complete. For the blasting operation, safety is ensured for transportation of explosives, storing them, blasting them, personnel safety, fire safety, and so on. Transportation of explosives should comply with all applicable federal, state, and local laws. Some common storage safety includes storing in an approved structure that is bullet resistant, weather resistant, and fire resistant, located in remote (out-of-sight) areas with restricted access, etc. All personnel responsible for handling explosives and present in and around blasting sites should be fully informed and trained in applicable safety precautions/procedures. Following detonation, the blasting area is to be inspected for undetonated or misfired explosives, hazards such as falling rock and rock slides. Special attention is to be given to preventing potential hazards in the blasting area resulting from flying rock, destabilized walls, structures, presence of low flying aircraft, dispersion of smoke and gases, etc. The presence of explosive materials on the project site could potentially increase the risk of fire during construction. Blasting has the potential to cause adverse impacts to sensitive environmental resources, including biological and cultural resources, wells and springs, and cause noise disturbances to nearby residents.

31.17 Multiple-Choice Questions

31.1 Which of the following blasting agent is primarily used in blasting operation in construction? [Select all that apply]

- A. Dynamite
- B. Ammonium nitrate
- C. Ammonium nitrate in fuel oil (ANFO)
- D. Slurries

31.2 Which of the following items are not prohibited in or near explosive magazines or while explosives are being handled, transported, or used?

- Smoking
- Firearms
- Drill machine
- Matches

31.3 What type of driver can drive and be in charge of transportation of explosives?

- A. Any licensed driver who works in a construction company
- B. Any licensed driver who is physically fit
- C. Any licensed driver with explosive experience
- D. Any licensed driver who is physically fit and familiar with regulation governing the transportation of explosives

31.4 Smoking and open flames shall not be permitted within _____ of explosives and detonator storage magazine.

- A. 25 feet
- B. 50 feet
- C. 75 feet
- D. 100 feet

31.5 Permanent underground magazines containing detonators shall not be located closer than _____ to any magazine containing other explosives or blasting agents.

- A. 25 feet
- B. 50 feet
- C. 75 feet
- D. 100 feet

31.6 What is the regulation for machines and tools not used for loading explosives into bore holes?

- A. Shall be removed from the immediate location of holes.
- B. Do not need to be removed from the immediate location of holes.
- C. Shall be moved at least 25 feet from the immediate location of holes.
- D. There is no regulations for unused tools.

31.7 What is the minimum size of the lettering in warning sign for blasting area?

- A. 4 inches
- B. 8 inches
- C. 12 inches
- D. 24 inches

31.8 What is the minimum length of safety fuse to be used in blasting?

A. Not be less than 20 inches
B. Not be less than 30 inches
C. Not be less than 40 inches
D. Not be less than 50 inches

31.9 When hand-lighting devices are used, what is the maximum number of fuses that shall be lighted by each blaster?

A. Not more than 4 fuses shall be lighted by each blaster.
B. Not more than 8 fuses shall be lighted by each blaster.
C. Not more than 12 fuses shall be lighted by each blaster.
D. Not more than 18 fuses shall be lighted by each blaster.

31.10 After the blasting, what is the minimum time that should be given before returning to shot area in tunnels?

A. Not less than 15 minutes
B. Not less than 30 minutes
C. Not less than 45 minutes
D. Not less than 60 minutes

31.18 Practice Problems

31.1 Discuss the concept of blasting.
31.2 List some general safety regulations for blasting operation.
31.3 What are the guidelines for the qualifications of blasters?
31.4 What are the guidelines for surface transportation of explosives?
31.5 What are the guidelines for underground transportation of explosives?
31.6 What are the guidelines for storage of explosives and blasting agents?
31.7 What are guidelines for the misfires of blasting?

31.19 Critical Thinking and Discussion Topic

You are in charge of a blasting operation. The technician wants to use cell phone to communicate with his team members. He also plans to switch off the phone before starting the blasting operation. However, you are concerned about the extraneous electricity due to the cell phone usage. As a site manager, what will you do in this situation?

(Photo courtesy of 123RF.)

CHAPTER **32**

Electric Power Transmission and Distribution

(Photo courtesy of www.bioprepper.)

32.1   General

Electric power transmission is a bulk movement of electrical energy from a generating site, such as a power plant, to an electrical substation. Interconnected lines that facilitate this movement are known as a transmission network. This is distinct from local wiring between high-voltage substations and customers, which is typically referred to as electric power distribution (Fig. 32.1).

477

Figure 32.1   Electric power transmission.
(Photo courtesy of Pok Rie from Pexels.)

An electric power supply has three main stages, including generation, transmission, and distribution. Each stage has its own unique production processes, activities, and hazards. In most cases, electricity is generated from 13,200 to 24,000 volts. There are several hazards in the electrical power generation process, some of which include burns and explosions resulting from equipment failure. Sometimes accidents in power plants may occur if proper lockout procedures are not followed. Depending on the job activity, examples of safety hazards that can harm workers in this industry include:

- Powerlines and other energized components that can burn, shock, or electrocute
- Falls from heights while working with overhead transmission system
- Confined spaces
- Hazardous materials and chemicals on the powerlines
- Dangerous moving mechanical parts on heavy machinery and equipment
- Falling trees, branches, poles on the powerlines
- Moving traffic and work vehicles on the powerlines
- Contact with wildlife
- Harsh weather (e.g., heat, cold, wind) may destroy the integrity of the powerlines

32.2 Avoiding Hazards

Following are some measures that can be adopted to avoid hazards for electric power transmission and distribution:

- Look for overhead powerlines and buried powerline indicators. Post warning signs.
- Contact utilities for buried powerline locations.
- Stay at least 10 feet away from overhead powerlines.
- Unless you know otherwise, assume that overhead lines are energized.
- Deenergize the groundlines when working near them. Other protective measures include guarding or insulating the lines.
- Use nonconductive wood or fiberglass ladders when working near powerlines.
- Use ground-fault circuit interrupters (GFCIs) on all 120-volt, single-phase, 15- and 20-ampere receptacles, or have an assured equipment grounding conductor program (AEGCP).
- Follow manufacturers' recommended testing procedure to ensure GFCI is working correctly.
- Use double-insulated tools and equipment, distinctively marked.
- Use tools and equipment according to the instructions included in their listing, labeling, or certification.
- Visually inspect all electrical equipment before use. Remove from service any equipment with frayed cords, missing ground prongs, cracked tool casings, etc. Apply a warning tag to any defective tool and do not use it until the problem has been corrected.

Example 32.1
Stay at least _____ feet away from overhead powerlines.

A. 5

B. 10

C. 15

D. 20

Answer: *B*

32.3 General Safety Regulations

According to OSHA, it is the obligation of employers to create an appropriate hazard control and prevention methodology that is designed to prevent workplace illnesses and injuries. As an employer, you are required to implement safe working practices by providing workers with the necessary training. With this training, your workers will know the necessary requirements for a safer working environment. Some major general safety regulations from Part 1926, Subpart V, Standard Number 1926.950 are as follows:

- The contract employer shall ensure that each of its employees is instructed in the hazardous conditions relevant to the employee's work that the contract employer is aware of as a result of information communicated to the contract employer by the host employer.

- Before work begins, the contract employer shall advise the host employer of any unique hazardous conditions presented by the contract employer's work.

- The contract employer shall advise the host employer of any unanticipated hazardous conditions found during the contract employer's work that the host employer did not mention. The contract employer shall provide this information to the host employer within 2 working days after discovering the hazardous condition.

32.4   Medical Services and First Aid

Part 1926, Subpart V, Standard Number 1926.951 states that the employer shall provide medical services and first aid as required. In addition, when employees are performing work on, or associated with, exposed lines or equipment energized at 50 volts or more, persons with first-aid training shall be available as follows:

- Field work: For field work involving two or more employees at a work location, at least two trained persons shall be available.

- Fixed work locations: For fixed work locations such as substations, the number of trained persons available shall be sufficient to ensure that each employee exposed to electric shock can be reached within 4 minutes by a trained person. However, where the existing number of employees is insufficient to meet this requirement (at a remote substation, for example), each employee at the work location shall be a trained employee.

Example 32.2
For field work involving two or more employees at a work location, at least _____ trained persons shall be available.

A. One

B. Two

C. Three

D. Four

Answer: *B*

32.5   Job Briefing

Part 1926, Subpart V, Standard Number 1926.952 states that in assigning an employee or a group of employees to perform a job, the employer shall provide the employee in charge of the job with all available information that relates to the determination of existing characteristics and conditions. The briefing shall cover at least the following subjects:

- Hazards associated with the job
- Work procedures involved
- Special precautions
- Energy-source controls
- Personal protective equipment requirements

A more extensive discussion shall be conducted if the work is complicated or particularly hazardous, or if the employee cannot be expected to recognize and avoid the hazards involved in the job.

32.6   Enclosed Spaces

Based on Part 1926, Subpart V, Standard Number 1926.953, some major regulations for enclosed spaces are as follows:

- The employer shall ensure the use of safe work practices for entry into, and work in, enclosed spaces and for rescue of employees from such spaces.
- Each employee who enters an enclosed space or who serves as an attendant shall be trained in the hazards of enclosed-space entry, in enclosed-space entry procedures, and in enclosed-space rescue procedures.
- Employers shall provide equipment to ensure the prompt and safe rescue of employees from the enclosed space.
- Before any entrance cover to an enclosed space is removed, the employer shall determine whether it is safe to do so by checking for the presence of any atmospheric pressure or temperature differences and by evaluating whether there might be a hazardous atmosphere in the space. Any conditions making it unsafe to remove the cover shall be eliminated before the cover is removed.

32.7   Protective Equipment and Tools

Part 1926, Subpart V, Standard Number 1926.954 specifies the following for general and fall protection:

- Personal fall arrest systems
- Work-positioning equipment
- Care and use of personal fall protection equipment

Part 1926, Subpart V, Standard Number 1926.955 specifies the following for the portable ladders and platforms:

- Portable platforms shall be capable of supporting without failure at least 2.5 times the maximum intended load.
- Portable ladders and platforms may not be loaded in excess of the working loads for which they are designed.

- Portable ladders and platforms shall be secured to prevent them from becoming dislodged.
- Portable ladders and platforms may be used only in applications for which they are designed.
- Portable metal ladders and other portable conductive ladders may not be used near exposed energized lines or equipment. However, in specialized high-voltage work, conductive ladders shall be used when the employer demonstrates that nonconductive ladders would present a greater hazard to employees than conductive ladders.

Part 1926, Subpart V, Standard Number 1926.956 specifies the following for the hand and portable power equipment:

- Cord- and plug-connected equipment
- Portable and vehicle-mounted generators
- Hydraulic and pneumatic tools

32.8   Materials Handling and Storage

Part 1926, Subpart V, Standard Number 1926.958 states that in areas to which access is not restricted to qualified persons only, materials or equipment may not be stored closer to energized lines or exposed energized parts of equipment than the following distances, plus a distance that provides for the maximum sag and side swing of all conductors and for the height and movement of material-handling equipment:

- For lines and equipment energized at 50 kilovolts or less, the distance is 3.05 meters (10 feet).
- For lines and equipment energized at more than 50 kilovolts, the distance is 3.05 meters (10 feet) plus 0.10 meter (4 inches) for every 10 kilovolts over 50 kilovolts.

In areas restricted to qualified employees, materials may not be stored within the working space about energized lines or equipment.

32.9   Grounding for the Protection of Employees

Based on Part 1926, Subpart V, Standard Number 1926.962, some major regulations for grounding regulations for the protection of employees are as follows:

General.   For any employee to work on transmission and distribution lines or equipment, the employer shall ensure that the lines or equipment are deenergized and shall ensure proper grounding of the lines or equipment. However, if the employer can demonstrate that installation of a ground is impracticable or that the conditions resulting from the installation of a ground would present greater hazards to employees than working without grounds, the lines and equipment may be treated as deenergized, provided that the employer establishes that all of the following conditions apply:

- No possibility of contact. There is no possibility of contact with another energized source.
- No induced voltage. The hazard of induced voltage is not present.

Equipotential zone. Temporary protective grounds shall be placed at such locations and arranged in such a manner that the employer can demonstrate will prevent each employee from being exposed to hazardous differences in electric potential.

Protective grounding equipment. Protective grounding equipment shall be capable of conducting the maximum fault current that could flow at the point of grounding for the time necessary to clear the fault. Protective grounding equipment shall have an ampacity greater than or equal to that of No. 2 AWG copper. Protective grounds shall have an impedance low enough so that they do not delay the operation of protective devices in case of accidental energizing of the lines or equipment.

Testing. The employer shall ensure that, unless a previously installed ground is present, employees test lines and equipment and verify the absence of nominal voltage before employees install any ground on those lines or that equipment.

Connecting and removing grounds. The employer shall ensure that, when an employee attaches a ground to a line or to equipment, the employee attaches the ground-end connection first and then attaches the other end by means of a live-line tool. For lines or equipment operating at 600 volts or less, the employer may permit the employee to use insulating equipment other than a live-line tool if the employer ensures that the line or equipment is not energized at the time the ground is connected or if the employer can demonstrate that each employee is protected from hazards that may develop if the line or equipment is energized.

Additional precautions. The employer shall ensure that, when an employee performs work on a cable at a location remote from the cable terminal, the cable is not grounded at the cable terminal if there is a possibility of hazardous transfer of potential, should a fault occur.

Removal of grounds for test. The employer may permit employees to remove grounds temporarily during tests. During the test procedure, the employer shall ensure that each employee uses insulating equipment, shall isolate each employee from any hazards involved, and shall implement any additional measures necessary to protect each exposed employee in case the previously grounded lines and equipment become energized.

Example 32.3
The employer _____ permit employees to remove grounds temporarily during tests.

    A. May
    B. Shall not
    C. May not
    D. Shall

Answer: A

32.10   Testing and Test Facilities

Based on Part 1926, Subpart V, Standard Number 1926.963, some major regulations for testing and test facilities are as follows:

Safe work practices.   The employer shall establish and enforce work practices for the protection of each worker from the hazards of high-voltage or high-power testing at all test areas, temporary and permanent. Such work practices shall include, as a minimum, test area safeguarding, grounding, the safe use of measuring and control circuits, and a means providing for periodic safety checks of field test areas.

Training.   The employer shall ensure that each employee, upon initial assignment to the test area, receives training in safe work practices. The employer shall provide safeguarding within test areas to control access to test equipment or to apparatus under test that could become energized as part of the testing by either direct or inductive coupling and to prevent accidental employee contact with energized parts.

32.11   Summary

Some of the common hazards that power plant employees face include burns, electrical shocks, explosions, boiler fires, flashes, direct contact with hazardous chemicals, and thermal burns.

Workers in the electric power industry are potentially exposed to a variety of serious hazards, such as arc flashes (which include arc flash burn and blast hazards), electric shock, falls, and thermal burn hazards that can cause injury and death. Employers are required to implement the safe work practices and worker training requirements of OSHA's Electric Power Generation, Transmission and Distribution Standard.

32.12   Multiple-Choice Questions

32.1   Which of the following is not an electric power supply main stages?

   A.   Generation
   B.   Storage
   C.   Transmission
   D.   Distribution

32.2   In most cases, electrical power is generated from:

   A.   13,200 to 24,000 volts
   B.   3,200 to 4,000 volts
   C.   1,200 to 2,000 volts
   D.   19,200 to 32,000 volts

32.3   As an employer, you are not required to:

   A.   Implement safe working practices by providing workers with the necessary training.
   B.   Ensure that employees are aware of the safety practices before being hired.
   C.   Create an appropriate hazard control and prevention methodology.
   D.   Provide necessary protective equipment and tools.

32.4 In areas to which access is not restricted to qualified persons only, materials or equipment may not be stored closer to 50 kilovolts energized lines:

A. 2.05 meters
B. 3.05 meters
C. 4.05 meters
D. 5.05 meters

32.5 Portable platforms for working for powerlines shall be capable of supporting without failure at least:

A. 1.5 times the maximum intended load
B. 2.5 times the maximum intended load
C. 3.0 times the maximum intended load
D. 3.5 times the maximum intended load

32.6 Protective grounding equipment shall have an ampacity greater than or equal to that of:

A. No. 1 AWG copper
B. No. 2 AWG copper
C. No. 3 AWG copper
D. No. 4 AWG copper

32.7 For fixed work locations such as substations, each employee exposed to electric shock can be reached within _____ minutes by a trained person.

A. 2
B. 4
C. 6
D. 8

32.13   Practice Problems

32.1 List some examples of safety hazards for electric power transmission and distribution that can harm workers in this industry.

32.2 List some measures that can be adopted to avoid safety hazards for electric power transmission and distribution.

32.3 List some general safety regulations for electric power transmission and distribution.

32.4 List some protective equipment regulations for electric power transmission and distribution.

32.5 List some materials handling and storage for electric power transmission and distribution.

32.6 List some grounding regulations for the protection of employees for electric power transmission and distribution.

32.14   Critical Thinking and Discussion Topic

For a repair work on a powerline, it was stated that powerline carries 50 kilovolts of electricity. After going at the site you found that the voltage is about the double. Do you need to do any change in your safety plan or PPE adoption?

(Photo courtesy of 123RF.)

CHAPTER 33

Cranes and Derricks
in Construction

(Photo courtesy by Maroš Markovič from Pexels.)

33.1 General

A crane (Fig. 33.1a) is a machine that uses a hoist rope, a wire rope, or a chain with a sheave (a pulley system) to lift heavy objects. A derrick (Fig. 33.1b) is a machine that uses at least one mast or self-supporting tower and a boom. To lift objects, the guy is adjusting the wires. A derrick is a lifting device composed of at least one mast, just like a gin pole, which can be articulated over a load by adjusting its guys. Most derricks have at least two components, either a guyed mast or a self-supporting tower, and a boom hinged for articulation at their base. Both are machines with a history that spans back as far as ancient times. Their designs worked by using human and animal effort and transitioned to technological advances such as the steam engine or hydraulics. Its size determines how much weight it can carry. A study conducted by OSHA showed that nearly 30% of work-related electrocutions involved cranes. Safety practice is thus important while using cranes in constructions. However, overloading is responsible for a relatively small portion of mobile crane accidents.

The main differences between these two machines are movement and setup. A derrick is usually set in a fixed spot. It carries objects from side to side by manipulating guy wires. There are different types of setups that play around with the mast and boom position, as well as the number of masts and booms. A quick list of different types of derricks might include the guy, shearleg, stiffleg, and gin pole. A derrick is most commonly seen on oil rigs or cargo ships and at ports.

The job of the crane is essentially to lift; their movement is up and down. For cranes, there are two categories: mobile and fixed. These categories account for crane's other features. They can transport items using a mobile-type pick-and-carry method. Even fixed cranes, which have relinquished their mobility to handle heavier loads, have their way of transporting material by adjusting the boom length. The crane is more mobile when compared with the derrick.

a) Crane
(Photo courtesy of PhotoMIX Ltd.)

b) Derrick
(Photo courtesy of Liebherr.)

FIGURE 33.1 Crane and derrick.

The following are examples of various crane hazards:

- Improper load rating
- Excessive speeds
- No, unclear, or improper hand signals
- Inadequate inspection and maintenance
- Unguarded parts
- Unguarded swing radius
- Working too close to powerlines
- Improper exhaust system
- Shattered windows
- No steps/guardrails walkways
- No boom angle indicator
- Not using outriggers

OSHA has identified the major causes of crane accidents to be:

- Boom or crane contact with energized powerlines
- Overturned cranes
- Dropped loads
- Boom collapse
- Crushing by the counterweight
- Outrigger use
- Falls
- Rigging failures

Example 33.1

Overloading is responsible for a relatively _____ portion of mobile crane accidents.

    A. Large
    B. Huge
    C. Big
    D. Small

Answer: D

33.2 Assembly/Disassembly

The equipment must not be assembled or used unless ground conditions are firm, drained, and graded to a sufficient extent so that, in conjunction (if necessary) with the use of supporting materials, the equipment manufacturer's specifications for adequate support and degree of level of the equipment are met. Ground condition means the ability of the ground to support the equipment (including slope, compaction, and firmness). The requirement for the ground to be drained does not apply to marshes/wetlands. Part 1926, Subpart CC, Standard Number 1926.1402 states that the controlling entity must inform the user of the equipment and the operator of the location of hazards beneath the equipment setup area (such as voids, tanks, utilities) if those hazards are identified in documents (such as site drawings, as-built drawings, and soil analyses) that are in the possession of the controlling entity (whether at the site or off-site) or the hazards are otherwise known to that controlling entity. Part 1926, Subpart CC, Standard Number 1926.1406 states that when using employer procedures instead of manufacturer procedures for assembly/disassembly, the employer must ensure that the procedures:

- Prevent unintended dangerous movement, and prevent collapse, of any part of the equipment.
- Provide adequate support and stability of all parts of the equipment.
- Position employees involved in the assembly/disassembly operation so that their exposure to unintended movement or collapse of part or all of the equipment is minimized.

33.3   Inspections

Part 1926, Subpart CC, Standard Number 1926.1412 states that equipment that has had modifications or additions which affect the safe operation of the equipment or capacity must be inspected as directed by the manufacturer and OSHA by a qualified person after such modifications/additions have been completed, prior to initial use. The inspection must include functional testing of the equipment. Safe operation of the equipment means the modifications or additions involving a safety device or operational aid, critical part of a control system, power plant, braking system, load-sustaining structural components, load hook, or in-use operating mechanism. The crane's main frame, crawler track, and/or outrigger supports, boom sections, and attachments are all considered part of the structural integrity of lifting.

Example 33.2
Cranes must be inspected as directed by the _____ and OSHA standards by competent persons who are familiar with them.

   A. Salesman

   B. OSHA inspector

   C. Manufacturer

   D. New hire

Answer: B

33.4 Wire Rope-Inspection

Part 1926, Subpart CC, Standard Number 1926.1413 states that a competent person must begin a visual inspection prior to each shift the equipment is used, which must be completed before or during that shift. The inspection must consist of observation of wire ropes (running and standing) that are likely to be in use during the shift for apparent deficiencies. Untwisting (opening) of wire rope or booming down is not required as part of this inspection. Master links, shackles, wire rope, and all other rigging hardware must be capable of supporting at least five times the maximum intended load without failure. Cranes and associated rigging equipment must be inspected regularly to identify any existing or potentially unsafe conditions. Regular inspections should be conducted before and during use.

Example 33.3
Master links, shackles, wire rope, and all other rigging hardware must be capable of supporting at least _____ times the maximum intended load without failure.

    A. 1
    B. 3
    C. 5
    D. 7

Answer: C

33.4.1 Deficiency Level

Depending on the severity level, the deficiency can be categorized as follows:

Category I. Apparent deficiencies in this category include the following:

- Significant distortion of the wire rope structure such as kinking, crushing, unstranding, birdcaging, signs of core failure or steel core protrusion between the outer strands
- Significant corrosion
- Electric arc damage (from a source other than powerlines) or heat damage
- Improperly applied end connections
- Significantly corroded, cracked, bent, or worn end connections (such as from severe service)

Category II. Apparent deficiencies in this category are:

- Visible broken wires, as follows:
  - In running wire ropes: Six randomly distributed broken wires in one rope lay or three broken wires in one strand in one rope lay, where a rope lay is the length along the rope in which one strand makes a complete revolution around the rope

- In rotation-resistant ropes: Two randomly distributed broken wires in six rope diameters or four randomly distributed broken wires in 30 rope diameters
- In pendants or standing wire ropes: More than two broken wires in one rope lay located in rope beyond end connections and/or more than one broken wire in a rope lay located at an end connection
- A diameter reduction of more than 5% from nominal diameter

Category III.   Apparent deficiencies in this category include the following:

- In rotation-resistant wire rope, core protrusion, or other distortion indicating core failure
- Prior electrical contact with a powerline
- A broken strand

Example 33.4
Cranes and associated rigging equipment must be inspected _____ to identify any existing or potentially unsafe conditions. Regular inspections should be conducted before and during use.

A. Seldom

B. Regularly

C. Occasionally

D. Never

Answer: B

33.4.2   Critical Review Items
The competent person must give particular attention to all of the following:

- Rotation-resistant wire rope in use.
- Wire rope being used for boom hoists and luffing hoists, particularly at reverse bends.
- Wire rope at flange points, crossover points, and repetitive pickup points on drums.
- Wire rope at or near terminal ends.
- Wire rope in contact with saddles, equalizer sheaves, or other sheaves where rope travel is limited. The boom hoisting sheave must have pitch diameters of no less than 18 times the nominal diameter of the rope used.

Example 33.5

The boom hoisting sheave must have pitch diameters of no less than _____ times the nominal diameter of the rope used.

A. 0

B. 5

C. 10

D. 18

Answer: D

33.5 Safety Devices

Part 1926, Subpart CC, Standard Number 1926.1415 states that the following safety devices are required on all equipment covered by this subpart, unless otherwise specified:

- Crane-level indicator.
  - The equipment must have a crane-level indicator that is either built into the equipment or is available on the equipment.
  - If a built-in crane-level indicator is not working properly, it must be tagged-out or removed. If a removable crane-level indicator is not working properly, it must be removed.
  - This requirement does not apply to portal cranes, derricks, floating cranes/derricks and land cranes/derricks on barges, pontoons, vessels, or other means of flotation.
- Boom stops, except for derricks and hydraulic booms.
- Jib stops (if a jib is attached), except for derricks.
- Equipment with foot pedal brakes must have locks.
- Hydraulic outrigger jacks and hydraulic stabilizer jacks must have an integral holding device/check valve.
- Equipment on rails must have rail clamps and rail stops, except for portal cranes.
- Horn:
  - The equipment must have a horn that is either built into the equipment or is on the equipment and immediately available to the operator.
  - If a built-in horn is not working properly, it must be tagged-out or removed. If a removable horn is not working properly, it must be removed.

Operations must not begin unless all of the devices listed in this section are in proper working order. If a device stops working properly during operations, the operator must safely stop operations. If any of the devices listed in this section are not in proper working order, the equipment must be taken out of service and operations must not resume until the device is again working properly.

33.6 Operation

Part 1926, Subpart CC, Standard Number 1926.1417 states that the employer must comply with all manufacturer procedures applicable to the operational functions of equipment, including its use with attachments.

33.6.1 Unavailable Operation Procedures

- Where the manufacturer procedures are unavailable, the employer must develop and ensure compliance with all procedures necessary for the safe operation of the equipment and attachments.
- Procedures for the operational controls must be developed by a qualified person.
- Procedures related to the capacity of the equipment must be developed and signed by a registered professional engineer familiar with the equipment.

33.6.2 Accessibility of Procedures

- The procedures applicable to the operation of the equipment, including rated capacities (load charts), recommended operating speeds, special hazard warnings, instructions, and operator's manual, must be readily available in the cab at all times for use by the operator.
- Where rated capacities are available in the cab only in electronic form: In the event of a failure which makes the rated capacities inaccessible, the operator must immediately cease operations or follow safe shut-down procedures until the rated capacities (in electronic or other form) are available.

The operator must not engage in any practice or activity that diverts his/her attention while actually engaged in operating the equipment, such as the use of cellular phones (other than when used for signal communications).

33.6.3 Leaving the Equipment Unattended

The operator must not leave the controls while the load is suspended, except where all of the following are met:

- The operator remains adjacent to the equipment and is not engaged in any other duties.
- The load is to be held suspended for a period of time exceeding normal lifting operations.

- The competent person determines that it is safe to do so and implements measures necessary to restrain the boom hoist and telescoping, load, swing, and outrigger or stabilizer functions.

- Barricades or caution lines, and notices, are erected to prevent all employees from entering the fall zone. No employees are permitted in the fall zone.

33.6.4   Tag-Out

Where the employer has taken the equipment out of service, a tag must be placed in the cab stating that the equipment is out of service and is not to be used. Where the employer has taken a function(s) out of service, a tag must be placed in a conspicuous position stating that the function is out of service and is not to be used. If there is a warning (tag-out or maintenance/do not operate) sign on the equipment or starting control, the operator must not activate the switch or start the equipment until the sign has been removed by a person authorized to remove it, or until the operator has verified that:

- No one is servicing, working on, or otherwise in a dangerous position on the machine.

- The equipment has been repaired and is working properly.

If there is a warning (tag-out or maintenance/do not operate) sign on any other switch or control, the operator must not activate that switch or control until the sign has been removed by a person authorized to remove it, or until the operator has verified that the requirements have been met. Before starting the engine, the operator must verify that all controls are in the proper starting position and that all personnel are in the clear.

33.6.5   Storm Warning

When a local storm warning has been issued, the competent person must determine whether it is necessary to implement manufacturer recommendations for securing the equipment. If equipment adjustments or repairs are necessary:

- The operator must, in writing, promptly inform the person designated by the employer to receive such information and, where there are successive shifts, to the next operator.

- The employer must notify all affected employees, at the beginning of each shift, of the necessary adjustments or repairs and all alternative measures.

Safety devices and operational aids must not be used as a substitute for the exercise of professional judgment by the operator. If the competent person determines that there is a slack rope condition requiring re-spooling of the rope, it must be verified (before starting to lift) that the rope is seated on the drum and in the sheaves as the slack is removed. The competent person must adjust the equipment and/or operations to address the effect of wind, ice, and snow on equipment stability and rated capacity.

Example 33.6
Do not hoist employees if weather conditions are bad, or if any other indication of impending danger exists.

- True
- False

Answer:
True. Do not hoist employees if weather conditions are bad, or if any other indication of impending danger exists. If employees are hoisted and a dangerous situation arises, they should be grounded immediately and safely.

33.6.6 Compliance with Rated Capacity

The equipment must not be operated in excess of its rated capacity.

- The operator must not be required to operate the equipment in a manner that would violate the regulations. The operator must verify that the load is within the rated capacity of the equipment by at least one of the following methods:
  - The weight of the load must be determined from a source recognized by the industry (such as the load's manufacturer), or by a calculation method recognized by the industry (such as calculating a steel beam from measured dimensions and a known per foot weight), or by other equally reliable means. In addition, when requested by the operator, this information must be provided to the operator prior to the lift.
  - The operator must begin hoisting the load to determine, using a load weighing device, load moment indicator, rated capacity indicator, or rated capacity limiter, if it exceeds 75% of the maximum rated capacity at the longest radius that will be used during the lift operation.
- The boom or other parts of the equipment must not contact any obstruction.
- The equipment must not be used to drag or pull loads sideways.
- On wheel-mounted equipment, no loads must be lifted over the front area, except as permitted by the manufacturer.
- The operator must test the brakes each time a load that is 90% or more of the maximum line pull is handled by lifting the load a few inches and applying the brakes. In duty cycle and repetitive lifts where each lift is 90% or more of the maximum line pull, this requirement applies to the first lift but not to successive lifts.
- Neither the load nor the boom must be lowered below the point where less than two full wraps of rope remain on their respective drums.

33.6.7 Traveling with a Load

- Traveling with a load is prohibited if the practice is prohibited by the manufacturer.
- Where traveling with a load, the employer must ensure that:
  - A competent person supervises the operation, determines if it is necessary to reduce rated capacity, and makes determinations regarding load position,

boom location, ground support, travel route, overhead obstructions, and speed of movement necessary to ensure safety.

- For equipment with tires, tire pressure specified by the manufacturer is maintained.
- Rotational speed of the equipment must be such that the load does not swing out beyond the radius at which it can be controlled.
- A tag or restraint line must be used if necessary to prevent rotation of the load that would be hazardous.
- The brakes must be adjusted in accordance with the manufacturer procedures to prevent unintended movement.
- The operator must obey a stop (or emergency stop) signal, irrespective of who gives it.
- A locomotive crane must not be swung into a position where railway cars on an adjacent track could strike it, until it is determined that cars are not being moved on the adjacent track and that proper flag protection has been established.

33.6.8 Counterweight/Ballast

The following applies to equipment other than tower cranes:

- Equipment must not be operated without the counterweight or ballast in place as specified by the manufacturer.
- The maximum counterweight or ballast specified by the manufacturer for the equipment must not be exceeded.

33.7 Work Area Control
33.7.1 Swing Radius Hazards

Cranes that have telescopic booms may be capable of lifting a heavier load when the boom length and radius is small. If the boom is extended, the load capacity decreases. Swing radius hazards means the hazards in which the equipment's rotating superstructure (whether permanently or temporarily mounted) poses a reasonably foreseeable risk of:

- Striking and injuring an employee
- Pinching/crushing an employee against another part of the equipment or another object

According to Part 1926, Subpart CC, Standard Number 1926.1424, to prevent employees from entering these hazard areas, the employer must train each employee assigned to work on or near the equipment (authorized personnel) in how to recognize struck-by and pinch/crush hazard areas posed by the rotating superstructure. Erect and maintain control lines, warning lines, railings, or similar barriers to mark the boundaries of the hazard areas. When the employer can demonstrate that it is neither feasible to erect such barriers on the ground nor on the equipment, the hazard areas must be clearly marked by a combination of warning signs (such as "Danger—Swing/ Crush Zone") and high visibility markings on the equipment that identify the hazard areas. In addition, the employer must train each employee to understand what these markings signify.

Example 33.7
Cranes that have telescopic booms may be capable of lifting a heavier load when the boom length and radius is small. If the boom is extended, the weights of the object can _____ the crane.

    A. Underload

    B. No issues

    C. Overhaul

    D. Overload

Answer: D

33.7.2 Protecting Employees in the Hazard Area

Before an employee goes to a location in the hazard area that is out of view of the operator, the employee (or someone instructed by the employee) must ensure that the operator is informed that he/she is going to that location. Where the operator knows that an employee went to a swing radius location, the operator must not rotate the superstructure until the operator is informed in accordance with a prearranged system of communication that the employee is in a safe position. Where any part of a crane/derrick is within the working radius of another crane/derrick, the controlling entity must institute a system to coordinate operations. If there is no controlling entity, the employer (if there is only one employer operating the multiple pieces of equipment), or employers, must institute such a system.

Example 33.8
There are no practical measures to keep the swing radius area free from equipment and personnel.

    • True

    • False

Answer:
False. It is advised that all employees stay out of the swing radius of the crane. A practical method of making sure radius is clearly visible is to erect barriers. OSHA determined that the preferred way to protect employees in these situations is to completely barricade the entire swing radius of the equipment and prevent employee access to the area.

33.8 Tower Cranes

In addition to the general requirements of crane and derricks, some special regulations for tower crane based on Part 1926, Subpart CC, Standard Number 1926.1435 are listed below.

33.8.1 Erecting, Climbing, and Dismantling

Dangerous areas (self-erecting tower cranes). In addition to the requirements in the previous section, for self-erecting tower cranes, the following applies: Employees must not be in or under the tower, jib, or rotating portion of the crane during erecting, climbing, and dismantling operations until the crane is secured in a locked position and the competent person in charge indicates it is safe to enter this area, unless the manufacturer's instructions direct otherwise and only the necessary personnel are permitted in this area.

Foundations and structural supports. Tower crane foundations and structural supports (including both the portions of the structure used for support and the means of attachment) must be designed by the manufacturer or a registered professional engineer.

Addressing specific hazards. In addition to the general requirement the supervisor must address the following:

- Foundations and structural supports. Tower crane foundations and structural supports are installed in accordance with their design.
- Loss of backward stability. Backward stability before swinging self-erecting cranes or cranes on traveling or static undercarriages.
- Wind speed. Wind must not exceed the speed recommended by the manufacturer or, where manufacturer does not specify this information, the speed determined by a qualified person.

Plumb tolerance. Towers must be erected plumb to the manufacturer's tolerance and verified by a qualified person. Where the manufacturer does not specify plumb tolerance, the crane tower must be plumb to a tolerance of at least 1:500 (approximately 1 inch in 40 feet).

Multiple tower crane jobsites. On jobsites where more than one fixed jib (hammerhead) tower crane is installed, the cranes must be located such that no crane can come in contact with the structure of another crane. Cranes are permitted to pass over one another.

Climbing procedures. Prior to, and during, all climbing procedures (including inside climbing and top climbing), the employer must:

- Comply with all manufacturer prohibitions.
- Have a registered professional engineer verify that the host structure is strong enough to sustain the forces imposed through the braces, brace anchorages, and supporting floors.

Counterweight/ballast. Equipment must not be erected, dismantled, or operated without the amount and position of counterweight and/or ballast in place as specified by the manufacturer or a registered professional engineer familiar with the equipment. The maximum counterweight and/or ballast specified by the manufacturer or registered professional engineer familiar with the equipment must not be exceeded.

Signs. The size and location of signs installed on tower cranes must be in accordance with manufacturer specifications. Where these are unavailable, a registered professional engineer familiar with the type of equipment involved must approve in writing the size and location of any signs.

33.8.2 Safety Devices

The following safety devices are required on all tower cranes unless otherwise specified:

- Boom stops on luffing boom-type tower cranes:
  - Jib stops on luffing boom-type tower cranes if equipped with a jib attachment
  - Travel rail end stops at both ends of travel rail
  - Travel rail clamps on all travel bogies
  - Integrally mounted check valves on all load-supporting hydraulic cylinders
  - Hydraulic system pressure-limiting device
  - The following brakes, which must automatically set in the event of pressure loss or power failure, are required:
    - A hoist brake on all hoists
    - Swing brake
    - Trolley brake
    - Rail travel brake
  - Deadman control or forced neutral return control (hand) levers
  - Emergency stop switch at the operator's station
  - Trolley end stops must be provided at both ends of travel of the trolley
- Proper operation required. Operations must not begin unless the devices listed in this section are in proper working order. If a device stops working properly during operations, the operator must safely stop operations. The equipment must be taken out of service, and operations must not resume until the device is again working properly.

33.8.3 Operational Aids

Operations must not begin unless the operational aids are in proper working order, except where the employer meets the specified temporary alternative measures. More protective alternative measures specified by the tower crane manufacturer, if any, must be followed.

If an operational aid stops working properly during operations, the operator must safely stop operations until the temporary alternative measures are implemented or the device is again working properly. If a replacement part is no longer available, the use of a substitute device that performs the same type of function is permitted and is not considered a modification

33.8.4 Inspections

All equipment must have the recommended operating speeds, rated load capacities, and special hazard warnings conspicuously posted in the manner that they are visible to operators in the control stations.

Pre-erection inspection. Before each crane component is erected, it must be inspected by a qualified person for damage or excessive wear.

- The qualified person must pay particular attention to components that will be difficult to inspect thoroughly during shift inspections.

- If the qualified person determines that a component is damaged or worn to the extent that it would create a safety hazard if used on the crane, that component must not be erected on the crane unless it is repaired and, upon reinspection by the qualified person, found to no longer create a safety hazard.

- If the qualified person determines that, though not presently a safety hazard, the component needs to be monitored, the employer must ensure that the component is checked in the monthly inspections. Any such determination must be documented, and the documentation must be available to any individual who conducts a monthly inspection.

Post-erection inspection. In addition to the other requirements, the following requirements must be met:

- A load test using certified weights, or scaled weights using a certified scale with a current certificate of calibration, must be conducted after each erection.

- The load test must be conducted in accordance with the manufacturer's instructions when available. Where these instructions are unavailable, the test must be conducted in accordance with written load test procedures developed by a registered professional engineer familiar with the type of equipment involved.

- Monthly. The following additional items must be included:

  - Tower (mast) bolts and other structural bolts (for loose or dislodged condition) from the base of the tower crane up or, if the crane is tied to or braced by the structure, those above the uppermost brace support.

  - The uppermost tie-in, braces, floor supports, and floor wedges where the tower crane is supported by the structure, for loose or dislodged components.

- Annual. In addition to the items that must be inspected, all turntable and tower bolts must be inspected for proper condition and torque.

Example 33.9

_____ equipment must have the recommended operating speeds, rated load capacities, and special hazard warnings conspicuously posted in the manner that they are visible to operators in the control stations.

A. All

B. Some

C. No

D. All of the above

Answer: A

33.9 Derricks

A derrick is powered equipment consisting of a mast or equivalent member that is held at or near the end by guys or braces, with or without a boom, and its hoisting mechanism. The mast/equivalent member and/or the load is moved by the hoisting mechanism (typically base-mounted) and operating ropes. Derricks include: A-frame, basket, breast, Chicago boom, gin pole (except gin poles used for erection of communication towers), guy, shearleg, stiffleg, and variations of such equipment. In addition to the general requirements of crane and derricks, some special regulations for derricks from Part 1926, Subpart CC, Standard Number 1926.1436 are listed below.

33.9.1 Load Chart Contents

Load charts must contain at least the following information:

- Rated capacity at corresponding ranges of boom angle or operating radii.
- Specific lengths of components to which the rated capacities apply.
- Required parts for hoist reeving.
- Size and construction of rope must be included on the load chart or in the operating manual.

33.9.2 Load Chart Location

- Permanent installations. For permanently installed derricks with fixed lengths of boom, guy, and mast, a load chart must be posted where it is visible to personnel responsible for the operation of the equipment.
- Nonpermanent installations. For derricks that are not permanently installed, the load chart must be readily available at the jobsite to personnel responsible for the operation of the equipment.

33.9.3 Construction

General requirements. Derricks must be constructed to meet all stresses imposed on members and components when installed and operated in accordance with the manufacturer's/builder's procedures and within its rated capacity. Welding of load-sustaining members must conform to recommended practices in ANSI/AWS.

Guy derricks:

- The minimum number of guys must be 6, with equal spacing, except where a qualified person or derrick manufacturer approves variations from these requirements and revises the rated capacity to compensate for such variations.
- Guy derricks must not be used unless the employer has the following guy information from the manufacturer or a qualified person, when not available from the manufacturer:
  - The number of guys.
  - The spacing around the mast.
  - The size, grade, and construction of rope to be used for each guy.

- For guy derricks manufactured after December 18, 1970, the employer must have the following guy information from the manufacturer or a qualified person, when not available from the manufacturer:
    - The amount of initial sag or tension.
    - The amount of tension in guy line rope at anchor.
- The mast base must permit the mast to rotate freely with allowance for slight tilting of the mast caused by guy slack.
- The mast cap must:
    - Permit the mast to rotate freely.
    - Withstand tilting and cramping caused by the guy loads.
    - Be secured to the mast to prevent disengagement during erection.
    - Be provided with means for attaching guy ropes.

Stiffleg derricks:

- The mast must be supported in the vertical position by at least two stifflegs; one end of each must be connected to the top of the mast and the other end securely anchored.
- The stifflegs must be capable of withstanding the loads imposed at any point of operation within the load chart range.
- The mast base must:
    - Permit the mast to rotate freely (when necessary).
    - Permit deflection of the mast without binding.
- The mast must be prevented from lifting out of its socket when the mast is in tension.
- The stiffleg connecting member at the top of the mast must:
    - Permit the mast to rotate freely (when necessary).
    - Withstand the loads imposed by the action of the stifflegs.
    - Be secured so as to oppose separating forces.

Gin pole derricks:

- Guy lines must be sized and spaced so as to make the gin pole stable in both boomed and vertical positions. Exception: Where the size and/or spacing of guy lines do not result in the gin pole being stable in both boomed and vertical positions, the employer must ensure that the derrick is not used in an unstable position.
- The base of the gin pole must permit movement of the pole (when necessary).
- The gin pole must be anchored at the base against horizontal forces (when such forces are present).

Chicago boom derricks. The fittings for stepping the boom and for attaching the topping lift must be arranged to:

- Permit the derrick to swing at all permitted operating radii and mounting heights between fittings.

- Accommodate attachment to the upright member of the host structure.
- Withstand the forces applied when configured and operated in accordance with the manufacturer's/builder's procedures and within its rated capacity.
- Prevent the boom or topping lift from lifting out under tensile forces.

33.9.4   Operational Aids

Boom angle aid.   A boom angle indicator is not required but if the derrick is not equipped with a functioning one, the employer must ensure that either:

- The boom hoist cable must be marked with caution and stop marks. The stop marks must correspond to maximum and minimum allowable boom angles. The caution and stop marks must be in view of the operator, or a spotter who is in direct communication with the operator; or
- An electronic or other device that signals the operator in time to prevent the boom from moving past its maximum and minimum angles, or automatically prevents such movement, is used.
- The outrigger must be placed on timbers or cribbed so as to spread the weight of the crane and the load over a large enough area.

Example 33.10
The _____ must be placed on timbers or cribbed so as to spread the weight of the crane and the load over a large enough area.

  A. In rigger
  B. Outrigger
  C. Side rigger
  D. Downrigger

Answer: *B*

33.9.5   Load Weight/Capacity Devices

Derricks manufactured more than one year after November 8, 2010 with a maximum rated capacity over 6000 pounds must have at least one of the following: load weighing device, load moment indicator, rated capacity indicator, or rated capacity limiter. The weight of the load must be determined from a source recognized by the industry (such as the load's manufacturer), or by a calculation method recognized by the industry (such as calculating a steel beam from measured dimensions and a known per foot weight), or by other equally reliable means. This information must be provided to the operator prior to the lift.

A load weight/capacity device that is not working properly must be repaired no later than 30 days after the deficiency occurs. If the employer documents that it has ordered the necessary parts within 7 days of the occurrence of the deficiency, and the part is not received in time to complete the repair in 30 days, the repair must be completed within 7 days of receipt of the parts.

33.9.6   Functional Test

Prior to initial use, new or reinstalled derricks must be tested by a competent person with no hook load to verify proper operation. This test must include:

- Lifting and lowering the hook(s) through the full range of hook travel
- Raising and lowering the boom through the full range of boom travel
- Swinging in each direction through the full range of swing
- Actuating the anti-two-block and boom hoist limit devices (if provided)
- Actuating locking, limiting, and indicating devices (if provided)

33.9.7   Load Test

Prior to initial use, new or reinstalled derricks must be load tested by a competent person. Test loads must be at least 100% and no more than 110% of the rated capacity, unless otherwise recommended by the manufacturer or qualified person, but in no event must the test load be less than the maximum anticipated load.

The test must consist of:

- Hoisting the test load a few inches and holding to verify that the load is supported by the derrick and held by the hoist brake(s)
- Swinging the derrick, if applicable, the full range of its swing, at the maximum allowable working radius for the test load
- Booming the derrick up and down within the allowable working radius for the test load

33.10   Floating Cranes/Derricks and Land Cranes/Derricks on Barges

This section contains supplemental requirements based on the Part 1926, Subpart CC, Standard Number 1926.1437 for floating cranes/derricks and land cranes/derricks on barges, pontoons, vessels, or other means of flotation (i.e., vessel/flotation device). The sections of this subpart apply to floating cranes/derricks and land cranes/derricks on barges, pontoons, vessels, or other means of flotation, unless specified otherwise. The requirements of this section do not apply when using jacked barges when the jacks are deployed to the river, lake, or sea bed and the barge is fully supported by the jacks. The requirements of this section apply to both floating cranes/derricks and land cranes/derricks on barges, pontoons, vessels, or other means of flotation.

33.10.1   Work Area Control

The employer must either:

- Erect and maintain control lines, warning lines, railings or similar barriers to mark the boundaries of the hazard areas; or
- Clearly mark the hazard areas by a combination of warning signs (such as "Danger-Swing/Crush Zone") and high visibility markings on the equipment that identify the hazard areas. In addition, the employer must train each employee to understand what these markings signify.

33.10.2   Operational Aids

An anti-two-block device is required only when hoisting personnel or hoisting over an occupied cofferdam or shaft. Rated capacities (load charts) are posted at the operator's station. If the operator's station is moveable (such as with pendant-controlled equipment), the load charts are posted on the equipment. Procedures applicable to the operation of the equipment (other than load charts), recommended operating speeds, special hazard warnings, instructions, and operators manual, must be readily available on board the vessel/flotation device.

Inspections.   In addition to meeting the general requirements for inspecting the crane/derrick, the employer must inspect the barge, pontoons, vessel, or other means of flotation used to support a floating crane/derrick or land crane/derrick, and ensure that:

Shift.   For each shift inspection, the means used to secure/attach the equipment to the vessel/flotation device is in proper condition, including wear, corrosion, loose or missing fasteners, defective welds, and (when applicable) insufficient tension.

Monthly inspections.   For each monthly inspection:

- The means used to secure/attach the equipment to the vessel/flotation device is in proper condition, including inspection for wear, corrosion, and, when applicable, insufficient tension.
- The vessel/flotation device is not taking on water.
- The deckload is properly secured.
- The vessel/flotation device is watertight based on the condition of the chain lockers, storage, fuel compartments, and hatches.
- The firefighting and lifesaving equipment is in place and functional.
- The shift and monthly inspections are conducted by a competent person.
- If any deficiency is identified, an immediate determination is made by a qualified person whether the deficiency constitutes a hazard.
- If the deficiency is determined to constitute a hazard, the vessel/flotation device is removed from service until the deficiency has been corrected.

Annual inspections.   For each annual inspection:

- The external portion of the barge, pontoons, vessel, or other means of flotation used is inspected annually by a qualified person who has expertise with respect to vessels/flotation devices and that the inspection includes the following items:
  - Cleats, bitts, chocks, fenders, capstans, ladders, and stanchions, for significant corrosion, wear, deterioration, or deformation that could impair the function of these items.
  - External evidence of leaks and structural damage; evidence of leaks and damage below the waterline may be determined through internal inspection of the vessel/flotation device.
  - Four-corner draft readings.
  - Firefighting equipment for serviceability.

- Rescue skiffs, lifelines, work vests, life preservers, and ring buoys are inspected for proper condition.
- If any deficiency is identified, an immediate determination is made by the qualified person whether the deficiency constitutes a hazard or, though not yet a hazard, needs to be monitored in the monthly inspections.
  - If the qualified person determines that the deficiency constitutes a hazard, the vessel/flotation device is removed from service until it has been corrected.
  - If the qualified person determines that, though not presently a hazard, the deficiency needs to be monitored, the deficiency is checked in the monthly inspections.

Four-year inspections.   Internal vessel/flotation device inspection. For each four-year inspection:

- A marine engineer, marine architect, licensed surveyor, or other qualified person who has expertise with respect to vessels/flotation devices surveys the internal portion of the barge, pontoons, vessel, or other means of flotation.
- If the surveyor identifies a deficiency, an immediate determination is made by the surveyor as to whether the deficiency constitutes a hazard or, though not yet a hazard, needs to be monitored in the monthly or annual inspections, as appropriate.
  - If the surveyor determines that the deficiency constitutes a hazard, the vessel/flotation device is removed from service until it has been corrected.
  - If the surveyor determines that, though not presently a hazard, the deficiency needs to be monitored, the deficiency is checked in the monthly or annual inspections, as appropriate.

Working with a diver.   The employer must meet the following additional requirements when working with a diver in the water:

- If a crane/derrick is used to get a diver into and out of the water, it must not be used for any other purpose until the diver is back on board. When used for more than one diver, it must not be used for any other purpose until all divers are back on board.
- The operator must remain at the controls of the crane/derrick at all times.
- The means used to secure the crane/derrick to the vessel/flotation device must not allow any amount of shifting in any direction.

33.10.3   Manufacturer's Specifications and Limitations

- The employer must ensure that the barge, pontoons, vessel, or other means of flotation must be capable of withstanding imposed environmental, operational, and in-transit loads when used in accordance with the manufacturer's specifications and limitations.
- The employer must ensure that the manufacturer's specifications and limitations with respect to environmental, operational, and in-transit loads for a barge, pontoon, vessel, or other means of flotation are not exceeded or violated.

- When the manufacturer's specifications and limitations are unavailable, the employer must ensure that the specifications and limitations established by a qualified person with respect to environmental, operational, and in-transit loads for the barge, pontoons, vessel, or other means of flotation are not exceeded or violated.

33.10.4 Floating Cranes/Derricks

For equipment designed by the manufacturer (or employer) for marine use by permanent attachment to barges, pontoons, vessels, or other means of flotation:

- The employer must not exceed the manufacturer load charts applicable to operations on water. When using these charts, the employer must comply with all parameters and limitations (such as dynamic and environmental parameters) applicable to the use of the charts.
- The employer must ensure that load charts take into consideration a minimum wind speed of 40 miles per hour.
- The employer must ensure that the requirements for maximum allowable list and maximum allowable trim as specified in Table 33.1 are met.

The employer must ensure that the equipment is stable under the conditions specified in Table 33.2 of this section. Freeboard is the vertical distance between the waterline and the main deck of the vessel. For backward stability of the boom, the high boom with no load at full back list must be stable at 90 mph.

Rated capacity	Maximum allowable list (degrees)	Maximum allowable trim (degrees)
Equipment designed for marine use by permanent attachment (other than derricks):		
25 tons or less	5	5
Over 25 tons	7	7
Derricks designed for marine use by permanent attachment:		
Any rated capacity	10	10

TABLE 33.1 Maximum Allowable List and Maximum Allowable Trim (OSHA Part 1926, Subpart CC, Standard Number, Table M1)

Operated at	Wind speed (mph)	Minimum freeboard (feet)
Rated capacity	60	2
Rated capacity plus 25%	60	1
High boom, no load	60	2

TABLE 33.2 Stability Condition under Wind Speed (OSHA Part 1926, Subpart CC, Standard Number, Table M2)

33.11   Summary

Cranes and associated rigging equipment must be inspected regularly to identify any existing or potentially unsafe conditions. Regular inspections should be conducted before and during use. If there are problems, necessary repairs must be completed before continuing work. Preventive maintenance must also be performed according to the crane manufacturer and/or the supplier specifications. Studies and analyses show that mechanical failures are frequently due to the result of a lack of preventive maintenance or adequate training, and/or experience on the part of the personnel involved.

Crane operators must know the load limits of the crane and the approximate weight of the load to be lifted. Load weights can often be determined by referring to shipping documentation that accompanies the load, and once the load weight is known, the operator must verify lift calculations to determine if the load is within the load rating of the crane.

Only necessary employees should occupy a personnel platform, and the platform must only be used for employee tools and materials necessary to perform the work. When employees are not being hoisted, the personnel platform should not be used for hoisting tools and materials. Additionally, a suspension system should be designed to minimize tipping the platform due to the movement of workers. Moreover, the personnel platform should be capable of supporting its own weight and at least five times the maximum intended load without any failure. The activity of hoisting a personnel platform should be performed in a controlled, slow, and cautious manner.

33.12   Multiple-Choice Questions

33.1   A machine that makes use of a least one guyed mast or self-supporting tower and a boom is called:

   A.   Crane
   B.   Derrick
   C.   Tower crane
   D.   Shearleg

32.2   Guy, shearleg, stifleg, and gin pole are the examples of:

   A.   Crane
   B.   Derrick
   C.   Tower crane
   D.   Shearleg

33.3   In crane and derrick community, ground conditions mean the:

   A.   Ability of the ground to support the equipment
   B.   Moisture condition and the soil type
   C.   Elevation of the ground surface
   D.   Composition of soil such as gravel, silt, or organic components

33.4   Significant distortion of the wire rope structure such as kinking, crushing, unstranding, birdcaging, signs of core failure, or steel core protrusion between the outer strands is known as:

   A.   Category I deficiency
   B.   Category II deficiency
   C.   Category III deficiency
   D.   Category IV deficiency

33.5 The operator must not leave the controls while the load is suspended, except:

A. The operator remains adjacent to the equipment and is not engaged in any other duties.

B. The load is to be held suspended for a period of time exceeding normal lifting operations.

C. The competent person determines that it is safe to do so and implements measures necessary to restrain the boom hoist and telescoping, load, swing, and outrigger or stabilizer functions.

D. All of the above.

33.6 Where the employer has taken the crane or derrick equipment out of service:

A. An out-of-service tag must be placed in the cab.

B. It must be stored in the warehouse garage.

C. It must be sent back to manufacturer immediately.

D. It must be repaired as soon as possible.

33.7 If there is a warning (tag-out or maintenance/do not operate) sign on the equipment or starting control, the operator must not activate the switch or start the equipment until:

A. The sign has been removed by a person authorized to remove it.

B. The operator has verified that no one is servicing, working on, or otherwise in a dangerous position on the machine.

C. The operator has verified that the equipment has been repaired and is working properly.

D. All of the above.

33.8 The operator must test the brakes each time a load that is _____ or more of the maximum line pull is handled by lifting the load a few inches.

A. 60%

B. 70%

C. 80%

D. 90%

33.9 If a load weight/capacity device does not work properly, it must be repaired no later than:

A. 30 days after the deficiency occurs

B. 60 days after the deficiency occurs

C. 90 days after the deficiency occurs

D. There is no such regulations

33.10 For floating crane/derrick, for backward stability of the boom, the high boom with no load at full back list must be stable at:

A. 75 mph

B. 90 mph

C. 120 mph

D. 155 mph

33.11 Who can operate a crane or derrick?

A. Only designated personnel

B. Any employee who needs something moved by the equipment

C. Any truck driver who needs to load his truck

D. Only the safety manager

33.13   Practice Problems

33.1   Distinguish between crane and derrick.

33.2   Discuss the wire rope inspection regulations for cranes and derricks.

33.3   What are the critical review items for rope inspection for crane and derricks?

33.4   What are the regulations for leaving the equipment unattended for crane and derricks?

33.5   What are the regulations for tagging out-of-service equipment for crane and derricks?

33.6   What are the regulations for travelling with a load for crane and derricks?

33.7   What are the regulations for protecting employees in the hazard area?

33.8   List the erecting, climbing, and dismantling regulations for tower cranes.

33.9   List the safety devices requirement for tower cranes.

33.10   List the load chart contents for derricks.

33.11   List the work area control regulations for floating cranes/derricks and land cranes/derricks on barges.

33.14   Critical Thinking and Discussion Topic

A crane operator said, "I am operating this crane during last 10 years. I do not need any training for it." As a supervisor, you said, "OSHA made regulations—every operator needs to take training every 2 years." The operator replied, "In my last 10 years, I had no accident nor any safety issue; please stay away of my road."

After this conversation, what will you tell to the operator or how will you work with the operator for the training?

(Photo courtesy of pexels-athena from pexels.)

CHAPTER 34

Ergonomics Program

(Photo courtesy of www.safetybydesigninc.)

34.1   General

Ergonomics, the human engineering, is the practice of designing machines, products, and places to better accommodate people. The principles of ergonomics are geared toward adapting the design and engineering of products and workplaces to people's sizes and shapes, physical strengths and limitations, biological needs, ability to handle information and make decisions, as well as their capacities for dealing with such psychological factors as isolation and stress.

Many workers across the United States have to carry out physically demanding tasks on a daily basis. If these tasks are not carried out in proper postures, they may cause fatigue or discomfort. Carrying out such tasks for prolonged periods may cause severe damage to muscles, nerves, tendons, ligaments, joints, cartilage, blood vessels, or spinal discs affecting the neck, shoulder, elbow, forearm, wrist, hand, abdomen (hernia only), back, knee, ankle, and foot. Such injuries are known as musculoskeletal disorders (MSDs). MSDs not only affect individual workers but also increase the cost of business in terms of workers' compensations, employee turnover, absenteeism, decreased efficiency, and finally, the overall productivity. In order to reduce the costs associated with MSDs and avoid the problem of work-related injuries, the OSHA has issued its long-awaited final rule for an Ergonomics Program Standard, 29 CFR Part 1910.900. According to the Bureau of Labor Statistics (BLS) in 2013, MSD cases accounted for 33% of all worker injury and illness cases in all areas. This standard contains stringent requirements to identify and abate MSDs. The costs to implement the ergonomics requirements may be significant and, in some cases, result in severe financial hardship. In the final rule are some surprising components that were not contained in the proposed rule. For instance, full implementation of the required program is now triggered by the activation of defined "action triggers" for at least one of five specified MSD "risk factors." Also, the final rule contains substantially expanded provisions addressing an employee's right to dispute the findings of the initially consulted healthcare professional (HCP). Employers now face the possibility and expense of as many as three separate HCP evaluations regarding an alleged MSD.

Employment covered by OSHA's standards for the construction, maritime, and agriculture industries, including directly related office management and support services, is expressly excluded, as are railroad operations. This is a substantial change from the proposed rule, which was limited to "manufacturing jobs" and "manual handling jobs," or any general industry job where an MSD actually occurs. Although the final standard has a broader application, the requirement for full implementation of the entire ergonomics program now depends entirely on the actual report of, or of the signs and symptoms of, an MSD. Until such an occurrence, the final rule only requires employers to provide all employees with initial, basic information addressing recognition and reporting of MSDs.

Example 34.1

Ergonomics attempts to adapt the design of products and workplaces to which of the following items? [Select all that apply]

    A. The costs of ergonomic solution to construction industries

    B. People's gender and race issues in construction sites

    C. People's emotional issues in construction sites

    D. People's medical histories and its effects in construction sites

    E. People's physical strength and limitations for work

    F. People's biological needs

    G. People's ability for dealing with information and make decisions

    H. People's capacities for dealing with isolation and stress

Answers: *E, F, G, and H*

34.2   Musculoskeletal Disorder (MSD)

The standard defines MSD as a disorder of the muscles, nerves, tendons, ligaments, joints, cartilage, blood vessels, or spinal discs affecting the neck, shoulder, elbow, forearm, wrist, hand, abdomen (hernia only), back, knee, ankle, and foot. Excluded are injuries arising from slips, trips, falls, motor vehicle accidents, or blunt trauma.

Examples of MSDs identified by OSHA include:

- Muscle strains and tears
- Ligament sprains
- Joint and tendon inflammation
- Pinched nerves
- Spinal disc degeneration
- Epicondylitis or tennis elbow
- Carpal tunnel syndrome
- Arthritis and other rheumatological disorders

MSD injuries may be manifested by medical diagnoses of low back pain, tension neck syndrome, carpal tunnel syndrome, rotator cuff syndrome, DeQuervain's syndrome, trigger finger, tarsal tunnel syndrome, sciatica, epicondylitis, tendinitis, Raynaud's phenomenon, hand-arm vibration syndrome, carpet layer's knee, and herniated spinal disc. MSDs most often develop as a result of repeated exposure to MSD "risk factors"; however, exposure to only one MSD risk factor may be enough to cause an MSD. The following types of employee behavior may indicate the presence of ergonomics-related problems:

- Employees shaking arms and hands or rolling shoulders due to discomfort
- Employees voluntarily modifying workstations and equipment to increase comfort
- Employees bringing in ergonomic products to the worksite (such as wrist braces)

34.3   Risk Factors for the MSD

There are certain aspects of tasks that can increase the risk of fatigue, MSD symptoms, and other types of issues. The associated risk factors can be divided into two categories:

1. Physical factor
2. Environmental factor

34.3.1   Physical Factors

The physical factors include:

a) Repetitive tasks or motions
b) Forceful exertions

c) Awkward postures

d) Pressure points

e) Recovery time

f) Vibration

Repetitive tasks or motions. Repetitive tasks require workers to perform the same task over and over again, using the same muscles, ligaments, and tendons. Repetitive motion may cause severe damage, injury, and discomfort to the worker. The risk of injury increases if the worker fails to take any breaks to relax his or her muscles.

Forceful exertions.   The amount of muscular effort required to perform a task is called force. Exerting more force than a body can sustain can cause severe damage to muscles and ligaments. The amount of force required for tools or machinery depends upon various factors, including those that follow.

- Load weight, shape, and bulkiness

- Grip type

- Amount of pressure required to accelerate, or decelerate, the load

The degree of risk generally increases with increasing force. Various parts of the body can be affected due to high force, including shoulders, neck, lower back, forearm, wrist, and hands.

Awkward postures.   The position of the body while performing a task is known as posture. The muscle groups used while performing a task are affected by the worker's posture. An awkward position can make the task more physically demanding by over-exerting small muscle groups and not using larger muscle groups. This can increase the likelihood of poor blood-flow, which can lead to fatigue and injury. Some examples of awkward postures include repeated or prolonged reaching, bending, twisting, kneeling, holding fixed positions, or squatting. Several areas of the body can be affected due to these postures including shoulders, arms, wrists, hands, knees, neck, and back. Awkward postures are often caused by poorly designed work areas, equipment, and tools along with poor work practices.

Postures to avoid are as follows:

- Prolonged or repetitive bending or extension of the wrist

- Prolonged or repetitive bending at the waist

- Prolonged standing or sitting without shifting your position

- Suspending an outstretched arm for extended periods of time

- Holding or turning your head consistently to one side

- Any posture that is held repeatedly or for a prolonged time

Motions to avoid are as follows:

- Repeated motion without periods of rest

- Repeated motion with little or no variation

- Repeated motions done with force
- Resting or compressing a body part on or against a surface
- Lifting heavy objects away from the body
- Frequent reaching or working above shoulder height

Pressure points.   Exerting pressure on different parts of the body by pressing them against hard or sharp surfaces may cause injury. Some body parts are at a greater risk as nerves, blood vessels, and tendons are present just under the skin in certain areas. Fingers, wrists, palms, elbows, forearms, and knees are examples of such body parts.

Recovery time.   Recovery time is the amount of time allocated to rest the muscles and tendons in any strained part of the body. It is very important for workers to take pauses between tasks that require forceful exertions. These breaks not only provide relief to workers but also enhance their performance. Employers must assess the duration of breaks according to the workload along with the risk factors present.

Vibration.   Exposure to continuous vibration can cause damage if uncontrolled. Exposure to vibration can occur with the use of vibrating tools such as sanders, chippers, chain saws, drills, grinders, routers, and impact guns. Vibrations can cause fatigue, numbness, and pain in the exposed area. It may also cause decreased sensitivity to touch and increased sensitivity to cold.

Example 34.2
Five technical terms are as follows:

a. Physical risk factors

b. Ergonomics

c. Musculoskeletal disorders

d. Force

e. Recovery time

Five explanatory sentences are as follows:

A. Amount of time needed to rest the muscles and tendons in any strained part of the body

B. Amount of muscular effort required to perform a task

C. Severe damage to muscles, ligaments, tendons, blood vessels, and nerves

D. Practice of designing machines, products, and place to better accommodate people

E. Interactions between work areas and workers such as awkward postures and repetitive motions

Match the technical terms with the explanatory sentences.

Answers: *Ea, Db, Cc, Bd, Ae*

34.3.2 Environmental Factors

Heat stress. Heat stress is the amount of heat that a body is exposed to while performing a task. Heat stress can be attributed to the worker's environment and also his or her own internal metabolism. Exposure to excessive heat can cause various disorders, including heat exhaustion, heat cramps, and heat stroke. The symptoms of heat stress may include headaches, thirst, nausea, muscle cramps, dizziness, and weakness. Due to the severity of the consequences of heat stress, employees must regularly monitor the workplace and take appropriate preventive measures.

Cold stress. Cold stress occurs when a worker is exposed to cold temperatures. Cold stress results in the decrease of the worker's body temperature and may cause shivering, unconsciousness, pain, and inadequate circulation of the blood. Cold stress may also cause the worker to lose the ability to grasp due to the decrease in body strength. Cold temperatures combined with the risk factors may increase the risk of MSDs.

Noise. Continuous sound at levels above 80–85 dB in the workplace can cause severe damage to a worker's hearing. Continued exposure to high noise levels may also result in impaired hearing or permanent deafness, tinnitus, or speech misperception. Furthermore, high levels of noise may also affect the worker's ability to concentrate on his or her work.

Lighting. Improper lighting in the workplace may cause eye fatigue and may result in headaches and a loss of focus. A worker's ability to perform task efficiently depends greatly on the proper lighting of the work area.

Example 34.3
There are certain aspects of tasks or workplace that can increase the risk of MSD symptoms and injuries. The physical factors are fundamentally the interaction between the work area and the worker. Mark the symptoms to the correct factor categories listed in Table 34.1.

Example 34.4
Determine if the program listed in Table 34.2 represents a physical factor or an environmental factor.

	Heat	Vibration	Sound
Thirst	X		
Nausea	X		
Sunburns	X		
Reduced tactile sensitivity		X	
Increased sensitivity to cold		X	
Numbness		X	
Tinnitus			X
Speech misperception			X
Permanent deafness			X

TABLE 34.1   Correlating the Symptoms to the Factor Categories

	Physical factor	Environmental factor
Records program		X
Audiogram program	X	
Audiometric program	X	
Noise monitoring program		X
Training program	X	
Hearing protection program		X
Hearing protector program	X	

TABLE 34.2   Identifying Physical Factors and Environmental Factors

34.4   Ergonomic Improvements by Employers

Employer can adopt various ergonomic improvements in their worksite to improve the working environment for their employees. Ergonomic improvements can be divided into three categories:

1. Engineering controls
2. Administration controls
3. Use of personal protective equipment (PPE)

34.4.1   Engineering Controls

Employees can make engineering controls in their workplaces by redesigning, rearranging, modifying, or replacing tools, equipment, workstations, products, or actions that increase the risks of injury. Implementing effective engineering controls can greatly reduce the risk factors. The following controls are particularly recommended:

- Install worktables with work surface that can be raised or lowered to the employee's body size and position. This can reduce bending, reaching, and awkward postures that can contribute to body damage.
- Cutout work surfaces may be appropriate to allow employees to adjust their distance from the worktable to reduce awkward postures.
- By allowing employees to reposition their work, their bending and reaching efforts can be reduced.
- Modifying the work surface according to the task can reduce the effort needed to complete the task. For example, to deliver a package from one area to another slider or roller can be used.
- Ladders, scaffolds, steps, or work platforms must be provided to employees to reach a high off surface.
- Loading and unloading jobs can be improved by implementing lifting devices, etc.
- All materials, products, and tools that are used frequently must be stored in a place that can be accessed easily.
- Materials should be transported around the workplace using mechanical aids when possible and appropriate.

- Unnecessary repetitive reaching, twisting, bending, and forceful exertions can be avoided by properly organizing the equipment and materials stored by grouping stored items by container size or shape.

- Installing proper lighting systems in the workplace, including all storage facilities, can help reduce eye strains and headaches.

- By utilizing good design and carrying out proper maintenance of all machinery and equipment, employees can ensure pressure points on the hands and wrists, awkward postures, and forceful exertions are minimized or are avoided.

- Exposure to vibration may be reduced by routine maintenance of vibrating equipment, covering handles with vibration-dampening wraps, operating the tool from a distance when possible, using vibration-dampening gloves, using alternate tools that produce less vibration, utilizing vibration isolators for workers who are seated, employing cushioned floor mats for tasks that have to be carried out while standing, or using vibrating tools at low speeds.

Example 34.5

How can an employer reduce the amount of bending and reaching the workers must do to complete their tasks?

   A. Prohibit them from repositioning their work.

   B. Allow them to reposition their work.

   C. Redesign workstation so they are all the same.

   D. Limit the use of mechanical aids.

Answer: *B*

Example 34.6

Ergonomic improvement can make the working environment safer for workers. One way to do this is to install engineering controls. Which of the following things are examples of engineering controls?
   [Select all that apply]
   A. Using a ladder to retrieve a container on the top shelf

   B. Working at a cutout work surface that can adjust your distance from the worktable

   C. Using work surfaces that can be raised or lowered to accommodate your body size and position

   D. Rearrange tools so workers don't have to bend, crouch, or stretch to reach them

   E. Taking short breaks to stretch

   F. Cleaning up after yourself and removing clutter from your work area

   G. Wearing slip-resistant boots

   H. Exercising regularly to avoid injury

Answers: *A, B, C*

Example 34.7

Suppose, you are the supervisor in your warehouse. Some of your workers have been complaining about fatigue and sore muscles. You decide to implement some engineering controls to reduce injuries. Five engineering control solutions available to you are as follows:

    a. Articulated arm

    b. Fork lift

    c. Work surface with rollers

    d. Cutout work surfaces

    e. Work surfaces that can be raised or lowered

Five ergonomic hazards present at your warehouse are as follows:

    A. Luis strains his back leaning to read the schematics on a computer screen.

    B. Christin is really tall and has to bend down to reach objects on a worktable.

    C. Abby has to lift heavy boxes and move them to the other side of the room.

    D. Kevin has to move palettes of containers to a work area 25 feet off the ground.

    E. Amber goes back and forth to the warehouse opening a large heavy door.

How will you use these five engineering control solutions to solve the five ergonomic hazards present at your warehouse?

Answers: Ea, Db, Cc, Ad, Be

34.4.2 Administrative Controls

Administrative controls involve developing work practices and methods that best protect the worker. These are often focused toward devising and implementing new practices and policies in order to allow employees to carry out their jobs effectively and efficiently and avoid any on-the-job injuries, illnesses, and accidents. Administrative controls rely on communication and training, as well as feedback from management and employees on the effectiveness of the controls. Administrative controls may include:

- Job rotation
- Adjusting work schedules and work pace
- Allowing more frequent breaks
- Modifying work practices
- Regular housekeeping and maintenance
- Encouraging regular exercise

Job rotation.   Adopting a job rotation system may be one effective measure to reduce damage caused to employees by using the same muscle groups every day. With a properly designed and implemented system, employees are rotated through different jobs, thus increasing job and muscle-use variety. Another system through which employees can increase job variety is through job enlargement. Through this system, employers combine two or more jobs or add different tasks to an existing job. These systems aim to prevent overuse and overexertion of muscles and body parts, by reducing the amount of repetition, altering the pace of work, reducing the physical exertion required, and controlling visual and mental demands.

Adjusting work schedules and work pace.   Employers must be careful not to assign too heavy of a workload to employees. They also must limit the amount of time that an employee spends performing a particularly challenging job in awkward positions even when physical improvements have been incorporated.

Allowing more frequent breaks.   Breaking work into smaller tasks allows employees to take adequate breaks between them. These breaks may help employees relax their muscles, thus preventing fatigue and injury.

Modifying work practices.   Supervisors and managers should regularly observe how workers perform their jobs. When employees perform all jobs while in a neutral posture, the body is less susceptible to injury. Employees may be able to adopt this posture by sitting or standing upright and not bending any joints into extreme positions; they should keep their necks, backs, arms, and wrists in a neutral position. Supervisors should encourage employees to work in a comfortable position and shift their positions or stretch often. Other work practices include the following:

- Minimize distances for carrying, pushing, and pulling.
- Manage an equal amount of weight in each hand.
- Avoid unnecessarily twisting of the body.
- Use smooth and even motions, avoiding jerking.
- Utilize legs to accomplish tasks rather than using the upper body or back.
- Ensure that all paths are free from obstacles and even-surfaced.
- Organize tasks to provide a gradual increase in the amount of force required.
- Ensure shoes worn are slip-resistant.

Regular housekeeping and maintenance.   Employers must devise a system to carry out regular housekeeping and maintenance of workspaces, equipment, machinery, and tools. There should be no cluttering in the workspace, as clutter can force employees to reach, bend, or twist their bodies while handling different objects. Additionally, employers must ensure that workspaces comply with the following points:

- All floor surfaces must be kept dry and free of any obstacles when possible. This can minimize hazards associated with slipping or tripping in the work area. Problems related to overexertion can often be minimized by carrying out regular maintenance of all tools and equipment.
- Ensure that handles and padding on vibrating tools are well maintained to help reduce vibration and awkward postures while tasks are being performed.
- All moving or mechanical parts on carts and pulleys are properly lubricated and maintained so as to reduce the amount of force required to move them.

Encouraging regular exercise.   Regular exercise is very important to one's well-being. It not only keeps the body fit, it reduces the risk of injury. Individuals who are in good physical condition are more productive and sustain fewer injuries. Employers may encourage their workers to increase their energy levels, coordination, and alertness by exercising regularly. Regular exercise can also increase the efficiency of their joints and improve blood circulation. Some organizations allow and encourage employees to warm up and engage in proper stretching before beginning work and while taking a break from work.

Example 34.8
Which of the solutions listed in Table 34.3 are considered engineering controls and which are administrative controls?

34.4.3   Use of Personal Protective Equipment

Personal protective equipment (PPE) includes all protective equipment, such as gloves, footwear, knee and elbow pads, eye protection, and other equipment that employees wear according to the type of task they are involved in.

Gloves.   Properly selected gloves help to protect hands from sustaining injuries, improve grip, and avoid contact with chemicals. However, if gloves do not fit properly or are not made of the proper materials, they can restrict hand movement and make it harder for employees to grip things.

Footwear.   Choosing the proper footwear according to the nature of the job can greatly reduce the risk of slipping. Some soles are designed to reduce fatigue for employees who are required to stand for long hours while performing a task.

Knee and elbow pads.   Knee and elbow pads can protect body parts that are pressed against hard or sharp surfaces. These aim to minimize the risk of negatively affecting pressure points until proper engineering improvements can be made.

Back belts.   OSHA does not recognize back belts as effective engineering controls to prevent back injury. While they may be accepted by individual workers because they feel as if they provide additional support, the effectiveness of back belts in the prevention of low back injuries has not been proven in the work environment.

OSHA's preferred approach to prevention of injuries and illnesses, including back injuries, is to eliminate the hazardous conditions in the workplace, primarily through engineering controls.

	Engineering control	Administrative control
Vibration isolators	X	
Job rotation		x
Adjustable equipment	X	
Job enlargement		X
Short breaks		X

TABLE 34.3   Identifying Engineering Controls and Administrative Controls

Example 34.9

Ergonomic improvements for the working environment can be divided into engineering controls and administrative controls. What is an example(s) of administrative controls? [Select all that apply]

A. Allowing short breaks

B. Modifying work practices

C. Installing mechanical lifting devices

D. Regular housekeeping and maintenance

E. Using rollers or a dolly to move heavy equipment

F. Adding rises to worktables to adjust table height

G. Encouraging regular exercise

Answers: A, B, D, G

Refresh your understanding of different types of controls using the examples of practices presented in Table 34.4.

34.5   Primary Elements of a Complete Ergonomics Program

Listed below are the primary elements of a complete ergonomics program:

- Management leadership
- Employee participation
- MSD management
- Job hazard analysis
- Hazard reduction and control
- Training

Activities	Engineering control	Administrative control	PPE
Wear back belt while lifting			X
Forklift and hand cart	X		
Job rotation system		X	
Proper material organization	X		
Distribute weight equally		X	
Properly fitting gloves			X
Workers use midrange posture		X	
Knee and elbow pads			X
Good lighting in the workroom	X		

TABLE 34.4   Different Types of Controls

The "Quick Fix" option may be available as an alternative to implementation of the full program under certain conditions. If the employer is eligible to utilize this option, the employer must take steps to eliminate or reduce the identified MSD hazard within 90 days according to specific criteria set out in the standard.

34.5.1   Management Leadership

This program element measures the employer's dedication to an effective program. To meet this obligation the employer must do the following:

- Assign and communicate the responsibilities for setting up and managing the program.
- Ensure each responsible person has the necessary authority, resources, and information to meet his assigned responsibilities.
- Ensure company practices encourage, and do not discourage, prompt reporting of MSDs and employee participation in the program.
- Periodically communicate with employees about the program and any employee MSD concerns.

34.5.2   Employee Participation

This program element reflects OSHA's firm belief that employee interest and involvement is vital to the success of the ergonomics program. It is based on the premise that employees have the most direct interest in their own safety and health on the job and that they have an in-depth knowledge of the operations and tasks they perform.

To comply with this program element, employers must establish mechanisms for employee reporting of MSDs, respond promptly to employee reports, and involve employees in the development, implementation, and evaluation of the program. (Such participation, however, must be consistent with limitations on the use of labor management committees imposed by the National Labor Relations Act, according to the Preamble to the proposed rule.) Employees must also receive a summary of the requirements of the standard and have ready access to the standard, the employer's program, and general MSD information.

34.5.3   MSD Management

The term "MSD management" refers to the employer's collective obligations to employees who have sustained a confirmed MSD injury. In addition to appropriate temporary work restrictions, this obligation includes ensuring employees receive prompt and effective medical evaluation and follow-up of their MSD and recovery by an HCP, all at no cost to the employee. The final rule provides an affected employee with the right to obtain a second and even a third HCP opinion, if necessary, regarding his or her condition, all at the employer's expense. This last provision was entirely absent from the proposed rule.

MSD management also includes the requirement for "work restriction protection" (WRP) for up to 90 days. Basically, WRP means maintaining 100% of wages for employees assigned to light duty or 90% of wages where total removal from the workplace occurs. WRP also guarantees no reduction in employment rights or benefits, including seniority, insurance programs, or retirement and savings plans.

34.5.4   Job Hazard Analysis

This element is the core of the ergonomics program. If an action trigger is activated, job hazard analysis is the step that determines whether the employer has a "problem job" on his hands. This determination is not a simple matter. OSHA has incorporated by reference several professional treatises, studies, and papers that address detailed methods for ergonomic analysis of work functions. These are referred to as "hazard identification tools."

34.5.5   Hazard Reduction and Control

This element refers to the employer's obligation to eliminate or materially reduce the identified hazards to the extent feasible, using engineering, administrative, and/or work practice controls, and as a last resort, personal protective equipment (excluding back belts/braces and waist braces/splints). In other words, this element constitutes the acceptable goals of an ergonomics program.

34.5.6   Training

All employees must receive basic MSD training. However, more detailed and specific training is required for employees in jobs where the action trigger has been activated, for their supervisors and team leaders, and for the employees involved in the set up and continuing management of the program. This focused training must include a re-emphasis of the MSD basic training, supplemented with information about specific MSD hazards and risk factors present in the relevant job. The training must also address the employer's ergonomics program and the roles the respective trainees play in the program. Finally, the training must include details about the employer's plan and time-table for correction of the hazards, as well as instruction regarding the employee's role in evaluating the effectiveness of the chosen controls.

34.6   Summary

Repetitive tasks require workers to perform the same task over and over again, using the same muscles, ligaments, and tendons. Repetitive motion may cause severe damage, injury, and discomfort to the worker. The risk of injury increases if the worker fails to take breaks to relax his or her muscles. Recovery time is the amount of time allocated to rest the muscles and tendons in any strained body part. It is very important for workers to take adequate pauses between tasks that require exertions. These breaks not only provide relief to workers but also enhance their performance. High or continuous vibration can cause fatigue, numbness, and pain in the exposed area. It may also cause decreased sensitivity to touch and increased sensitivity to cold. To prevent the consequences of heat stress, employers must regularly monitor the workplace and take the appropriate measures to ensure that all employees are adequately protected. In short, in order to reduce the costs associated with MSDs and to avoid the problem of work-related injuries, employers must implement principles of ergonomics in their workplaces.

All supervisors required to administer an ergonomics program should be provided with special training focusing on how to effectively make the workplace safe by adopting sound ergonomic principles and practices. Apart from ergonomic principles, the training must also include the risk potential of damage to the body and injuries that can result from the failure to adopt proper ergonomic practices. Employers must provide formal training to all employees who could be exposed to hazards to inform them about the hazards associated with their jobs and the tools, machinery, and equipment they use. New employees and those assigned new tasks must be made aware of the specific risks associated with a particular job before they start their work.

An ergonomics training program must include:

- All employees who are exposed to different risk factors
- Supervisors
- Managers
- All engineers and maintenance personnel

An effective training program includes a mix of both theoretical and practical ways in which employees can develop their skills to work safely. They must manage the amount of time that an employee spends performing a particularly challenging job.

34.7 Multiple-Choice Questions

34.1 In order for an employee who suffered muscle strain to heal, the employer must provide and the employee must take advantage of:

A. Strength training
B. OSHA training
C. Exercise programs
D. Recovery time

34.2 Risk factors that lead to MSD can be divided into the following categories:

A. Administrative factors and environmental factors
B. Physical factors and environmental factors
C. Engineering factors and physical factors
D. All of the above

34.3 Job rotations, modified work pace, and frequent breaks are all examples of _____ controls.

A. Administrative
B. Engineering
C. Safety
D. Design

34.4 Those areas of the body where nerves, blood vessels, and/or tendons lie just beneath the skin are highly susceptible to _____ injuries.

A. Pressure point
B. Repetitive motion
C. Forceful exertion
D. Vibration

34.5 Willard runs an industrial, hand-operated floor sander 4 days a week. Although he wears his PPE religiously, he is still at risk for all these potential injuries, except:

A. Hearing loss
B. Hand numbness
C. Muscle pain
D. Muscle fatigue

34.6 Which personal protective equipment can reduce the risk of pressure points?

    A. Padded, leather gloves
    B. Rubber-soled boots
    C. Knee and elbow pads
    D. Weight belts

34.7 All of the following awkward postures must be avoided, except:

    A. Suspending an outstretched arm for extended periods of time
    B. Prolonged standing while shifting your position
    C. Prolonged or repetitive bending at the waist
    D. Turning your head consistently to one side

34.8 How often might ergonomic training be offered in the workplace?

    A. Biennially
    B. Biannually
    C. Annually
    D. With each new hire

34.9 What is the ergonomically optimum body posture for performing all physical jobs?

    A. Erect
    B. Midrange
    C. Active
    D. Prone

34.10 Higher workers' compensation premiums, increased employee turnover, absenteeism, and decreased efficiency are common results of:

    A. Musculoskeletal disorders
    B. Ergonomic improvements
    C. Engineering controls
    D. Extensive safety training

34.8 Practice Problems

34.1 Define the term "Ergonomics." Why is it important in construction industry?

34.2 Define musculoskeletal disorder. Why is it important in construction industry?

34.3 List the employee behavior indicating the presence of ergonomics-related problems.

34.4 List the physical factors of MSD.

34.5 List the environmental factors of MSD.

34.6 Discuss the engineering controls for the MSD.

34.7 Discuss the administrative controls for the MSD.

34.8 Discuss the PPE controls for the MSD.

34.9 List the primary elements of a complete ergonomics for the MSD.

34.9   Critical Thinking and Discussion Topic

The first day back at work after the holidays is usually a pain in the neck, and today is no different. Everybody's playing catch up. We are all a little behind and very busy, so everything isn't set up efficiently. There is a group of people assembling parts in the east workroom, but all of the components aren't there. Adam will have to go get them from boxes located on the top shelves in the warehouse. For most of the day, warehouse crews will be in the yard sorting the newly delivered containers of solvent and other substances. John will have to go from the warehouse out to the loading docks several times to check on them. There is this huge metal door that he'll have to raise and lower every time he goes out there.

What sorts of things will these workers need to do their jobs safely?

(Photo courtesy of tiger-lily from pexels.)

OSHA.com Construction Safety Training Programs

The OSHA 30-hour Construction Industry Outreach Training course is a comprehensive safety program designed for anyone involved in the construction industry. This online course is for construction foremen, supervisors, and safety directors, or anyone who works in construction and requires additional training beyond the basics. You'll receive a comprehensive overview of the policies, procedures, and best practices in OSHA's 29 CFR 1926 standards for construction. Upon completion of the course, you will get your laminated official Department of Labor (DOL) card in the mail within 2 weeks.

Topics covered in OSHA 30-hour construction industry outreach training are as follows:

Lesson 1: Introduction to OSHA
Module 2: Managing Safety and Health
Module 3: OSHA Focus Four Hazards
Module 4: Personal Protective Equipment
Module 5: Health Hazards in Construction
Module 6: Stairways and Ladders
Module 7: Concrete and Masonry Construction
Module 8: Confined Spaces
Module 9: Cranes, Derricks, Hoists, Elevators and Conveyors
Module 10: Ergonomics
Module 11: Excavations
Module 12: Fire Protection and Prevention
Module 13: Materials Handling, Use and Disposal
Module 14: Motor Vehicles, Mechanized Equipment and Marine Operations; Rollover Protective Structures and Overhead Protection; and Signs, Signals and Barricades
Module 15: Safety and Health Programs
Module 16: Scaffolds
Module 17: Tools—Hand and Power
Module 18: Welding and Cutting
Module 19: Silica Exposure
Module 20: Lead Exposure
Module 21: Asbestos Exposure

A 10-hour Construction Safety Training Program is also available and is designed for construction workers as an ideal orientation to those who are new to the industry and as a reminder to those who have been working in the industry to the hazards associated with their work. For more information, please visit the site listed in the reference.

Reference: https://www.osha.com/courses/outreach-construction.html. Accessed November 22, 2020.

References

Goetsch, D. (2018). *Construction Safety and the OSHA standards*, Second Edition. Upper Saddle River, NJ: Pearson.

Heinrich, H. W. (1931). *Industrial Accident Prevention: A Scientific Approach*. New York, NY: McGraw-Hill.

Levine, D. I., Toffel, M. W., Johnson, M. S. (2012). "Randomized government safety inspections reduce worker injuries with no detectable job loss." *Science*, Vol. 336, No. 6083, pp. 907–911.

Mahmood, W. Y., Mohammed, A. H. (2006). A Conceptual Framework for the Development of Quality Culture in the Construction Industry. Association of Researchers in Construction Management, ARCOM 2007—Proceedings of the 23rd Annual Conference.

OSHA 2020. PART 1926—Safety and Health Regulations for Construction. Washington, DC: Occupational Safety and Health Administration (OSHA), US Department of Labor. https://www.osha.gov/laws-regs/regulations/standardnumber/1926. Accessed November 8, 2020.

Reason, J. T. (1991). "Too little and too late: a commentary on accident and incident reporting." In: Van Der Schaaf T. W., Lucas D. A., Hale A. R. (eds.). *Near Miss Reporting as a Safety Tool*. Oxford, UK: Butterworth-Heinemann, pp. 9–26.

U.S. Bureau of Labor Statistics (BLS). News Release, USDL-19-2194, December 17, 2019. https://www.bls.gov/news.release/pdf/cfoi.pdf

Index

Note: Page numbers followed by f indicate figure and by t indicate table.